POPULATION GEOGRAPHY

Second Edition

Huw Jones is Head of the Department of Geography at the University of Dundee. He has undertaken spatial demographic research in several parts of the developed and less developed world including Barbados, Canada, France, Malta, Mauritius, Scotland and Wales. He is a former Secretary and Chairman of the Population Geography Study Group of the Institute of British Geographers.

Population Geography

Second Edition

Huw Jones

First published 1990
Paul Chapman Publishing Ltd
144 Liverpool Road
London
N1 1LA

British Library Cataloguing in Publication Data

Jones, Huw R. (Huw Roland)
Population geography. – 2nd ed.
1. Population. Geographical aspects
I. Title
304.6

ISBN 1 85396 071–3

Typeset by Inforum Typesetting, Portsmouth
Printed by St Edmundsbury Press Ltd, Bury St Edmunds, Suffolk
Bound by W H Ware & Sons Ltd, Avon

E F G H I 8 7 6 5 4 3

CONTENTS

PREFACE

Population geography has moved on – and forward – in the ten years from the first edition of this text. Consequently, this is a completely rewritten and extended work, not simply a revised one. It reinforces my earlier belief that research, policy formulation and student interest in population geography have become concentrated in process-orientated population dynamics. Emphasis in this edition continues, therefore, to focus on evolving patterns of fertility, mortality and migration, the spatio-temporal processes that fashion them, and their interaction expressed in population growth, population problems and population policies.

A concern for disciplinary boundaries is now unfashionable in the social sciences, so that this text unashamedly roams the interface between social demography, development studies and human geography. Inevitably, this provides a broader interpretation of population geography than is usually conveyed by advocates of spatial demography. Moreover, there is a commitment here to draw widely on illustrative material from both the developed and less developed worlds; after all, it is curiosity about other places and other peoples that makes geographers in the first place.

Finally, I am much indebted in the preparation of this edition to Pat Michie and Jim Ford for their assiduous and professional secretarial and cartographic work.

ACKNOWLEDGEMENTS

The author and publishers would like to thank the following Publishers who have kindly given permission for use of copyright material:

Edward Arnold, for Tables 3.1 and 3.2 from McKeown, T., *The Modern Rise of Population*, 1976.

Harcourt Brace Jovanovich, Inc, for Figure 33.2 from Baumol, W. J. and Blinder, A. S., *Economics: Principles and Policy*, 1979.

Heinemann, for diagrams from Johnson, B. L., *South Asia*, 1969.

Macmillan Publishing Company for Figure 2.7 from Boughey, A., *Ecology of Populations 2/e*. Copyright © 1975 Arthur S. Boughey.

Population Reference Bureau, Inc., Washington D.C., for *1992 World Population Data Sheet*, April 1992.

The Population Council for selected data from Tables 9 and 19, from Ming-Cheng Chang, Ronald Freedman and Te-Hsiung Sun, Trends in Fertility, Family Size Preferences, and Family Planning Practice: Taiwan 1961–85, *Studies in Family Planning* Vol. 18 no. 6 Nov/Dec 1987; and Table 4 from Robert J. Lapham and W. Parker Mauldin, Contraceptive Prevalence: The Influence of Organized Family Planning Programs, *Studies in Family Planning* Vol. 16 no. 3 May/June 1985.

The University of Chicago Press, for Figure 10 from Hauser, P. and Duncan, O., *The Study of Population: An Inventory and Appraisal,* 1959; Table 3 from Friedlander, D. *Economic and Cultural Change*, Vol. 22, pp. 39–51, 1973–4.

Universe Books, N.Y., for material from Donella H. Meadows, Dennis L. Meadows, Jorgen Randers, William W. Behrens, III. *The Limits to Growth: A Report for The Club of Rome's Project on the Predicament of Mankind,* 1972. A Potomac Associates book, graphics by Potomac Associates.

The Population Council for Figures 6.5, 7.6 and 7.7 and Tables 6.4 and 11.1.

1

THE NATURE OF POPULATION GEOGRAPHY

As a response to the stresses imposed on societies and their habitats throughout the world by unprecedented increases in human numbers and consumption in this century, there has been ever-growing interest by academics, general public and governments in 'the population question' at all levels of scale from global to local. Major international organizations like the United Nations and World Bank have concerned themselves more and more with the study of global population problems, especially with a view to incorporating appropriate population policies in overall development strategies in the disadvantaged Third World. Meanwhile, the governments of many Western nations feel they can no longer shelter behind their shield of affluence and ignore the population dimension. Several such governments have formed special advisory commissions or panels on population, although because population tends to be viewed in crisis terms rather than as a critical element of an ever-evolving social scene, these initiatives have been poorly funded, half-heartedly supported and sometimes aborted (Nam, 1979).

By the 1980s the more hysterical outbursts of concern about population growth had abated somewhat as the more gloomy scenarios for the future interaction of population and resources had been discredited by critical examination of their assumptions. Moreover, the growth rate in world population has fallen from about 2.0 per cent per annum in the 1960s to 1.7 per cent in the 1980s, essentially due to fertility falls in all developed countries and in many Third World countries, including most notably and dramatically the population giant of China. Yet such is the inbuilt momentum for absolute growth in population numbers globally and such are the uncertainties about the world's ability to provide supportive resources that population questions will remain a fundamental concern of humanity for decades to come. Even if developed countries myopically consider that their modest fertility levels cushion them from a survival crisis, they will certainly be concerned with a whole host of economic, social and physical planning problems stemming from the changing age structure of their populations. In the Third World, too, there is a broadening of population issues away from simple population growth in relation to environmental resources towards consideration of the critical role in development strategies of human resources based particularly on qualities of health and education.

The study of human populations embraces such a wide academic spectrum that it is important to assess how distinctive is the approach of population geography.

The roots of population geography

The nature of contemporary population geography owes more to the radical changes in approach and methods operating throughout human geography from the 1960s than to the particular emergence and development of population geography as a recognized branch of geography in the 1950s; yet these early roots (Kosiński, 1984) must be acknowledged and traced briefly.

In authoritative reviews of the evolution and nature of geography, Hartshorne (1939) and Wooldridge and East (1951) make not a single reference to population geography, indicating its late entry into the recognized systematic branches of the subject. Zelinsky (1966) attributes this to the late arrival and modest impact on the academic scene of demography, compared with, for example, the role of economics in stimulating economic geography. Another factor would have been the strong interests of many of the leading human geographers like Brunhes and Demangeon in settlement geography, where the distribution of population was studied through what was thought to be the more geographically acceptable medium of the cultural landscape. Population *per se* was relegated to consideration in the more sterile forms of regional geography as part of a place–work–people chain, with a naïve implicit assumption of unidirectional causation in that the physical environment was thought to influence strongly economic activities, which in turn controlled population patterns.

It was in the early 1950s that population geography emerged as a systematic branch of geography in the sense that it dealt with a recognizably distinct group of phenomena and systematically related processes, the study of which involved a particular form of training. Much of the credit, at least in the English-speaking world, must go to Trewartha (1953), who used the platform of his presidential address to the Association of American Geographers to make a powerful case for, and an outline of, population geography. He argued (1953, p. 97) that population was 'the pivotal element in geography, and the one around which all the others are oriented, and the one from which they all derive their meaning'. Other important formative statements were made at this time by P. George (1951) and James (1954), and such was the growing interest in population geography that several synthesizing introductory texts appeared in the 1960s (P. George, 1959; Clarke, 1965; Zelinsky, 1966; Beaujeu-Garnier, 1966; M. G. Wilson, 1968; Trewartha, 1969). Illustrative of growth in the subject at this time, papers in population geography increased from 3 per cent in 1962 to 13 per cent in 1972 of papers presented at the annual meetings of the Association of American Geographers and from 5 per cent to 12 per cent at the same time of papers in the leading American geographical journals (Hansen and Kosińksi, 1973, p. 12).

Spatial distribution and areal differentiation of population attributes were clearly the unifying threads within population geography at this time. Thus Trewartha (1953, p. 87) saw its purpose as 'an understanding of the regional differences in the earth's covering of peoples', while James (1954, p. 108) thought 'the objective of population geography is to define and to bring forth the significance of differences from place to place in the number and kind of human inhabitants'. The role of the population geographer was regarded by Zelinsky (1966, p. 5) as studying 'the spatial aspects of population in the context of the aggregate nature of places', and by Beaujeu-Garnier (1966, p. 3) as describing 'the demographic facts in their present environmental context'. More explicitly, Clarke (1965, p. 12) stated that population geography 'is concerned with demonstrating how spatial variations in the distribution, composition, migrations and growth of populations are related to spatial variations in the nature of places'.

In these traditional approaches a focus on spatial distribution was thought sufficient to distinguish population geography from demography, which is much more concerned with the intrinsic nature and universal attributes of populations and with a temporal, rather than spatial, dimension. But a problem had always existed in specifying population attributes appropriate for direct study by the population geographer. There is agreement on a core comprising distribution, density, age-, sex- and marital-composition, fertility, mortality and migration, but opinions differ widely on the inclusion of attributes like occupation, religion, language and ethnicity. The stance taken here is that the demographically more marginal attributes of population like occupational composition should be excluded from a central position in population geography; otherwise there would be little to distinguish it within the broad field of human geography. One rule-of-thumb way of circumscribing the field, suggested by Zelinsky (1966, p. 7), is that attention should be confined to those human characteristics 'appearing in the census enumeration schedules and vital registration systems of the more statistically advanced nations', but inevitably there are data recording variations between such nations, and, more critically, there is no theoretical justification for the suggestion.

Two questions have dominated the traditional approach of the geographer to population study: Where? and Why there? The first has been responsible for considerable work in observing, identifying and, above all, depicting patterns of spatial distribution. Mapping of cross-sectional patterns dominated the early work in population geography. This was well exemplified by James (1954) devoting more than three-quarters of his discussion of research frontiers in population geography to problems of mapping, while the first five years (1963–8) of what has become the dominant forum and co-ordinating body of British population geography, the Population Geography Study Group of the Institute of British Geographers, were given over entirely to the mapping of census data. Various commissions of the International Geographical Union have been active in this area (Kosiński, 1980), so that many geographers have developed considerable expertise in population mapping.

Why there? takes the population geographer's approach a step further into an essentially ecological field, since 'the areal facts of population are so closely orchestrated with the totality of geographic reality' (Zelinsky, 1966, p. 127). Consequently, the analysis and explanation of 'complex inter-relationships between physical and human environments on the one hand, and population on the other . . . is the real substance of population geography' (Clarke, 1965, p. 2). Figure 1.1 schematizes the interactions between the fundamental dimensions of the ecological complex from a functional point of view. Emphasis is given to the way in which a population organizes and equips itself for survival in a particular habitat. There is no unidirectional causation since the lines in the diagram represent linkages of functional interdependence. It follows that if dislocating forces affect any one dimension (e.g. concentrated mortality reduction, technological innovation or rapid environmental change), a chain reaction of modifications is set up throughout the complex as a new equilibrium is sought.

Modern population geography: its changing emphasis

The spatial and ecological approach to population phenomena remains the dominant and distinctive dimension of population geography, but the practice of the subject has changed appreciably in response to the winds of conceptual and quantitative change which have swept through human geography in the last three decades. Population geographers themselves may not have figured prominently among the stormtroopers of methodological transformation, yet their take-up of the new outlooks and methods grounded in positivism, behaviouralism and structuralism (Johnston, 1986) has been rapid and substantial.

The several strands of transformation in modern population geography can be summarized by reference to selected changing procedures and purposes – in other words, changing means and ends.

New procedures

One of the most important enabling developments in population geography has been the emergence of Geographical Information Systems (GIS). As in all areas of demographic inquiry, the extent, quality and accessibility of data are of paramount importance. Study of the detailed spatial structure of population in contemporary Britain, for example, has been greatly enhanced by comprehensive standard sets of data being made available from censuses for small areas (parishes, wards, enumeration districts, grid squares and postcodes), comprising what has been termed the largest multivariate spatial series in human geography. GIS are beginning to enable routine computerized matching, by users' areal specification, of these data with a host of other spatially referenced data files (housing, planning, electoral registration, health, etc.), so that important demographic relationships can be pursued – at an ecological or cross-sectional level of analysis in the first place.

Paralleling the greater availability of small area demographic data has been the greater analytical rigour now being adopted in population geography. First, expertise derived from the dominant 1960s positivist paradigm of human geography as spatial science has ensured that cross-sectional multivariate studies comprise one field of population analysis for which modern geographers are particularly well equipped. This is essentially because the pitfalls of routine multivarite analysis of spatial series have now been widely understood after their often uncritical and sometimes inappropriate adoption in the heyday of geography's positivist era.

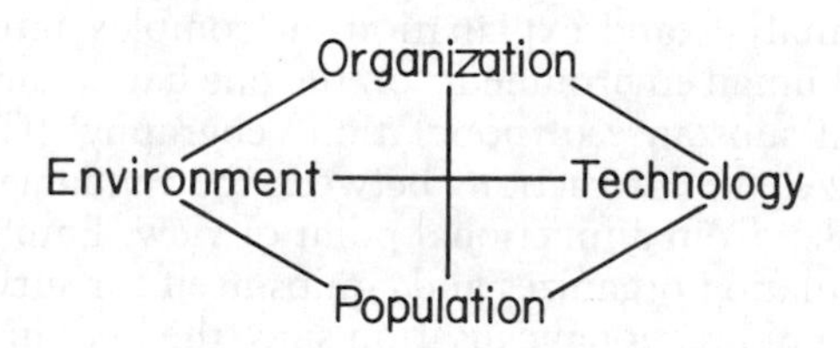

Figure 1.1
The ecological complex
Source: Duncan (1959), figure 10. By permission of The University of Chicago Press © 1959 by The University of Chicago.

Second, there have been determined attempts by Woods (1979, 1982a) to incorporate in population geography a wide range of analytical methods and theories from formal demography. Allied to this have been developments in spatial demographic accounting to estimate future regional population structures on the basis of area- as well as age- and sex-specific estimates of mortality, fertility and migration. Population projections are thus being applied in an explicitly spatial manner (Rees and Wilson, 1977; Woods and Rees, 1986; Congdon and Batey, 1989). Similarly, the geographer's interest in optimization or 'best location' has important applications in the population field. Work by Robertson (1972, 1974) is an example of the application of statistical and cartographic methods to a matrix of population data on a regular grid to give precise yet flexible answers to problems of facility location when maximum accessibility of population to facilities is desired.

New purposes

Analytical work in population geography has been moving away from *ad hoc* empirical investigation of particular demographic situations to work which is focused increasingly on the pursuit of two central aims: to seek demographic patterns, regularities and order in space; and to explain these patterns by the processes operating differentially in space which create, modify, destroy and replace such patterns. There is no implication here that specifically spatial processes are at work; indeed, it can be argued that 'there are no such things as purely spatial processes; there are only particular social processes operating over space' (Massey, 1985, p. 11).

Demographic regularity and demographic regimes can be identified and analysed at very different levels of scale (Clarke, 1976) – from continents to local communities. The basis of such demographic patterning within space has normatively been regarded as economically based (on modes of production, levels of wealth, etc.), but there is growing appreciation of specifically culture-driven demographic differentiation. This is exemplified by the fundamental macro-demographic division between northern and southern India that is now thought to be based on very different kinship systems (Dyson and Moore, 1983) and by the Caldwells' (1987, 1989) insistence on the distinctiveness of the sub-Saharan African social and demographic system.

Turning now to explanatory processes, there has been a growing awareness in human geography that too much attention in teaching and research has been given to the observation and identification of spatial patterns and too little concern for the processes which create and subsequently modify such patterns. More and more it is being appreciated that form and structure – the statics – are dependent on process and spatial interaction – the dynamics. Indeed,

> in proper perspective, the distinctions we make between spatial process and spatial structure disappear because they are based upon a limited time perspective . . . Process and structure are, in essence, the same thing . . . When we distinguish spatial process from spatial structure we are merely recognizing a difference in relative rapidity of change . . . Properly considered, the spatial structure of a distribution is viewed as an index of the present state of an ongoing process.
>
> (Abler, Adams and Gould, 1971, p. 61)

Accordingly, geographers have become reluctant to infer process from structure, but ever more anxious to research the nature of the formative processes, not simply to explain past and present patterns but also to provide the basis of sound forecasting. Moreover, an important logistical advantage is that

> our fundamental shift away from the study of differentiation toward that of process carries with it liberation from any blanket-like constraint of scale in defining a geographical problem. The appropriate level of resolution becomes that at which the relevant process may best be recognized and analysed.
>
> (Brookfield, 1973, p. 15)

Heenan (1967, p. 714) argues that 'population geographers have hitherto tended to concentrate their interest very heavily upon the end product or summation of change in preference to the study of those dynamic processes whereby change is wrought'. In sympathy with that perceptively early view, the structure of this book has been designed specifically to recognize the prime explanatory importance in contemporary population geography of the dynamic components – fertility, mortality and migration – and the processes that fashion them; it is their spatial and temporal interaction which produces changes in population numbers, distribution and composition. Interaction is all-important, and none of the dynamic components and their associated processes can be studied effectively in isolation. Consider, for example, how appreciable emigration (selective in terms of young male adults) from congested, famine-ridden Ireland in the nineteenth century promoted changes in the age-, sex- and marital-structure of Ireland, which in turn had repercussions on the level of birth, death and marriage rates and on a wide range of social and economic structures.

It has been the spur of process study that has stimulated the theoretical, model-based and simulation-oriented work on the spatial diffusion of ideas, behaviour and technology which arguably has been one of the very few significant strands of indigenous theory developed within geography, as opposed to theory derived from other sciences. Diffusion studies of the type pioneered by Hagerstrand (1967) have a clear relevance to an understanding of many of the key demographic processes. It is possible, for example, to consider, although not necessarily to accept, that spatial patterns of both fertility and mortality reduction reflect the spatial diffusion of innovations like contraception knowledge and public health technology, with diffusion from metropolitan innovation centres being controlled by spatial and urban hierarchical proximity.

The way forward

There is no such thing as a geography of population or indeed a geography of anything, unless there are those who still believe in a very restricted and intellectually unsatisfying equivalence of geography with simple spatial distribution. Accordingly, population geography will achieve full academic respectability only if it sheds its traditional unambitious concern for pattern description and inferential interpretation and embraces unequivocally those geographical perspectives and procedures which can contribute fresh insights into demographic problems which are studied also in several other disciplines, notably demography, sociology, social and economic history, and sociology. As Woods (1982b, p. 247) has argued, the primary objective of population geography is 'that of providing the spatial perspective to the wide field of which it is but one part, namely population studies'.

It is encouraging to observe three modern developments in demography which are empathetic to human geography. First, there has been some unease about the mechanistic and almost exclusively macro-analytical nature of much of formal demography, reflected in the International Union for the Scientific Study of Population (IUSSP) establishing in 1982 a Working Group on the Micro-Approach to Demograpahic Research with workshops in Canberra in 1984 and London in 1985. The micro-analytical approach comprises

> an intensive examination, employing methods akin to those of anthropology, in a few chosen localities in the expectation that participatory observation along with repeated, intimate, and discursive discussions would yield more than rapidly conducted surveys could ever hope to achieve.
>
> (Caldwell *et al.*, 1987, p. 6)

This is very much in line with the long tradition of detailed community studies within human geography and human ecology.

Second, demographers are increasingly using territorially disaggregated data to throw light on the dynamic relations between demographic change, especially fertility decline, and socio-economic change, particularly when their preferred time-series data are lacking. There is no justification, therefore, for viewing spatial analysis of population data as the exclusive preserve of population geography, but geographers do have valuable conceptual and technical skills in assembling and analysing spatial series that other social scientists often lack. In population geography, 'space' is certainly our distinctive, although not exclusive, dimension for contributing understanding to demographic relationships within human societies; but 'space' for us should be only a means to a greater interpretative end.

Third, although demographers would claim that their work has always been informed by an appreciation that demographic events and processes are embedded in their socio-economic and institutional settings, it is only recently that they have began to attach more than 'background' causal status to such structural setting. But there is now a growing appreciation of the need systematically and holistically to link structural change in societies with demographic behaviour, particularly through applying many of the theoretical and conceptual advances in the sociology of development and underdevelopment to studies of differential fertility, mortality and migration. There is, of course, the danger of excessively course reductionism and determinism in which distinctive features of countries and regions (i.e. geography) simply disappear under a blanketing of global core, periphery, semi-periphery and the like.

Yet a recent book, *The Demography of Inequality in Brazil* (Wood and Carvalho, 1988), demonstrates a particularly fruitful way forward in its blend of demography, development sociology and geography. Its basic methodology is to decompose aggregate measures of fertility, mortality and migration to correspond with key dimensions of geographic and socio-economic stratification (it finds, for example, a huge 23-year difference in life expectancy at birth between the lowest income group in the north-east and the highest income group in the south). It then interprets this demographic mosaic within a broader model of transformation in the country's political economy. There is a full appreciation that capitalist development within Brazil has been uneven, in that the process has been neither linear nor homogeneous. This study, then, is a good, path-finding example of a qualified historical materialist model of socio-economic, political and demographic interaction which respects the uniqueness of a particular country's historic development path and its internal geographical and social diversity.

2

INTRODUCTION TO POPULATION GROWTH AND REGULATION

The purpose of this chapter is to provide a broad overview of the nature of population growth as the essential context for detailed examination of the major components of population change in subsequent chapters.

The fascinating story of world population growth is how a few thousand wanderers a million or so years ago grew into today's billions of inhabitants of cities, towns and countryside. *Homo sapiens* is thought to have become distinct from its hominid predecessors about a million years ago. Between then and the beginning of agriculture and domestication of animals about 10,000 BC, the average rate of population growth can be very roughly estimated at about 0.0001 per cent per annum, so that in this 99 per cent of man's history the population had increased to a mere 5–10 million (the number which anthropologists estimate as being capable of support by hunting and gathering cultures). After the establishment of agriculture and sedentary communities, the world's population grew to about 300 million by AD 1 and 800 million by 1750, but still the average growth rate was well below 0.1 per cent per annum (Dumond, 1975).

The modern period of rapid population growth may be regarded as starting about 1750. Average annual growth rates climbed to about 0.5 per cent between 1750 and 1900, 0.8 per cent in the first half of this century and 1.7 per cent in the second half. In this minute period of human history the world's population has sextupled, from 0.8 billion in 1750 to 5.3 billion in 1990. Thus in less than 0.1 per cent of man's history has occurred more than 80 per cent of the increase in human numbers.

The modern period of population growth is clearly a transitory episode. If the growth rate of the 1980s were to be maintained, the population would double every 40 years or so, so that in 500 years there would be standing room only on the earth's surface (Figure 2.1). Simple arithmetic, therefore, shows that a return to a growth rate much nearer zero than 2 per cent per annum is inevitable, but whether this will be achieved through lower birth rates or higher death rates, or some combination of both, is the great imponderable. The most plausible scenarios must surely be based on an understanding of the past interaction of mortality, fertility, resources and culture. In effect, the past must serve as our laboratory.

Stepped growth

The first aid to understanding must be an appreciation that all figures for world population levels and growth rates, even for recent times, are estimates. The first censuses of national populations occurred in only the eighteenth century (in Scandinavian countries) and were delayed until the nineteenth century for most of Europe and well into the twentieth century for Third World countries; national registration systems of births and deaths were even later and more incomplete. However, despite the paucity of firm data, a modern consensus view has been emerging that the traditional representation of world population growth by a single inflection curve of the type shown in Figure 2.2 is misleading. Because the arithmetic scale of such representation has difficulty in coping with a huge range of time and population, the impression given is of an extremely low and constant rate of increase from 'the beginning' until the eighteenth century, followed by an abrupt surge to the present day.

When logarithmic scales are used, thereby extending the space available for depicting variations in distant times, a very different pattern emerges (Figure 2.3). Emphasis is now given to the long periods of population stability interrupted by relatively short spurts of growth, providing a stepped evolution of population size that is entirely concealed in graphs with arithmetic scales. The stepped profile of Figure 2.3 is intended not to be accurate in detail, but to provide a better understanding of the nature of population growth. It puts the growth of the last two

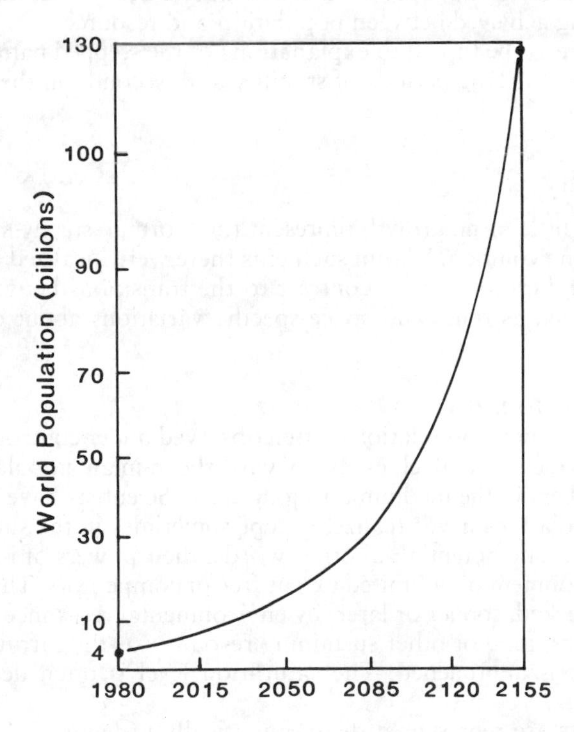

Figure 2.1
Projected growth of the world's human population at a constant growth rate of 2 per cent per year, 1980–2155
Source: Baumol and Blinder (1979), figure 33–2.

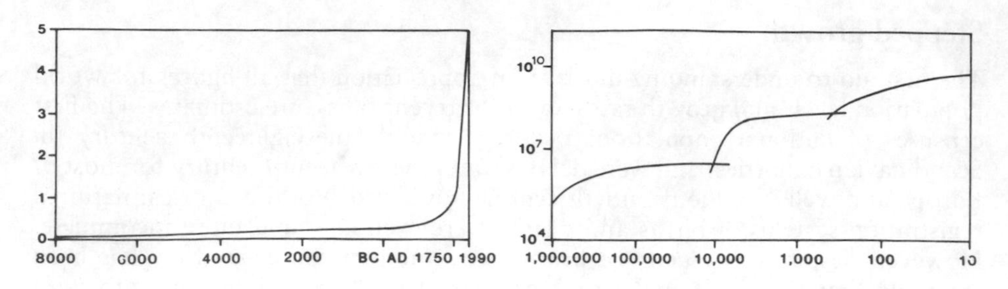

Figure 2.2
Growth of world's human population in billions

Figure 2.3
Use of logarithmic scales to represent the likely growth of the world's human population (vertical axis) in terms of years ago (horizontal axis)
Source: Deevey (1960), p. 198.

centuries into proper perspective as only the third in a sequence of major surges that reflect, chronologically, the tool-making revolution associated with the emergence of *Homo sapiens* and its dispersion over most of the globe, the agricultural or Neolithic Revolution from about 10,000 BC, and the modern industrial–scientific revolution. It also suggests, unlike Figure 2.2, the likelihood of a future equilibrium being achieved between population and resources.

Attention can now be turned to explanations of the stepped pattern of growth, focusing, first, on the long periods of stability and, second, on the revolutionary surges of growth.

Eras of stability

Long periods of little or no growth represent the 'normal' steady-state pattern of world population evolution. Within such eras there were localized fluctuations of varying size and duration but, in contrast to the transitions between eras, these should be regarded as time- and space-specific variations about or towards an equilibrium.

Animal populations

The behaviour of animal populations, often observed under controlled laboratory conditions, provides several clues as to why the human population tends to stability of numbers in the medium and long term. Scientists have long observed that animal populations never realize, except sometimes in the short term, their often enormous biotic potential; in other words, their powers of reproduction in an optimal environment of unlimited extent free of competitors. The biotic potential is always checked, sooner or later, by environmental resistance in the form of shortages of space, food or other sustaining resources as the carrying capacity of the environment is approached. The saturation level is often described as the population ceiling.

These concepts are represented diagrammatically in Figure 2.4, which shows the sigmoid or S-shaped pattern of growth observed among many populations of animals, insects and micro-organisms. Initially, under optimal conditions, the population expands at the exponential rate of its particular biotic potential. At

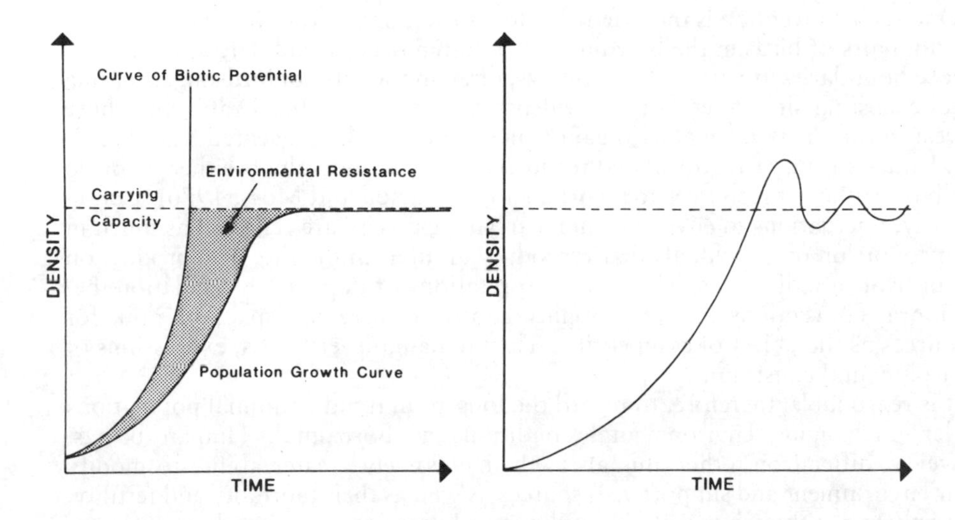

Figure 2.4
Theoretical relationships between biotic potential, environmental resistance and population growth
Source: Boughey (1968), figure 2.7.

Figure 2.5
Population overshoot and adjustment

some time, however, the population begins to fill up its jar, cage, island or whatever, so that under these more crowded conditions the growth rate diminishes and eventually ceases, accomplishing the state described by ecologists as equilibrium, in which births will just balance deaths.

Another, and perhaps more realistic, representation of this form of growth is given in Figure 2.5. Here the population overshoots the carrying capacity, crashes, and then recovers to oscillate around the ceiling. Such regulated populations are maintained in equilibrium by the balance of two opposing forces: the inherent growth potential of the population and the limits to population growth imposed by the environment. Explanation is now required of the mechanisms that sooner or later limit population growth. Such density-dependent mechanisms must obviously embrace increased mortality or reduced fertility or some combination of both.

Increases in mortality at critical levels of increasing density are readily understood, involving starvation, predation and the spread of infectious diseases as hosts become more crowded and undernourished. Density-dependent reduction in fertility among animal populations is more complex. Certainly, under conditions of extreme crowding, direct physiological restraints on fertility can be observed. Female fruit flies, for example, lay fewer eggs when jostled by one another; flour beetles eat their own eggs and pupae; and crowded rodents, through anxiety, suppress spermatogenesis and ovulation (Calhoun, 1963). But, in addition, behavioural conventions have been observed in some animal populations to regulate fertility so that excessive population pressure on limited resources, and therefore the extreme rigours of Darwinian natural selection, are avoided (Wynne-Edwards, 1962, 1965).

One such convention is the defence of territories, or territoriality. For example, mating pairs of birds in the breeding season often occupy and defend a territory whose boundaries are marked by displays, often in the form of birdsong. Securing a good nesting site or territory is evidence of high hierarchical rank, but those lowest in rank may be unable to gain a niche and thus be prevented from breeding. Under strong hierarchical and territorial systems, individuals either do or do not have sufficient resources for reproduction (Watson and Moss, 1970). Consequently, fluctuations in environmental carrying capacity are reflected as much in the proportion of individuals that reproduce at all as in the rate of reproduction or survival of individuals. Thus, in interpretations of population regulation, behavioural conventions like territoriality substitute breeding space or rank for resources as the object of competition. They remain, nevertheless, expressions of environmental constraint.

It is reasonable, therefore, to regard the long-term trend of animal populations under unchanging environmental conditions as horizontal. Human beings, however, differ from other animals in their massively greater ability to modify their environment and supportive resources as well as their mortality and fertility. Nevertheless, although population ceilings and carrying capacities have little relevance to humanity in the last two centuries of rapid technological advancement, it still remains true that when humanity has been restricted to a particular level of material culture for long periods (and this is the norm in human history) population regulation has been very similar to that in other animal populations. This applied particularly in the long Palaeolithic era of hunting and gathering economies and also among agricultural societies between the Neolithic and Industrial Revolutions. There were some modest advances in environmental exploitation, particularly within the agricultural era, but the concept of a ceiling to population growth, inching up certainly at times, remains applicable and appropriate to such societies (Wrigley, 1967).

We now need to consider the long-term stabilizing influences on human, pre-industrial populations, bearing in mind that in the short term there were often appreciable fluctuations in births and especially deaths, and even sometimes in population totals.

Human mortality checks

Traditionally it has been thought that population was regulated and maintained at sub-ceiling levels by density-dependent mortality. Early Malthusian theory is of this type. Malthus, in his first *Essay on the Principle of Population* (1798), argues that the power of the population to increase is greater than that of the earth to produce subsistence. He illustrated the differential by suggesting that population, when unchecked, tends to grow in a geometrical ratio (1, 2, 4, 8, 16, 32), doubling perhaps every 25 years, whereas food supply at best increases arithmetically (1, 2, 3, 4, 5, 6). Therefore, to maintain a balance between population and food supply, population growth is inevitably checked by:

1. 'misery' – the positive checks of famine, disease and war;
2. 'vice' – abortion, sexual perversion and infanticide;
3. 'moral restraint' – the preventive checks of sexual abstinence and late marriage.

Only in later editions of his essay did Malthus think that 'moral restraint' (broadened now to include, at least implicitly, contraception) could be in any way effective. His earlier views, which constitute classical Malthusian theory, posit the mortality-inducing positive checks as the fundamental regulators of population growth. They are exemplified by the so-called demographic crises or catastrophes that historical demographers have identified in pre-industrial Europe.

It is important to appreciate that not all mortality fluctuations in pre-industrial populations were density-dependent. For example, although epidemic disease is often promoted by population crowding and nutritional inadequacy, it is sometimes exogenously determined by density-independent factors unrelated to resource availability. These can be illustrated by the decimation of indigenous populations in the Americas and the Pacific after contact in the colonial era with European diseases to which they had no immunity.

Human fertility checks

Now recognized as a myth is the traditional view of unbridled fertility in pre-industrial populations and its corollary that the necessary regulation of population size was accomplished solely by density-dependent mortality. Rather than a culturally unregulated surrender to sex, hunger and death, there is now thought to have been some form of optimizing effort in reproductive behaviour engaged in by individuals and groups to maintain or enhance well-being (Harris and Ross, 1987).

There is no evidence of any physiological suppression of fertility with crowding among humans as there is among animals observed under laboratory conditions of high density, presumably because no equivalent levels of stress, specifically through crowding, have been reached among humans (J. Freedman, 1980). However, we have now accumulated enough demographic evidence from pre-industrial societies to know that a combination of behavioural and nutritionally-related physiological factors have sufficiently suppressed fertility to have had a major stabilizing effect on world population growth up to the nineteenth century.

Particularly rich and revealing demographic information has been collected from a population of !Kung Bushmen who were still practising a hunting and gathering economy in the 1960s in the Kalahari Desert of southern Africa (Lee and DeVore, 1976; Howell, 1979, 1986). Expectancy of life at birth has been calculated from childbirth histories to be about 33 years. If this had been accompanied by natality levels common in today's less developed countries, there would have been a population explosion. Instead, the average completed family size of women surviving to the end of the child-bearing period was only five, three of whom would typically survive into their reproductive years, a figure just sufficient to produce a stationary or very slowly growing population.

How was this modest family size achieved in a technologically primitive society practising no contraception as we know it today? In a review of evidence from many primitive hunting and gathering societies, R. Short (1976) has shown conclusively that their overall fecundity (the physiological ability to conceive and bear children) was significantly lower than in almost all contemporary societies due to delayed age at menarche (first menstruation), to adolescent infertility (the delay of ovulation for some years after onset of menstruation) and to the lactational amenorrhoea (suppression of menstruation and ovulation during breast-feeding) that is widespread among mammals. Thus, among the !Kung Bushmen, 16 years was the average age at menarche, coincident with marriage; adolescent sterility delayed the birth of the first child until 19; and lactational amenorrhoea kept births four years apart until the relatively early arrival of menopause (cessation of menstruation) at about 40. Lactational amenorrhoea is thought to be of particular importance among short, thin people like the !Kung because the severe calorie drain of prolonged breastfeeding provides insufficient body fat for ovulation to take place (Frisch, 1978). The number of Venuses and other obese female figures found in prehistoric deposits suggests that hunters and gatherers were well aware of the crucial role of body fat in fertility.

It is also thought that there may have been some deliberate child-spacing among societies like the nomadic !Kung in order to accommodate activities necessary for hunting and gathering peoples, and to provide for the nutrition of the very young. Since mothers need to be mobile, not only between camps but also in daily food gathering, they are unable to carry more than a single child on hip or back. Moreover, foods soft enough to permit the early weaning of infants are often unavailable to hunting and gathering peoples, so that infants often have to be breastfed for three to four years. The needs for appreciable birth intervals are clear, but the methods used are not. Possibilities include post-birth sexual abstinence, coitus interruptus and induced abortion, although these seem little practised by the !Kung. The low fertility of such societies certainly contradicts traditional theories that the size of hunter–gatherer populations was limited solely by high mortality rates. !Kung population size remained stable essentially because there were so few children born. When the !Kung became sedentary, their birth intervals drop by a third. There is no longer the need to 'travel light' in terms of material possessions and young children, and infants can be weaned much sooner by giving them grain meal and cows' milk.

Infanticide was also practised widely in hunter–gatherer societies for child spacing (Hausfater and Hrdy, 1984). The regulatory role of infanticide has been described by Firth in his concern about the upsetting of population balance in the Pacific island of Tikopia by the spread of Christianity:

> the so-called sanctity of human life is not an end in itself but the means to an end, to the preservation of society . . . life can be taken to preserve life, so in Tikopia infants just born might be allowed to have their faces turned down, and to be debarred from the world they have merely glimpsed, in order that the economic equilibrium might be preserved, and the society maintain its balanced existence.
>
> (Firth, 1958, p. 376)

The interpretation of past processes in terms of present phenomena must be cautious, but studies of recent tribal populations like the !Kung have provided by inference new perspectives on demographically stabilizing influences in the Palaeolithic era that, in length, has dominated man's occupation of the globe.

For pre-industrial agricultural populations with stable or only slowly increasing population ceilings, a more direct study of demographic evidence is possible, largely from the work of historical demographers on European burial, baptism and marriage records from the sixteenth to the eighteenth centuries. The evidence, discussed in Chapter 5, reveals the adoption of behavioural restrictions on fertility (like delayed marriage), as supplemental nutrition for mother and infant, as well as new work patterns, reduced the normal duration of breastfeeding and lactational amenorrhoea. From their monumental reconstruction of English pre-industrial demography, Wrigley and Schofield (1981, p. 451) conclude: 'An accommodation between population and resources was secured not by sudden, sharp mortality spasms, but by wide, quiet fluctuations in fertility, which in their downward phase reduced fertility levels to the point where population growth ceased.'

A homeostatic regime

Homeostasis may be defined as the tendency for the internal environment of the body to remain constant in spite of varying external conditions. It has been argued convincingly (D. S. Smith, 1977; Corsini, 1977; Lesthaeghe, 1980; Coleman, 1986; Lee, 1987) that homeostatic demographic patterns characterize pre-industrial populations, in that the demographic variables tend to check and compensate one another to maintain an equilibrium. Such a population equilibrium, at or some way below a ceiling level, could be maintained in pre-industrial societies by quite different combinations of fertility and mortality (Figure 2.6). Any

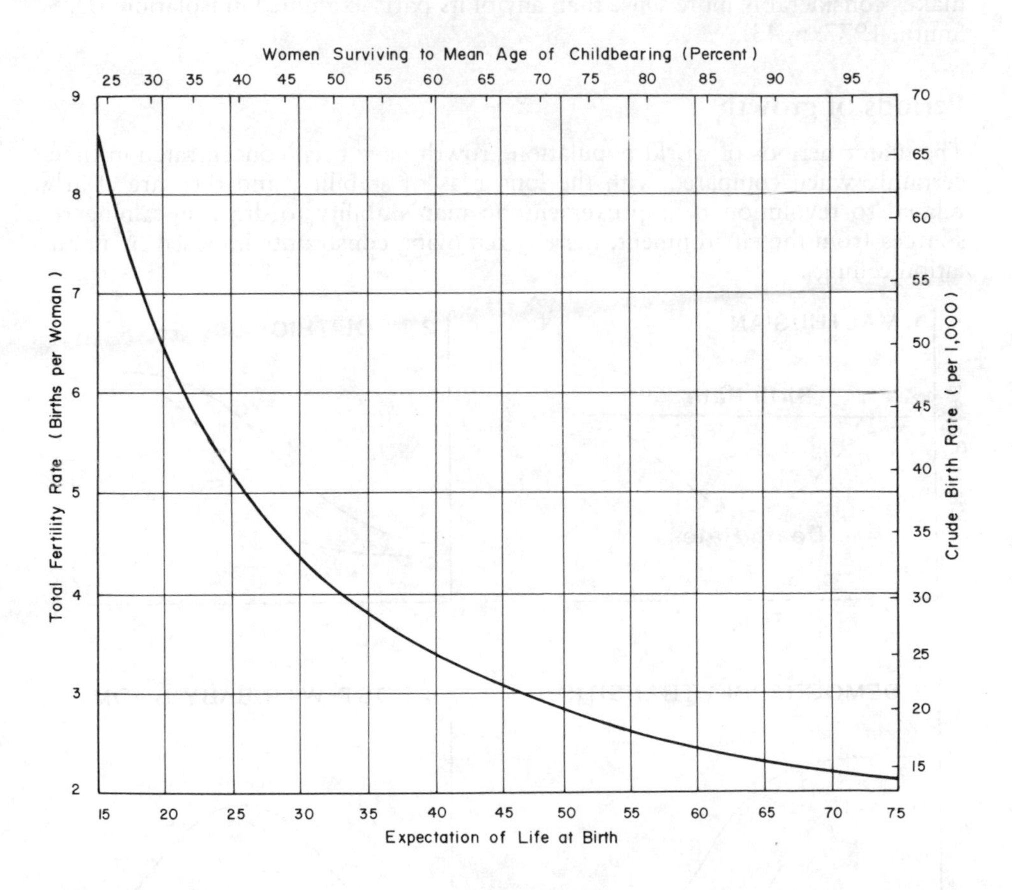

Figure 2.6
The maintenance of a non-migrating population at a constant size by different combinations of fertility and mortality. If, for example, the birth rate was 50 per 1,000, then life expectancy at birth would have been about 20 years, one-third of all women would have survived to mean child-bearing age, and these survivors would have averaged some 6.5 births
Source: Coale (1974), p. 45. Copyright © (1974) by Scientific American, Inc.

disequilibrium between population growth and resources tends to generate a correcting or homeostatic response. Classic Malthusian theory describes such a response in terms of positive checks on population growth, but we have seen that fertility fluctuations can play the same regulatory role. At local and even regional levels, migration can also be expected to have been an important element in population homeostasis, being capable of a quicker, more flexible response than fertility.

If one accepts the homeostatic theory of counterbalancing demographic forces, then the demographic diversity among pre-industrial populations becomes easier to reconcile. Differences in mortality, marital structure and marital fertility did not generate equivalent differences in population increase. The rate of natural increase was clearly more constant than the individual components of population change, so that 'the whole of the population engine of early modern Europe makes considerably more sense than any of its parts examined in isolation' (D. S. Smith, 1977, p. 43).

Periods of growth

The major periods of world population growth have been concentrated in time, certainly when compared with the long eras of stability, and they are clearly related to revolutionary improvements in man's ability to draw sustaining resources from the environment, thereby removing constraints imposed by population ceilings.

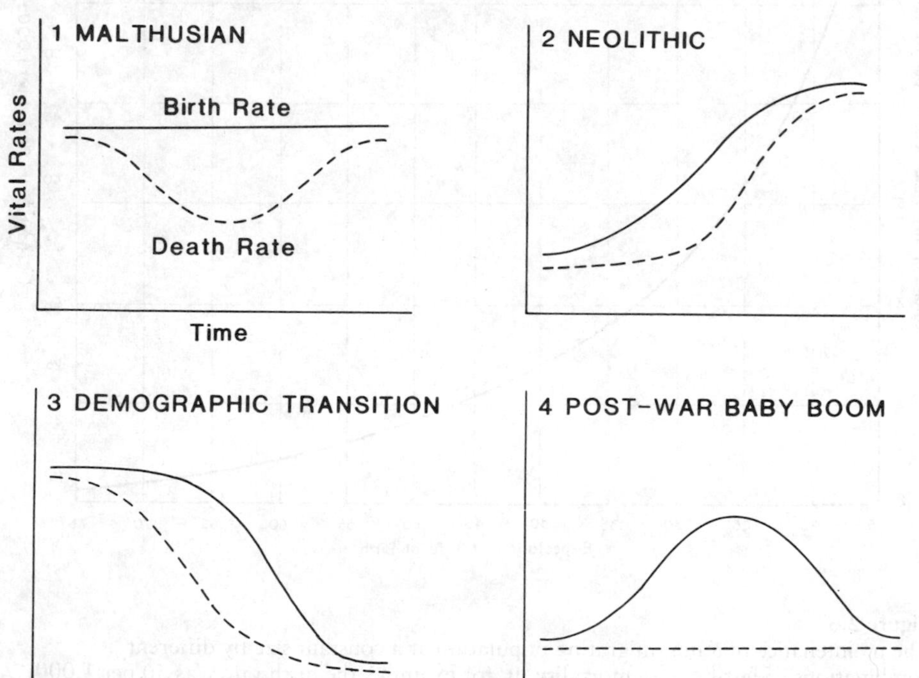

Figure 2.7
Population growth cycles
Source: Cowgill (1949).

There has been a tendency for the populations of individual countries, and indeed the world as a whole, to grow in a series of cycles which conform broadly to sigmoid or extended S-shaped curves (Figures 2.3 and 2.4), of which the logistic curve is one form. Cowgill (1949) has suggested four major ways in which such growth cycles can occur in closed populations (Figure 2.7).

Cycle 1

It has been argued by some, following Malthus, that this cycle characterizes pre-industrial populations. Whenever environmental conditions improve, bringing better harvests, more plentiful fish shoals or fewer epidemics than usual, the death rate would fall and, in association with a relatively constant birth rate, cause a population increase; the birth rate could actually increase somewhat because of the stimulus given by the improved conditions to earlier marriage. But sooner or later, because the improvements were only temporary, the growing population would press inexorably on limited resources so that the death rate would rise again to restore population stability.

Cycle 2

Here, the death rate lags behind the birth rate while both increase. Although Cowgill discounted this theoretical cycle as an improbability in demographic history, it may well have been the form of population growth associated with the emergence of agriculture in Neolithic times around 12,000 years ago (Figure 2.3), although somewhat later in the New World. There is no question that growth was stimulated by the increased carrying capacity of the habitat as man switched from being a predator and secondary and tertiary consumer in food webs to being dominantly a herbivore and primary consumer. But the traditional view that population growth was accomplished largely by diminished mortality is now discredited. Concentration of population in villages and later in cities could be expected to promote transmission of diseases (Coale, 1974; Cohen and Armelagos, 1984), and there was always vulnerability to crop failure. There may well, therefore, have been increases in mortality while almost certainly there were increases in fertility associated with reduced duration of breastfeeding and the labour needs of families in primitive agricultural societies (Roth, 1985).

One should also note that the causative relationship between the Neolithic Revolution and population growth is a matter of some controversy. Traditionally it has been assumed (e.g. Childe, 1936) that the adoption of an agricultural economy led to subsequent population growth, but some (notably Cohen, 1977) argue that population growth came first, and was the cause, not the result, of the Neolithic. More generally, C. Clark (1967) and Boserup (1981) argue that population growth is the stimulating cause of agricultural change, following 'a long-established if minor tradition in economic writing which argues that not only does population growth cause economic improvement but that without such a spur human society would remain culturally and economically stagnant' (Grigg, 1979, p. 64).

Cycle 3

This embraces a lag of the birth rate behind the death rate while both are declining. It is the form of growth that has characterized the most modern surge in the world's population (Figures 2.2 and 2.3). In developed countries the rapid economic development and associated social modernization from the eighteenth century impacted initially on mortality and then, about a century later, on fertility.

What distinguishes this cycle from Cycle 1, where a fall in mortality was also the initiator of growth, is that 'modern man is able to foresee demographic catastrophe long before it arrives and takes adaptive action long before it is forced on him by the brute forces of nature' (Bogue, 1969, p. 53). Emphasis is thus placed on the human's unique position as a highly flexible and adaptable creature, capable, in particular, of curtailing fertility when mortality reduction (a universally valued goal) causes population to grow faster than the environment's ability to sustain it.

In such circumstances a society has to regain control of its growth, so that the regulation of fertility becomes identified with group, as well as individual, welfare. It is not a straightforward automatic response, but embraces the entire process of social change whereby a solution in terms of behavioural norms must be devised, diffused and widely adopted. Consequently there may well be appreciable population growth during the adjustment because of cultural inertia.

The broad conformity of developed countries to this cycle of growth in the last 200 years has formed the basis of the demographic transition model.

Cycle 4

The most obvious case of population growth through a short-term increase in the birth rate while the death rate remains stable is the 1945–65 period in developed countries. Then, a baby-boom was caused by a trend towards younger marriage and by the coincidence of economic prosperity with low participation of married females in the labour force.

The demographic transition

Demographic transition theory provides a general description, although not a complete explanation, of the changing rates of mortality and fertility in the more developed countries since the eighteenth century. It has been aptly described by Demeny (1972, p. 153) as 'the central pre-occupation of modern demography', not simply for its intrinsic intellectual interest, but also because it has brought about the global population explosion which threatens the preservation of social systems. The theory posits a particular pattern of demographic change as accompanying a nation's progression from a largely rural, agrarian and illiterate society to a dominantly urban, industrial and literate one. During the course of this progression, which is regarded to all intents and purposes as irreversible, there are major reductions in both fertility and mortality, so much so that in the past two centuries average life expectancy in developed countries has doubled and fertility halved.

If fertility and mortality declines had been concurrent, population growth in developed countries would have been fairly modest, as was uniquely the case in France where there had been exceptionally early fertility decline in a dominantly agrarian society. But elsewhere in the developed world mortality decline preceded fertility decline by about a century, leading to the replacement of a demographic steady state by the population explosion of nineteenth-century Europe and North America. In some cases there may have been a slight increase of fertility at the beginning of industrialization, further accentuating population growth in the transition.

Figure 2.8 provides a stylized representation of the standard demographic transition. Note how in pre-industrial societies the birth rate is depicted as relatively

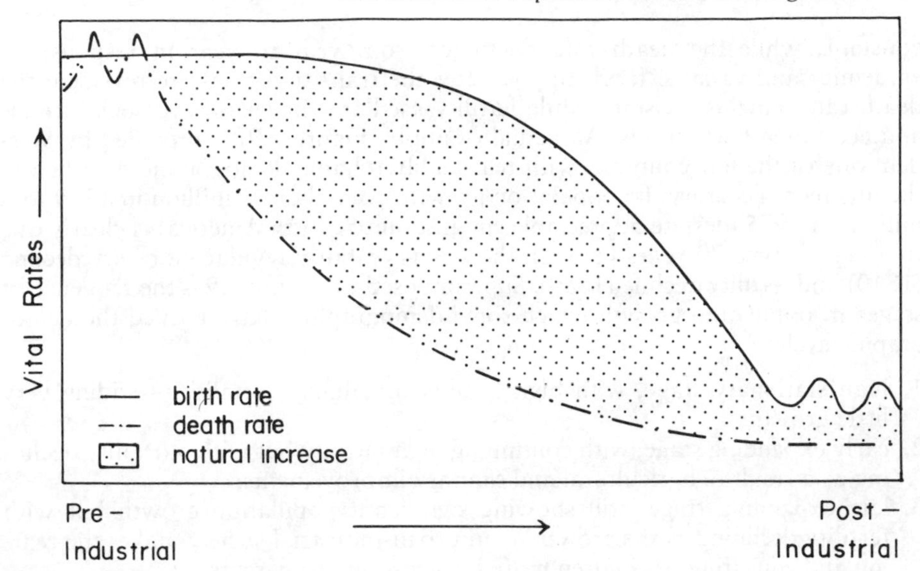

Figure 2.8
The standard demographic transition

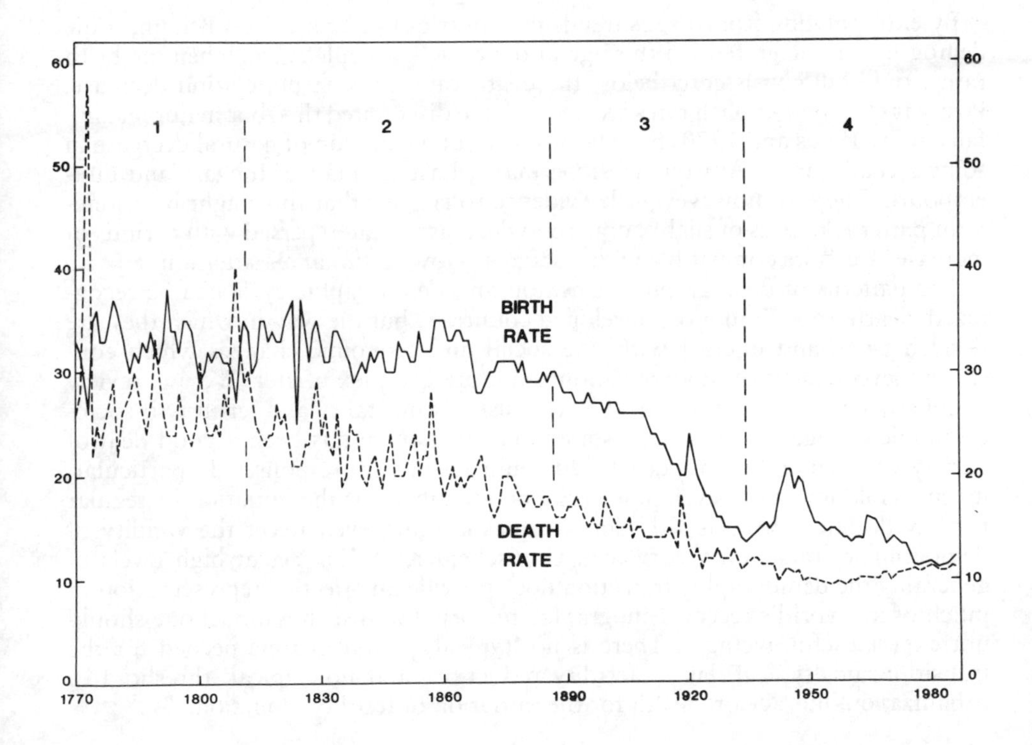

Figure 2.9
Crude birth rates and death rates (per 1,000) in Sweden, 1770–1986, and stages of the demographic cycle
Source: Data from D. S. Thomas (1941), Table 3, and UN *Demographic Year Books*.

constant, while the death rate fluctuates from year to year in response to epidemics and variable food supply. After the transition roles are reversed: the death rate remains constant while fertility oscillates in relation to social trends and economic fluctuations. An actual example (Figure 2.9) is provided by Sweden, one of the few countries with reasonably reliable demographic data before the nineteenth century. Its population growth from about 2 million in 1770 to 8 million in 1975 (despite appreciable emigration to North America) is clearly due to a lag of some 70 years between the onsets of fairly regular mortality decline (1810) and fertility decline (1880). Superimposed on Figure 2.9 is the sequence of stages in population growth which some demographers have termed the demographic cycle:

1. high stationary stage, with high fertility and high mortality providing very little growth;
2. early expanding stage, with continuing high fertility but with mortality declining as a result of agricultural and sanitary improvements;
3. late expanding stage, still showing significant population growth but with fertility declining as the growth of an urban-industrial society makes the rearing and educating of children more burdensome for parents;
4. low stationary stage, with fertility and mortality levelling out to re-establish a fairly stationary population, thus completing the cycle.

By extrapolating from 1930s trends in countries like France and Britain, some demographers suggested a fifth stage in the cycle – a senile stage, when the birth rate would fall consistently below the death rate, causing population decrease. Postwar elevation of birth rates seemed to have discounted this, but major fertility falls in the 1960s and 1970s have brought about a situation of natural decrease in some recent years in Austria, West Germany, East Germany, Hungary and Luxembourg. There is, however, little evidence to suggest that this might be a long-term pattern. Periods of slight population decrease, if interspersed with periods of increase, are quite compatible with a stage 4 – low stationary – situation.

The patterns of demographic transition and demographic cycle can be recognized clearly enough in most developed countries, but the way in which they are derived from, and interact with, the social and economic changes which constitute development or modernization is a more complex matter. We will review in subsequent chapters the separate responses of mortality and fertility to socio-economic change. But it can be stated here that there has been a good deal of variety between and within developed countries in such responses. In particular, no threshold levels of development can be identified for the initiation of regular fertility decline. This has led some to question and even reject the validity of demographic transition theory (e.g. Goldscheider, 1971). Yet at high levels of generality the demographic transition does provide an effective representation of much of the world's recent demographic history. From such a model one should not expect useful averages. There is no 'typical' period of time needed for the transition, no 'typical' lag in fertility reduction, and no 'typical' threshold of urbanization, literacy or wealth for the initiation of fertility reduction.

Modes of production

The most fundamental explanation of the demographic transition stems from the statement of Marx (1976 edn, p. 784) that 'every special historic mode of production has its own special laws of population'. Thus Caldwell (1978), Woods

(1982a) and Seccombe (1983) conceive of the demographic transition as a transition between demographic regimes associated with specific modes of production and their social formations. They attempt to relate European demographic change to the complex, prolonged transition from feudalism to industrial capitalism, and not simply to the rather vague notion of economic development and social modernization embodied in traditional demographic transition theory.

During the early phases of the capitalist mode of production a restructuring of agriculture leads to a greater food supply and therefore declining mortality. This is enhanced by public health developments stimulated by the needs of capital for labour reproduction. The reduction in fertility occurs at a later stage when the bourgeoisie and associated classes, recognizing that the family is no longer the basic unit of production, lead the way in investing through education in the future human capital of their offspring. This concern for quality, rather than quantity, of children subsequently spreads to the working class.

Less developed countries

For some time it has been fashionable to deny the applicability of demographic transition theory to the Third World. It is emphasized that imported medical and public health technology has drastically cut mortality but has not often been accompanied by the modernization changes necessary to reduce fertility from the high levels permitted by a pattern of early and near-universal marriage never experienced in pre-industrial Europe. It is shown that the resultant population growth rates of 3.5 to 2.0 per cent per annum, with populations doubling every 20–35 years, far exceeded growth in any of Europe's nineteenth-century transitions.

Yet there are clear relationships in the Third World between levels of development or modernization and levels of mortality and fertility, as Chapters 4 and 6 will demonstrate. The more developed countries have, on the whole, experienced the greatest mortality declines and the beginnings of fertility reduction. It has been observed that in Latin America 'as in Europe, the broadest implications of transition theory do materialize: birth rates are falling in the countries that are relatively developed and have low mortality . . . also as in Europe, natality tends to fluctuate in the predecline period' (S. E. Beaver, 1975, p. 146).

But although the early stages of demographic transition may be observed in the Third World, there is no assurance that later stages will replicate European experience and achieve, through fertility regulation, environmentally sustainable population levels. The other regulation method widely used in Europe – mass emigration to the New World – is unrealistic in today's socio-political climate.

Proponents of demographic regulation as a universally applicable theory suggest optimistically that the necessary rapid fertility regulation can be achieved in the Third World. They point out that latecomers to the transition process, notably Germany, southern and eastern Europe and Japan, all experienced a concentrated and accelerated fertility decline; that, contrary to some views, the pace of modernization in many Third World countries is actually more rapid than in nineteenth-century Europe; and that governments are now prepared and have the capability to promote policies of fertility regulation. There are others who maintain that high fertility is so institutionally interwoven with the entire cultural fabric of less developed countries that its reduction is an intractable problem. The demographic history of the next generation will therefore be of crucial importance and, to scholars, great fascination.

3

MORTALITY: INTERNATIONAL VARIATIONS

Both fertility and mortality rates are high in traditional, less developed societies and both are low in modernized, economically advanced societies. Population growth through natural increase is consequently low at both ends of the conventional development/modernization spectrum, but in between it is considerable as mortality invariably falls in advance of fertility through its earlier response to modern economic and social influences. Thus, although in individual life-cycles birth obviously occurs before death, it is mortality that merits initial consideration in any systematic treatment of demographic components and processes.

Measures of mortality

To study spatial and temporal variations in mortality, it is necessary to use comparative measures or indices of mortality for the periods and areas under consideration. The basic measures are essentially rates of incidence, relating the number of deaths to a unit of population, commonly 1,000, in a particular interval of time. The interval is invariably one year, to avoid the complicating effect of seasonal mortality variations on comparability of rates.

Crude death rate

This is the simplest measure, indicating the number of deaths in a particular year per 1,000 of the population:

$$\frac{D}{P} \times 1{,}000$$

The mid-year population is used, ideally, as the best available approximation to the average number of people exposed to death, or 'at risk', during the year. The term 'crude' is used advisedly, since no allowance is made for often considerable differences between communities in those aspects of population composition, particularly age and sex, which strongly influence mortality probabilities. It will be shown that mortality rates increase with age from late childhood, and that at nearly all ages mortality is higher for males than females. Consequently, age and sex structure plays an important role in influencing crude death rates, regardless of factors like living standards and health programmes. This is why care has to be taken in interpreting international differences in crude death rates. There are some less developed countries (e.g. Costa Rica, Guyana, Jamaica, Malaysia, Sri Lanka) which have lower crude death rates (about 5–7 per 1,000 in the late 1980s) than the economically and medically advanced countries of North America and north-western Europe (about 9–12 per 1,000). Clearly this is due to much younger age structures in less developed countries, resulting from their high fertility in the immediate past.

Age- and sex-specific death rates

A more critical appreciation of mortality level can be derived from a table of mortality rates disaggregated by age and sex. An age- and sex-specific death rate is the number of deaths during a year of persons of a given age and sex per 1,000 people of that age and sex category:

$$\frac{D_{as}}{P_{as}} \times 1{,}000$$

It is, in essence, a crude death rate for a particular age and sex group. It is conventional to use five-year or ten-year age groups, as a compromise between unnecessarily detailed and tediously compiled one-year groups and excessively generalized and often internally heterogeneous larger age groups. However, because of the importance of infant deaths, separate rates are usually calculated for the under one and 1–4 age groups.

Examples of age- and sex-specific death rates are given in Table 3.1 for a developed and less developed country. Although levels of mortality differ between England and Wales and Mauritius, the overall pattern of mortality by age and sex is similar. Mortality rates fall from relatively high levels in the first year of life to a minimum in the 5–14 age group and then rise in successive age groups, although the mortality level in the first year is not exceeded until old age. The lower mortality of females is found in all the England and Wales age groups and almost all in Mauritius. Traditional explanations include the greater exposure of men, in

Table 3.1 Age-specific death rates (per 1,000), England and Wales (1987) and Mauritius (1986)

	England and Wales		Mauritius	
	Male	Female	Male	Female
Under 1	10.4	7.9	30.8	23.1
1–4	0.5	0.4	1.2	1.3
5–14	0.2	0.1	0.5	0.4
15–24	0.7	0.3	1.1	1.0
25–34	0.9	0.5	2.1	1.4
35–44	1.7	1.1	4.5	2.3
45–54	5.0	3.2	11.7	5.0
55–64	15.7	9.0	27.8	13.7
65–74	41.7	22.8	57.3	39.6
75–84	98.8	60.7	123.0	83.5
85+	212.7	168.6	249.5	192.5

Source: *Population Trends*, Vol. 53, Table 19; *Demographic Yearbook* 1987, Central Statistical Office, Mauritius, Table 5.6.

Table 3.2 Calculation of standardized death rate, Mauritius and England and Wales

	Standard population Japan 1985 ('000)		Mauritius: Age-specific death rates 1986 (per 1,000)		Mauritius: Expected deaths in standard population		England and Wales: Age-specific death rates 1987 (per 1,000)		England and Wales: Expected deaths in standard population	
	M	F	M	F	M	F	M	F	M	F
Under 1	732	698	30.8	23.1	22,546	16,124	10.4	7.9	7,613	5,514
1–4	3,087	2,942	1.2	1.3	3,704	3,825	0.5	0.4	1,544	1,177
5–14	9,520	9,054	0.5	0.4	4,760	3,622	0.2	0.1	1,904	905
15–24	8,766	8,414	1.1	1.0	9,643	8,414	0.7	0.3	6,136	2,524
25–34	8,507	8,371	2.1	1.4	17,865	11,719	0.9	0.5	7,656	4,186
35–44	9,950	9,923	4.5	2.3	44,775	22,823	1.7	1.1	16,915	10,915
45–54	8,019	8,151	11.7	5.0	93,822	40,755	5.0	3.2	40,095	26,083
55–64	5,789	6,616	27.8	13.7	160,934	90,639	15.7	9.0	90,887	59,544
65–74	3,285	4,472	57.3	39.6	188,231	177,091	41.7	22.8	136,985	101,962
75–84	1,560	2,367	123.0	83.5	191,880	197,645	98.8	60.7	154,128	143,677
85+	256	529	249.5	192.5	63,872	101,833	212.7	168.6	54,451	89,189
	121,008				802,032	674,490			518,314	445,676

Standardized death rate:

Mauritius

$$\frac{(802{,}032 + 674{,}490)}{121{,}008{,}000} \times 1{,}000 = 12.20 \text{ per } 1{,}000$$

England and Wales

$$\frac{(518{,}314 + 445{,}676)}{121{,}008{,}000} \times 1{,}000 = 7.97 \text{ per } 1{,}000$$

general, to warfare, occupational and recreational hazards and to occupational stress, but the excess male mortality at infant and even foetal stages suggests that a biological difference is the decisive factor – the gentler sex being the stronger sex, at least for survival (Hart, 1988).

Sets of age- and sex-specific death rates represent refined measures of mortality, but they are cumbersome and difficult to assimilate. It is desirable, therefore, to have a summary measure which epitomizes a whole set of rates but which avoids the crudity of the crude death rate. Demographers have used their technical skills to produce several such measures and two of the most widely used will now be discussed to illustrate the procedure known as standardization. By this procedure death rates are adjusted or standardized for the age and sex composition of the population. Standardization can be regarded as a means of holding age and sex constant, in a not dissimilar way to the holding of altitude constant in the calculation of temperature or pressure 'reduced to sea level'.

Standardized death rate

This is a hypothetical figure which indicates what the crude death rate would be if the population being studied had the same age and sex composition as a population (any population, real or assumed) which is used as the standard. It is calculated (in the so-called direct method of standardization) by applying the age- and sex-specific death rates for each population under consideration to the age and sex composition of the standard population:

$$\frac{\Sigma(P_{as} D_{as})}{P} \times 1{,}000$$

P = standard population
P_{as} = number in standard population of age and sex category *as*
D_{as} = specific annual death rate of age and sex category *as* in the population under study

It is a weighted average of the age- and sex-specific death rates in a population, using as weights the age and sex distribution of the standard population.

Table 3.3 *Extract from life-table of England and Wales, males, 1980–2*

Age interval	No. living at beginning of interval	No. dying during interval	Average remaining lifetime in years at beginning of interval
x	l*x*	d*x*	e*x*
0–1	100,000	1,271	71.04
1–2	98,729	84	70.96
2–3	98,645	51	70.02
3–4	98,594	37	69.05
–			
–			
–			
109–110	1	1	0.94

Source: OPCS (1987) *English Life Tables* 14, HMSO, p. 8, Crown Copyright.

Worked examples for England and Wales and Mauritius are shown in Table 3.2, using the population of another country, Japan (an arbitrary choice), as the standard. The standardized death rates, calculated on this basis, confirm a theoretically reasonable expectation that mortality is somewhat lower in England and Wales than in Mauritius, despite crude death rates for the years in question of 11.3 per 1,000 in England and Wales and a mere 6.7 in Mauritius. It should be emphasized that standardized death rates have little intrinsic meaning; they are meaningful only in relation to each other and to the selected standard population.

Expectation of life at birth

This is the average number of years that would be lived by a group of persons born in the same year, assuming (unrealistically from the point of view of forecasting) that the age-specific death rates of that year would be maintained throughout the life history of the cohort. It is a measure of mortality in a particular year, and that year only, and is normally derived from a national life-table. An extract from an England and Wales life-table is shown in Table 3.3. It adopts the standard practice of setting up a hypothetical birth cohort of 100,000 persons and then plotting its diminution over time on the basis of age-specific mortality rates applying in the year of calculation, an average of 1980–2 in this particular example. In essence, it provides a cross-section or snapshot of mortality at one particular time.

The first figure in the final column (e_x) gives the expectation of life at birth, calculated in the following manner. Consider from column d_x that 1,271 persons die in the first year of life. They can be regarded as having lived, on average, about half a year, so that they contribute 1,271 × 0.5 years to the total years lived by the original 100,000 birth cohort. The 84 dying in the second year live, on average, 84 × 1.5 years. Thus the average length of life for the birth cohort can be regarded as:

$$\frac{(1{,}271 \times \tfrac{1}{2}) + (84 \times 1\tfrac{1}{2}) + \ldots + (1 \times 109\tfrac{1}{2})}{100{,}000} = 71.04$$

The formula, using the notations of Table 3.3, is:

$$e_0 = \frac{\sum_{x=0}^{x=109} d_x\,(x + \tfrac{1}{2})}{100{,}000}$$

As refined measures of mortality, expectation of life at birth and the standardized death rate should be expected to give broadly consistent results in comparative analyses. In the case of England and Wales and Mauritius in 1986–7, standardized death rate calculations (Table 3.2) have given figures of 7.97 and 12.20 per 1,000 respectively, which are consistent with the life expectancy at birth figures for those countries at that time, of 76 and 68 years respectively.

Infant mortality rate

This is a widely used indicator of health conditions and general living standards which recognizes the concentration of deaths in the first year of life (Tables 3.1–3.3). The rate is conventionally defined as the number of deaths of children under one year of age in a particular year per 1,000 live births in that year:

$$\frac{D_0}{B} \times 1{,}000$$

It is almost, but not quite, the same as the specific death rate in the under one-year-old age group. In the late 1980s its values ranged from 5–10 per 1,000 in developed countries to 140–170 per 1,000 in the poorest Third World countries.

Data quality

A word of warning is necessary at this stage, since there is a danger that pre-occupation with sophisticated statistical methods might blind one's eyes to the often unreliable primary data on deaths and ages. Clearly, any statistical analysis is only as sound as the quality of the data it uses.

The basic data on deaths are normally obtained from a country's vital statistics registration system. The legal requirement for registration of births, deaths and marriages is far from universal, and when it is absent registration will inevitably be incomplete. Many less developed countries collect death statistics by registration in only a part of the national territory, because of problems like nomadism, civil disorder and inadequate administrative systems. In India, for example, the Sample Registration System covers 6 million people spread over 6,000 sample areas in a country of over 800 million population. Enumerators make monthly rounds of households to record births and deaths, but informed judgements (Visaria and Visaria, 1981) suggest that both births and deaths are under-reported by 5–10 per cent. Another source of under-registration, common in many less developed countries, is that babies dying before the end of the legal registration period for births are not registered as births or deaths. Such is the extent of under-registration that a United Nations analysis in 1951–5 suggested that only about 33 per cent of the world's deaths were then being registered, the proportion varying regionally from some 7 per cent in East Asia to almost 100 per cent in Europe and North America (Shryock and Siegel, 1976).

In many of the poorest countries even sample registration is absent, and here demographers have used their technical ingenuity, particularly involving stationary and stable population theory (Woods, 1982a, ch. 2; Coale and Demeny, 1983), to estimate vital statistics rates. Such estimates do, however, require reliable information on age structure, generally from national censuses, for at least two points in time. Infant and child mortality rates are often estimated retrospectively from census and survey questions to women on the number of children ever born and their subsequent survival history.

Even in some developed countries, the establishment of comprehensive death recording has been achieved only fairly recently. In the United States the registration area at the beginning of this century embraced only 40 per cent of the country's population. Complete territorial coverage was not achieved until 1933, and even then a completeness of registration of only 90 per cent of deaths was the condition for joining the national death registration area. A useful, but by no means infallible, indication of the completeness of current national mortality data is given in the UN *Demographic Yearbooks*.

Most mortality rates require for their computation the appropriate age and sex distribution of the population from a census or sample survey. Errors in the recording of age are thought to be widespread and may be due to a host of causes: ignorance of correct age; the reckoning of age in parts of Asia by the so-called Chinese system and not by completed solar years since birth; deliberate misrepresentation for personal and sometimes practical reasons; and a tendency to give age in figures ending in certain digits, particularly 0, 5 and even numbers.

Demographers have evolved methods of adjustment and smoothing to minimize such data deficiencies (Brass *et al.*, 1968) but it is well to be aware that for much of the world the data on mortality still rest essentially on estimates of varying authority. The accuracy suggested by the common use of decimal points in much reported data is spurious.

Mortality and development

The modern period of world mortality decline and consequent population growth, dating essentially from the eighteenth century, was initiated and consolidated in the economically advanced areas of Europe and North America before spreading to less developed parts of the world, reaching the bulk of the world's population in Asia, Africa and Latin America only in the last half century. However, the functional relationship between economic development and mortality is far from straightforward.

In pre-industrial Europe, when disease was essentially density-dependent, mortality was actually highest in the wealthiest, most urbanized areas (de Vries, 1984) and often lowest in remote, fairly poor and sparsely peopled rural areas (Flinn, 1981) away from the trade routes, migrants and urban crowding that spread disease. For much of the nineteenth century, too, the high mortality rates of congested and insanitary cities and industrial areas retarded the establishment of the inverse relationship between development and mortality that seems so plausible. Only in this century has that relationship become well founded, being identifiable at local, regional and national scales (Mosk and Johansson, 1986).

Aggregate data analysis

To test the current relationship between national levels of mortality and development, Table 3.4 presents a correlation matrix, based on data from 99 countries in the mid-1980s, of the relationships between three measures of mortality and two measures of development.

The first row of Table 3.4 shows a clear relationship between national levels of GNP per capita and mortality. Predictably the relationship is stronger for the age-standardized measures than for the crude death rate, bearing in mind the young age structure of less developed countries.

But a scattergraph (Figure 3.1a) of GNP and expectation of life at birth indicates that the relationship is very far from a linear one. Life expectancy increases

*Table 3.4 Coefficients of linear correlation between indicators of development and of mortality, 99 countries,** c. *1985*

	CDR	LEB	IMR
GNP per capita	–0.33	0.69	–0.67
log GNP per capita	–0.59	0.87	–0.85
SEC EDUC	–0.65	0.89	–0.88
CDR		–0.83	0.80
LEB			–0.97

CDR = Crude death rate; LEB = Life expectancy at birth; IMR = Infant mortality rate; SEC EDUC = % of 12–17 age group in secondary school.

* Excluded are countries of less than 1 million population, countries with no data on at least one variable, and five high-income, oil-exporting countries.

Source: Data from Population Reference Bureau, *World Population Data Sheet* 1988 and World Bank, *World Development Report* 1986.

rapidly at first as GNP rises, but tends to level off in the richer group of countries at just a few years above the biblical allotment of three score years and ten. Thus at higher levels of economic development substantial increases in GNP per capita bring little or no reduction in mortality level. It is now widely recognized that such a pattern of strongly diminishing returns is common in the development process, in that increasing wealth and investment add progressively less economic and social benefit.

In order to transform the curvilinear graphical pattern into a linear one, log GNP values have been used in the second row of Table 3.4 and in a second scattergraph (Figure 3.1b). The linear correlation coefficient values increase accordingly, although the hint of curvilinearity still found in Figure 3.1b at higher GNP levels indicates that the relationship between GNP and mortality is a very tenuous one at high levels of economic advancement. It is almost certainly influenced by the variable *distribution* of income and other benefits within national societies. In the United States, for example, mortality reduction to the very lowest levels predicted on the basis of GNP may well be prevented by the heterogeneity of its population, a sizeable segment belonging to groups which are still severely disadvantaged. An obvious contrast is provided by Scandinavian countries.

The relationship between mortality and education revealed by the third row of Table 3.4 and by Figure 3.1c is much more straightforward, being essentially linear and very impressive statistically.

Spatial pattern

Figure 3.2 depicts the international pattern of survival based on actual or estimated 1986 age-specific death rates. A zonal pattern, dependent on development differences, is the dominant feature, with the two ends of the survival spectrum being the temperate-latitude developed countries, on the one hand, and the hard-core disadvantaged countries of inter-tropical Africa and parts of southern Asia on the other.

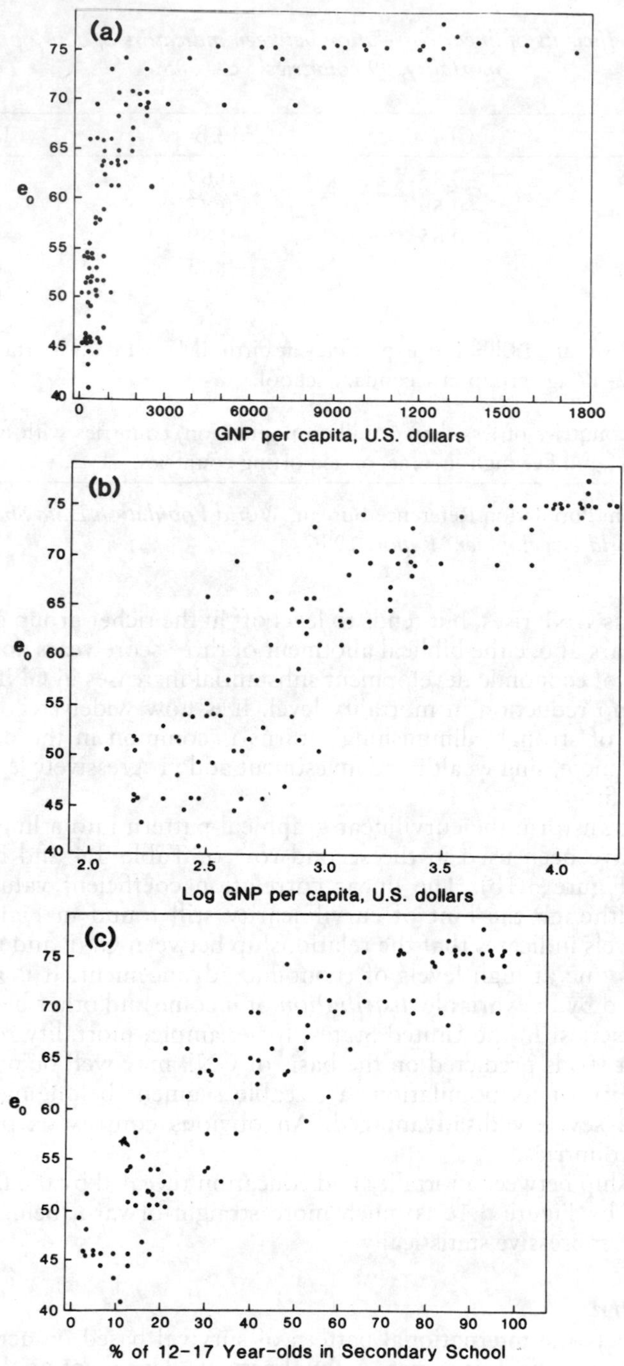

Figure 3.1
Expectation of life at birth (in years) by (a) gross national product per capita, (b) log GNP per capita, (c) percentage of 12–17 year-olds in secondary school, 99 countries, *c.* 1985
Source: See Table 3.4.

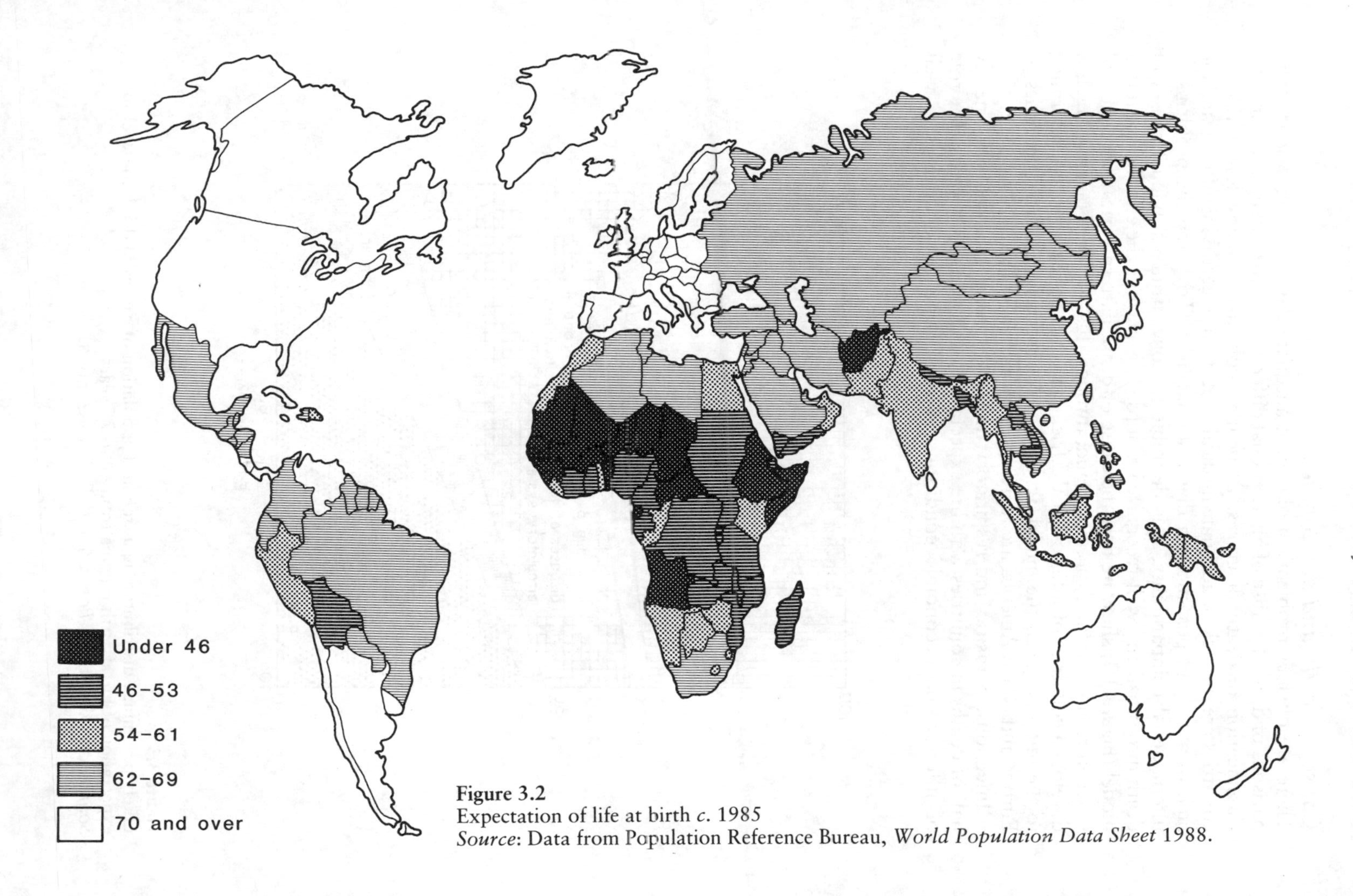

Figure 3.2
Expectation of life at birth *c.* 1985
Source: Data from Population Reference Bureau, *World Population Data Sheet* 1988.

Cause of death, and development

There are obvious problems in obtaining adequate data for international comparisons of death, because of international differences in the extent, accuracy and classification of death diagnosis. Particular problems in many less developed countries are the lack of adequate medical education of certifying officers and the lack of ante- and post-mortem diagnostic facilities. To lessen the problems of international comparability, it is desirable to concentrate individual causes of death into broad and well-recognized groups. This permits the establishment of a model (Figure 3.3) showing how the cause-of-death pattern varies over time and space in relation to life expectancy (and, through this, to development). The less developed countries of inter-tropical Africa may thus be regarded as conforming to the initial stage of the model, while the economically advanced countries of Europe and North America are represented by its final stages.

Infectious, parasitic and respiratory diseases (Group 1) account for nearly half of deaths in countries with life expectancies as low as forty years. The role of inferior socio-economic conditions in promoting these diseases is well

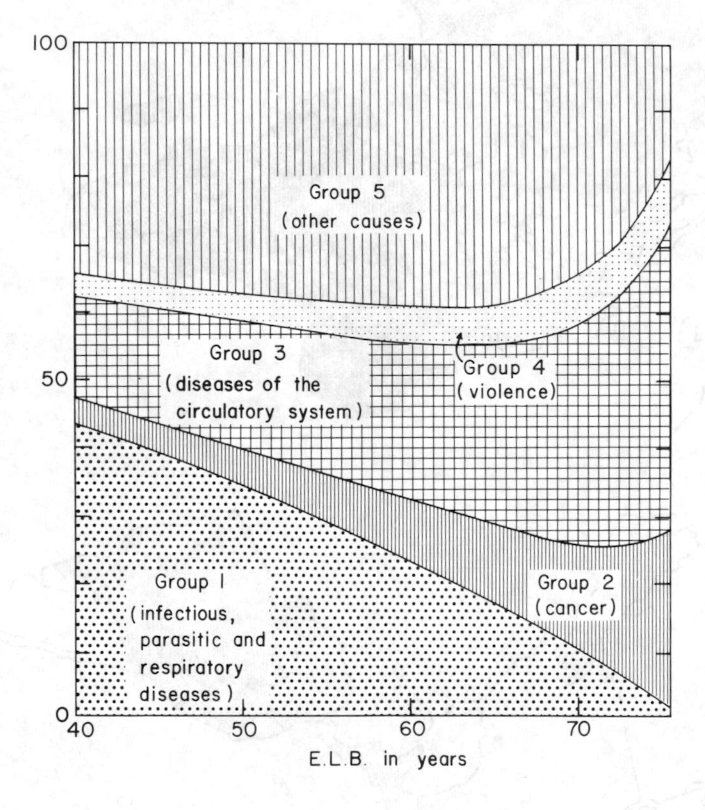

Figure 3.3
Likely percentage distribution in a standard population of causes of death for levels of expectation of life at birth ranging from 40 to 76 years
Source: United Nations, *Population Bulletin* 1962, no. 6.

known, but it should be noted that the great majority of less developed countries with very low life expectancies are currently in low latitudes, where a warm humid climate is ideal for the propagation of infective mircro-organisms or germs and their transmission by animal vectors like flies, mosquitoes and snails. Tropical and sub-tropical areas are thus affected by infectious diseases like yellow fever, malaria and bilharzia which did not occur or were much less serious in the early development stages of Western nations. Group 1 diseases are now of trivial significance in developed countries, except for pneumonia and influenza among the elderly.

Group 2 and 3 diseases, largely cancers, heart disease and stroke, are essentially the chronic or deteriorative diseases associated with older adulthood. They have thus become the dominant causes of death in developed countries, largely as a function of relatively old age structures. But even when age structure is held constant (as in Figure 3.3), these diseases still dominate the cause-of-death pattern. It is not so much that the age-specific rates of these diseases have increased with development or modernization. Some may well have at times, but the overall evidence from several developed countries is that these rates have remained fairly stable in recent decades. The essential point is that the incidence of these diseases has not declined significantly with development, setting them apart from other causes of death. There are, however, some interesting distinctions emerging among developed countries. In particular, there have been major declines (more than 25 per cent) in heart disease mortality among males 30–69 between the mid-1970s and the mid-1980s in the United States, Canada, Australia and Japan, whereas in northern and western Europe, including the countries with the very highest rates (UK and Finland), the declines have been much more modest (World Health Organization, 1987).

Deaths due to violence (Group 4) make a fairly constant proportional contribution to the mortality pattern at different development levels. Decline is prevented by the importance of motor vehicle accidents. It is difficult to comment meaningfully on the residual 'other causes of death' (Group 5) because of its diversity. But diarrhoeal diseases are prominent and these 'can neither be prevented nor cured by injections or other direct remedies readily dispensed through public health programmes. Even purification of water supplies has had limited benefits. Diarrhoeal disease is prompted by poverty and ignorance' (Preston and Nelson, 1974, p. 43). This partly explains the retarded reduction of Group 5 mortality in Figure 3.3 as well as the retention of high infant mortality in most Third World countries.

Sex differentials and development

Almost universally, mortality rates are lower for females at nearly every age group, thereby establishing the well-known demographic feature that life expectancies at birth are substantially higher for females than males. Sex ratios are directly affected and, in turn, marriage possibilities, expected length of widowhood and a host of other demographic and social variables. But the extent of female superiority in survival varies by development level, being comparatively modest under conditions of high mortality in less developed countries and appreciable in economically advanced societies. The 1986 *Demographic Year Book* indicates that the great majority of less developed countries record a female superiority in life expectancy at birth of 2–4 years, with a very small minority (India, Pakistan, Bangladesh, Nepal, Bhutan, Maldives and Iran) showing a very slight male superiority; but in the developed countries of Europe and North America the female superiority amounts to 6–7 years.

Given the appreciable evidence for the innate biological basis of the sex differential in mortality, it remains to consider what factors reduce its impact in less developed societies and enhance it in the developed world. Significant factors in less developed societies appear to be the subordinate position of women, the relative aversion to and neglect of female children, and the maternal mortality rate under conditions of poor medical care. The role of such factors has been discussed for the Indian sub-continent by D'Souza and Chen (1980), Barbara (1981), Chen, Hue and D'Souza (1981), Bairagi (1986), Das Gupta (1987) and Basu (1989). These studies show that food intake and health care are biased in favour of male children in rural families. In patriarchal, patrilineal societies a boy is considered an economic asset, because he will become an economically productive member of the family, whereas a girl, particularly of later parities, is often considered an economic liability since she will join another family after marriage.

Illustrations of how the sex differential evolves in response to development are provided by England and Wales and Mauritius (Figure 3.4). In England and Wales there are now two distinct concentrations of excess male mortality. In the post-50 age groups, the critical factor has been the greater vulnerability of males

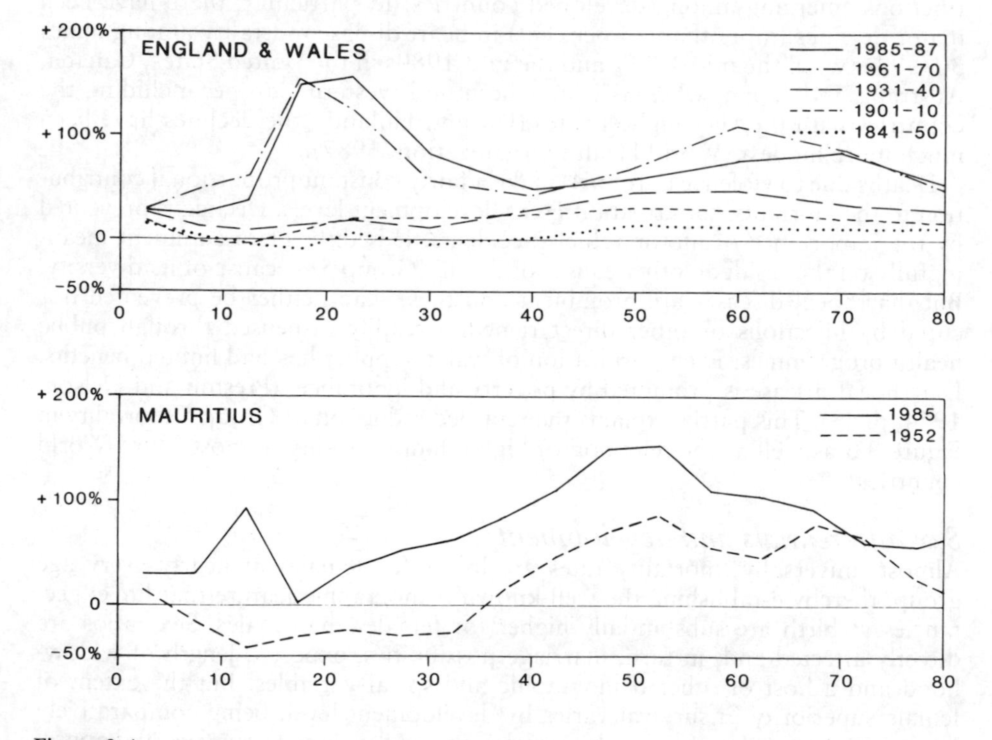

Figure 3.4
Percentage excess of male over female death rate by age, England and Wales, 1841–1987, and Mauritius, 1952 and 1985
Sources: Office of Population Censuses and Surveys (1978a), figure 1.4 Crown Copyright; *Demographic Year Book* 1987, Central Statistical Office, Mauritius, Table 5.6.

to cancers (above all, lung cancer) and heart diseases. There are signs, however, that the sex mortality differential may have reached its maximum in the most modernized societies, as the life-styles of the sexes increasingly merge.

The temporal and spatial processes which have created the current world pattern of mortality can now be explored through a consideration of groups of countries distinguished on the basis of their differing onset, pace and pattern of mortality decline.

North-western Europe

The nature of population growth from about the mid-seventeenth to the mid-nineteenth century in the north-western European core area of world population expansion is still a matter of controversy, although the scale and extent of growth are not disputed. There are those who argue that associated revolutions in agriculture, industry and transport directly stimulated population growth, although the traditional view that this was achieved through lowered mortality rather than by increased fertility has increasingly been challenged (Krause, 1958; Langer, 1963; Habbakuk, 1971; Wrigley and Schofield, 1981). Others argue for the primacy of population growth as an independent variable, promoting in its wake a series of economic changes (Boserup, 1981). Consider, for example, the population growth in Ireland from 3 million at the beginning of the eighteenth century to just over 8 million at the 1841 census immediately preceding the Great Famine. This growth is often attributed, in an entirely derivative fashion, to the widespread adoption of the potato, to subdivision of large farms and to reclamation of bogland, all of which could be expected to lower mortality and promote earlier marriage and hence higher fertility (Connell, 1950). But it can also be argued that it was population pressure, however caused, that was the primary stimulus during this period to Irish agricultural change (Grigg, 1976).

Data

Deficiency of data is clearly the factor preventing a definitive interpretation of demographic trends before the nineteenth century. Civil registers, maintained by state officials, of births, deaths and marriages, became common only in the nineteenth century (instituted, for example, in 1806 in France, 1837 in England and Wales and 1855 in Scotland), and only a minority of European countries began national censuses of population before well into the nineteenth century (Iceland 1703, Sweden 1749, Denmark–Norway 1769, France 1801, England and Wales 1801). A variety of sources has therefore been used to reconstruct past demographic patterns. Among the most bizarre is a scrutiny of Roman tombstone inscriptions, which led Durand (1960) to estimate the expectation of life at birth in the Roman Empire during the first and second centuries as 25–35 years. Such methods illustrate the risk of superficial generalization from unrepresentative data, since there can be little doubt that the provision of burial memorials varies by age, sex, class and other factors. For example, Henry (1957) found less than 10 per cent of funerary inscriptions at a Lyon cemetery in the early nineteenth century were for ages below fifteen, whereas over 40 per cent of all deaths in France were in this age group.

The most widely used demographic sources for the pre-civil registration era are the ecclesiastical registers of births, deaths and marriages. The extraction, analysis and interpretation of their data form the core of the widely practised, interdisciplinary field of historical demography, whose methods are described in Wrigley (1966), Henry (1967) and Willigan and Lynch (1982). At various times from the fifteenth to the eighteenth centuries it became common for European clergy to keep local registers of vital events, often stimulated initially by diocesan initiative but later by royal requirement. In England, for example, the keeping of parish registers dates from Thomas Cromwell's Ordinance of 1538. Invaluable as the registers are as raw demographic data banks, and ingenious as are some of the extractive methods adopted, they are subject nevertheless to major problems of error, bias and uncertainty, which can readily contaminate historical demographic data; it is also rare for complete sets of registers to survive.

The most critical deficiency of the registers is under-enumeration of vital events (Krause, 1965). There was rarely any check on an individual's diligence in register keeping, and Hollingsworth (1972, p. 79) notes that, for the clergyman, 'keeping a parish register may have been like keeping his diary. He remembered to write the event down at once perhaps 80 per cent of the time; sometimes he forgot and never wrote it up afterwards.' Growth of anti-clericalism and nonconformity clearly promoted under-enumeration, since the events recorded in the registers by the established church were essentially religious events – baptism and burials rather than births and deaths. In Britain, it is widely recognized that growth of dissenting congregations diminishes greatly the demographic value of parish registers from the late eighteenth century, although another factor was the inability of the ecclesiastical recording system to cope with rapidly increasing urban populations. Yet another defect relates specifically to infant deaths in the period between birth and baptism, which averaged almost one month in late eighteenth-century England (Berry and Schofield, 1971). Invariably both birth and death were unrecorded in such cases. It has been estimated that between one-third and one-half of infant deaths were unrecorded in English parish registers, leading to total deaths being under-counted by some 10 per cent (R. E. Jones, 1976).

The most outstanding work of demographic reconstruction from the European ecclesiastical registers has been by Wrigley and Schofield (1981) on the English parish registers. From counts in 404 parishes of burials and baptisms, statistically adjusted to take account of the above defects, they have estimated national totals of births and deaths annually from 1541. These are then converted to annual measures of fertility and mortality by relating the numbers of vital events to estimated population totals and age structures. These estimates are derived from what is called back-projection, in which the starting point is a reliable national population total and age–sex structure from a nineteenth-century census. Chronologically backward adjustments to this structure are made sequentially for 5-year periods, taking into account known births and deaths. Although a few reviewers have baulked at the cumulative, error-inducing impact of the necessary adjustments and assumptions, the resulting series are widely regarded as authoritative and credible. Indeed, one review title enthused: 'History Will Never Be The Same Again' (*Times Higher Education Supplement*, 5 February 1982).

Pre-decline pattern

Explanations for mortality variability over time and space in pre-industrial Europe can be discussed in terms of long-term and short-term fluctuations.

A speculative but plausible case has been made out by Galloway (1986) for the demographic impact of long-term variations in climate operating through environmental carrying capacity. Climatic historians have shown that temperatures have varied over the long run in close synchrony over the middle latitudes of the northern hemisphere, and Galloway has attempted to match these variations to demographic changes. In particular, he shows how the cold period of the seventeenth century, known as the Maunder Minimum or the Little Ice Age, is matched by poor yields, rising mortality and falling nuptiality and fertility.

Relationships between harvests and mortality have also dominated discussion of short-term fluctuations, although the random impact of epidemics and wars is now more favoured as the critical controlling mechanism. The traditional view is that of a Malthusian 'boom and bust' model (Figure 2.7) in which population expansion is thwarted by recurrent crises or mortality surges, constituting the positive checks which contain population growth within the environment's carrying capacity. Since these are often termed 'subsistence crises', the implication is that rises in mortality were caused by famine and the spread of disease in malnourished, vulnerable populations. Recent empirical assessment, however, fails to show any obvious positive relationship between short-term fluctuations in grain price and mortality. Thus, in a series of regressions using data from several countries of pre-industrial Europe, Galloway (1988) finds that price fluctuations typically 'explain' only between 20 and 40 per cent of the variance in mortality. The English evidence is particularly discouraging, with Figure 3.5 showing little obvious relationship in the short term between life expectancy and an index of real wages that is dominated by the price of food. It has also been shown (Livi-Bacci, 1983) that death rates of élite groups like Italian Jesuits and English aristocrats, who were never short of food, closely matched those of the population at large.

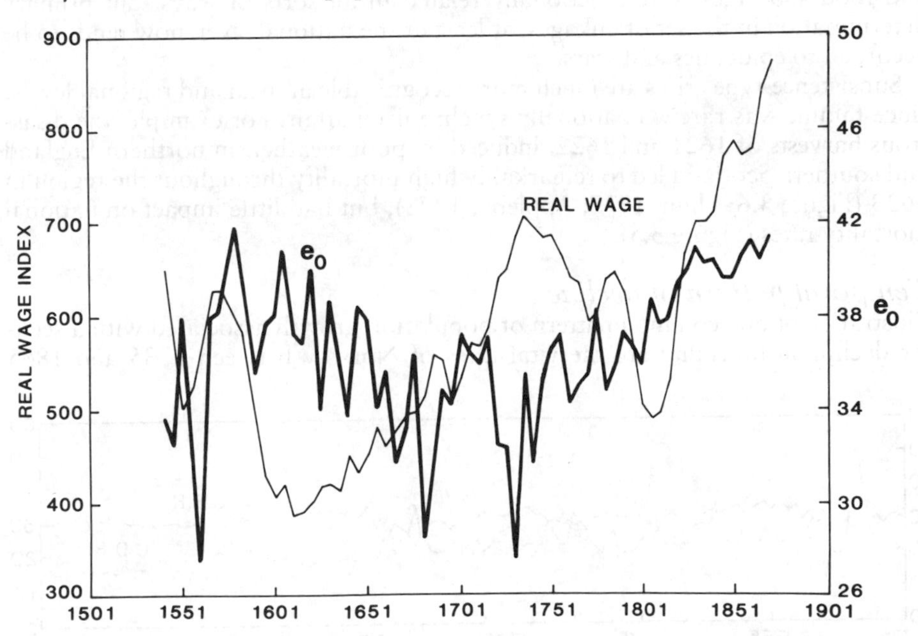

Figure 3.5
Expectation of life at birth (five-year periods) compared with an eleven-year moving average of a real-wage index, England
Source: Wrigley and Schofield (1981), figure 10.5.

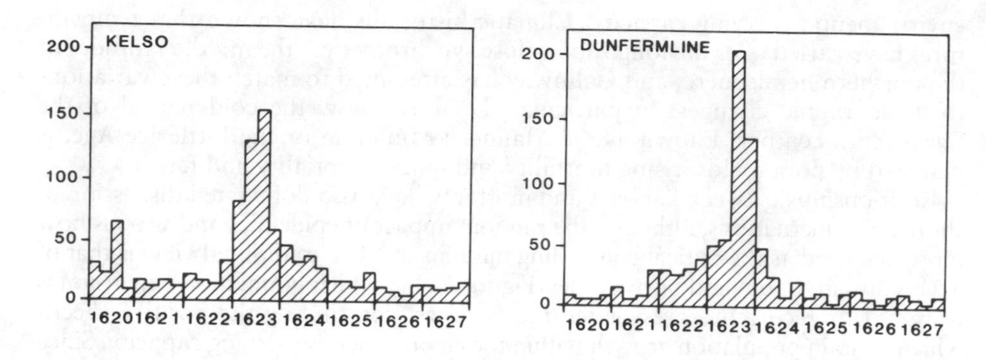

Figure 3.6
Quarterly burials at Kelso and Dunfermline, Scotland, 1620–7
Source: Flinn (1977), figure 3.2.2. By permission of Cambridge University Press.

The weight of modern opinion (Wrigley and Schofield, 1981; Boserup, 1987) is that short-term mortality variations in pre-industrial Europe owe much more to the unpredictable waxing and waning of epidemics and wars unrelated to man–land ratios and economic conditions. Bubonic plague, alone, dominated the pattern of mortality variation from its European debut in the 1340s to its almost complete disappearance after 1670, when smallpox epidemics may well have assumed a similar determining role (Mercer, 1985). Of course, epidemics, wars and food shortages were functionally related in all sorts of ways, but primary determination in the set of linkages, at least at the national level, now tends to be accorded to epidemics and wars.

Subsistence-type crises are much more recognizable at local and regional levels, since famine was rarely a nationally synchronized affair. For example, the disastrous harvests of 1621 and 1622, induced by poor weather, in northern England and southern Scotland led to remarkably high mortality throughout the region in 1623 (Figure 3.6; Flinn, 1977; Appleby, 1978), but had little impact on national mortality rates (Figure 3.5).

Temporal pattern of decline

Illustrative of one common pattern of population growth associated with a secular decline in mortality are the vital rates of Norway between 1735 and 1865

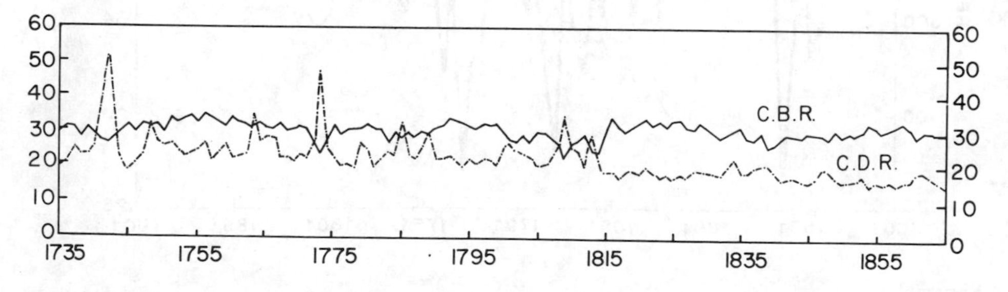

Figure 3.7
Crude birth rates and crude death rates per 1,000, Norway, 1735–1865
Source: Drake (1969), figure 2. By permission of Cambridge University Press.

(Figure 3.7), derived by Drake from the ecclesiastical registers and population censuses of the time. He observes that

> the evidence of the vital statistics and census returns support each other, suggesting a high degree of diligence and skill on the part of the clergy of the Norwegian state church who were responsible for their collection. The clergy were aided by cooperation and homogeneity of the public – nonconformity was not a problem – and by the efficiency and continued interest of the central administration. Fortunately too, they were spared the rapid growth of urban areas which bedevilled English vital registration in the late eighteenth and nineteenth centuries
>
> (Drake, 1969, p. 150)

The pattern of Norwegian population growth revealed by Figure 3.7 is consistent with that found in Sweden (Figure 2.9) and elsewhere in north-western Europe, although not in England or the Netherlands (de Vries, 1986). For the eighteenth and presumably previous centuries growth in Norway is restricted not simply by a high level of average mortality, but even more by the intermittent peaks in mortality commonly referred to as demographic crises or catastrophes. In the very nature of things, the theoretical range of fluctuation for the death rate – from no one to everyone dying in a particular year – is much greater than for the birth rate, and it is clear that the demographic crises were an immediate response to the onset of variable mixes of famine, epidemics and war. Thus of the three major peaks in Norwegian mortality revealed by Figure 3.7, that of the early 1740s is related to climatically induced harvest failures and to pandemics of influenza and typhus which raged throughout Europe in those years (Helleiner, 1957), that of the early 1770s to the harvest failures and acute distress widely present then in Scandinavia (Gille, 1949–1950), and that of the very early nineteenth century to the collapse of the herring and cod fisheries along the west coast of Norway, to failures in the grain harvest and to the disruption of shipping and timber trades during the Napoleonic Wars (Drake, 1969). An interesting demographic feature revealed by Figures 3.7 and 2.9 and confirmed by parish register data in Scotland (Flinn, 1977) is that in the years immediately following a demographic crisis, mortality rates generally fell to very low levels, which suggests that disease and famine tended to weed out the physiologically weak, leaving a relatively tough and resistant population. But of greater impact is the elimination in Norway of demographic crises in the nineteenth century, so that population growth was not terminated, as in previous epochs, by catastrophe.

The secular downward trend of mortality throughout north-western Europe in the eighteenth and nineteenth centuries can be explained by the interaction of several processes, most but not all of which are associated, directly or indirectly, with economic development.

Improved nutrition

To Malthus and many other students of population the fundamental resource constraint on population growth is food supply, but there was considerable expansion of agricultural output throughout north-western Europe from the early eighteenth century. Langer (1975) attributes this largely to the widespread adoption of the new wonder crop from America, the potato, with its enormous calorific yield per acre. Drake (1969) attaches great demographic importance to the introduction of the potato to Norway in the 1750s, since it rapidly proved to be a more reliable provider of subsistence in agriculturally marginal climates than traditional grain crops. More generally, the increased European food supply can be attributed to a series of agricultural advances which, in the case of Britain, have collectively been termed the Agricultural Revolution. Major advances in Britain included the enclosure of open fields and the reclamation of fenland and moorland; the introduction of new crops, notably turnips and clover, providing winter fodder for animals that formerly would have been killed in autumn; the development of balanced crop rotations in place of the traditional grain–fallow rotation; the conservation of soil fertility by marling, liming and mixed farming. In addition, regional famines were eliminated by the growing economic integration of national territory promoted by road and canal developments in the eighteenth century and railways in the nineteenth. In Scotland more emphasis has been placed (Flinn, 1977) on the emergence of a reasonably efficient system of famine relief, involving Poor Law authorities, landlords and central government, so that the decimation by blight of the potato harvests in the western Highlands and Islands in 1836–7 and 1845–6 did not have the same devastating human toll as in Ireland.

By these means Britain was able to feed, almost entirely from its own resources, a population which trebled between 1700 and 1850, although Razzell (1974) argues that per capita food consumption may not have increased during this period, and the modern consensus view among historians (A. J. Taylor, 1975) is that real wages and mass living standards in Britain and western Europe did not increase significantly, if at all, until well into the nineteenth century. Later in the nineteenth century substantial imports of food became available, above all from the newly opened-up wheatlands of the North American interior.

The more substantial and varied food supply lowered mortality rates by reducing the ravages of infanticide, starvation and infectious diseases. There is no direct statistical evidence on infant mortality through wilful neglect or violence, but there is appreciable documentary evidence for its wide practice until this century in Europe and for its basic cause being the inability of parents to feed their growing families (Langer, 1974; Sauer, 1978). Contemporary observers noted that parents were frequently indifferent to the loss of a child and that smallpox was commonly referred to in Britain as 'the poor man's friend'. Illustrative of parental inability to cope with large families were the foundling hospitals or orphanages present in all large cities of nineteenth-century Europe. Langer suggests that in the 1817–20 period about one-third of children born in Paris were abandoned to these institutions. He shows that many contemporary writers denounced the practice as legalized infanticide, one even suggesting that foundling hospitals should erect a sign reading 'Children killed at government expense'.

Reliable figures on deaths wholly attributable to starvation are just as difficult to obtain, but the weight of opinion is that starvation deaths *per se* have often been exaggerated in accounts of demographic catastrophes in pre-industrial Europe, as indeed they have been in the famines of the 1970s and 1980s in the Sahel, Ethiopia, Sudan and Bangladesh (Watkins and Menken, 1985). Much more emphasis is now placed on the way in which malnutrition induces death through the medium of infectious diseases like typhus, tuberculosis and measles. Among the most explicit statements is that

> Extensive experience in developing countries leaves no doubt about the profound effect of nutritional state on response to micro-organisms; malnourished populations have

> higher infection rates and are more likely to die when infected. The predominance of infectious disease in pre- and early industrial societies was due largely to malnutrition, and an improvement in nutrition was a necessary condition for a substantial and prolonged reduction of mortality and growth of population.
>
> (McKeown, 1976, pp. 128–9)

We have seen, however, that this orthodox view of cause and effect is being questioned by the recent appreciation that there is no clear temporal relationship in pre-industrial Europe between harvests and mortality.

Increased manufacturing output

The change from domestic handcraft to factory mass-production systems, which comprises the Industrial Revolution, greatly increased the amount and variety of goods, many of which – like soap, iron bedsteads, washable cotton underclothes and heavy winter clothing – had a direct role in warding off illness and death (for example in ridding bodies of lice, the vectors of typhus). More indirectly, many of the technologically based advances in agriculture and public health in the second half of the nineteenth century were dependent on an abundant production of commodities like agricultural machinery, iron pipes and pumping systems.

Medical advances

The views of Griffith (1926), attributing much of the eighteenth-century fall in mortality to medical developments, were widely accepted until the 1950s. He had been impressed by the expansion of hospital, dispensary and midwifery services (the number of hospitals in England increasing from 2 in 1700 to over 50 in 1800), by advances in the understanding of anatomy and physiology, and by the introduction of a specific protective measure, inoculation against smallpox. At least one modern historian (Razzell, 1977) supports the important demographic effect of inoculation, a crude method of immunization introduced from the Middle East into Europe at a time when vaccination was unknown. The method consisted of inoculating, i.e. infecting, a healthy person with pus from a mild case of the disease, thus conferring protection against a more serious attack.

These conclusions on eighteenth-century mortality decline have been severely questioned by McKeown and Brown (1955) and others. They emphasize that effective medicines were non-existent, that hospitals furthered rather than controlled infectious disease (hence the well-known statement by Florence Nightingale that the first requirement of a hospital is that it should do the sick no harm), and that the crude and dangerous measure of inoculation is thought likely by modern virologists to have spread rather than limited smallpox; but Cherry (1980) provides a less pessimistic view of the role of early hospitals.

Even for most of the nineteenth century, medical developments are likely to have had little or no influence on falling mortality rates, essentially because until the work of Pasteur and Koch in the 1870s and 1880s there was no appreciation of the existence of germs, their manner of reproduction and transmission, and their specificity in causing disease. Similarly, society had to await Lister's work in the 1880s before a start could be made on aseptic and antiseptic surgery through the sterilization of instruments, the use of masks, and the scrubbing of operating theatres with carbolic acid. Even as late as 1870–1, 10,000 out of 13,000 amputations performed by French army surgeons in the Franco-Prussian War proved fatal (Newsholme, 1929). Effective medical therapy, acting directly on the infective micro-organism, was delayed until the introduction of chemotherapeutic agents, particularly sulphonamides and antibiotics, from the 1930s.

Probably the only medical measure to contribute significantly to mortality reduction before this century was vaccination against smallpox by the cowpox vaccine developed by Jenner in 1798, but opinions differ on its contribution to overall mortality decline. In Britain, where vaccination was made compulsory for infants in 1854 although not legally enforced until 1871, data from McKeown (1976) suggest that mortality decline from smallpox accounted for about 5 per cent of total mortality decline in the second half of the nineteenth century. But a review of the wider European evidence (Mercer, 1985) suggests a more important demographic impact. Thus in Norway, Drake (1969) attributes to vaccination a role second only to the potato in overall mortality reduction. He shows that vaccination was made compulsory in 1810 and was enforced by the refusal of the church to confirm or marry unvaccinated persons, so that by the 1850s the annual number of vaccinations amounted to 82 per cent of live births. Similarly, in Scotland vaccination was not only extolled by the General Assembly of the Church of Scotland, but was carried out extensively by its parish ministers from 1803. The effect is shown clearly in the much reduced death toll from smallpox recorded in the bills of mortality issued at that time by several urban parishes (Flinn, 1977).

A strong statement of the very limited role of medicine in bringing about mortality decline, even in the twentieth century, is provided by the view of McKeown (1976, p. 162) that 'the health of man is determined essentially by his behaviour, his food and the nature of the world around him, and is only marginally influenced by personal medical care'. However, there are those (e.g. Woodward, 1984) who argue that McKeown's opinion is based on a rather narrow view of medicine. Adopting a wider perspective, one can note that exposure to infection has been considerably reduced in this century by medically stimulated measures like pasteurization and bottling of milk, which M. W. Beaver (1973) regards as a major factor in reducing infant mortality in Britain. There is also the important role of health education, whereby people become aware of the need for cleanliness in body, in food preparation and in domestic waste disposal. Razzell (1974) has argued that there were significant improvements in this field in Britain in the first half of the nineteenth century – before, in fact, the germ theory of disease was appreciated.

Public health developments

Drinking water, contaminated with human faeces or urine, was clearly the major medium for transmission of intestinal diseases (typhoid, cholera, dysentery) in crowded and unsanitary urban environments before the gradual provision of public utility systems during the nineteenth century. London provides an illustration of this provision, although as capital of the most materially advanced country in the world at that time its dates of provision were atypically early. In 1829 a start was made on water purification by the use of sand filters for Thames river water (although chlorination had to await the turn of the century); in 1848 public agencies for refuse removal began operation; and in 1865 a network of sewers was completed, to take the place of open ditches and cesspools.

A striking example of how public utility provision contributed to mortality decline in nineteenth-century Europe has been provided by Preston and Van de Walle (1978). They present graphs (Figure 3.8) of life expectancy at birth for the

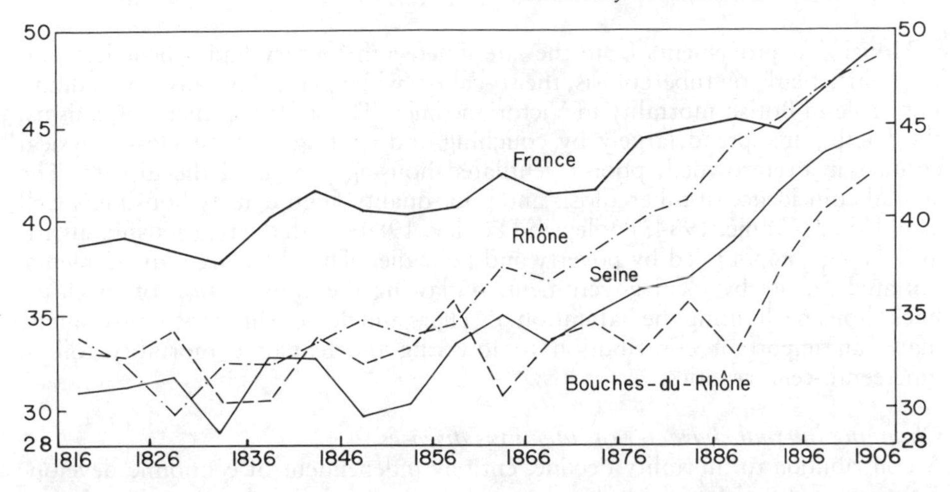

Figure 3.8
Nineteenth-century trends in life expectancy at birth (in years) in France and the *départements* of Seine, Rhône and Bouches-du-Rhône
Source: Preston and Van de Walle (1978), figure 2.

three *départements* which contain the largest cities in France: Seine (Paris), Rhône (Lyon) and Bouches-du-Rhône (Marseille). The fact that life expectancy is much lower in all three urban *départements* than in France as a whole reflects an important urban–rural differential in mortality, which will be discussed in the next chapter. What demands immediate attention from Figure 3.8 is the striking divergence in mortality experience between the three *départements* from the middle of the nineteenth century. This is attributed to their differing progress in provision of public health utilities. In Lyon the provision of filtered, piped water and a sewerage system occurred in the 1850s. In Paris water and sewerage improvements were less decisive and were spread more gradually throughout the 1850–1900 period. Cause-of-death data (Table 3.5) show that by far the greater part of overall mortality decline in Paris at this time was due to the reduced impact of water- and food-borne diseases, particularly cholera. The most retarded development was in Bouches-du-Rhône, where Marseille, almost up to the end of the century, had sewers only in a few privileged neighbourhoods and derived its water from the Durance by an uncovered 83-km canal which passed through several contaminating settlements. It took a severe cholera epidemic in 1884–5 to sting city authorities into providing appropriate utilities in the 1890s.

Table 3.5 Crude death rate (per 1,000) by cause in nineteenth-century Paris

	1854–6	1887–9
Airborne diseases	7.2	6.9
Water- and food-borne diseases	8.6	2.2
Typhoid	2.0	0.5
Diarrhoea, gastritis, enteritis	3.7	1.7
Cholera	2.9	0.0
Other causes	14.9	13.4
All causes	30.7	22.5

Source: Preston and Van de Walle (1978), table 3.

Housing improvements from the late nineteenth century had a beneficial impact particularly on tuberculosis, the so-called 'white plague' because of its dominant role in British mortality in Victorian times. The infective micro-organisms, the bacilli, are spread largely by coughing and spitting, so that close physical contact in overcrowded, poorly ventilated housing promoted the disease. The spatial coincidence of tuberculosis and poor-quality, high-density housing is well established (Cronjé, 1984; Pooley and Pooley, 1984), with the relationship invariably being compounded by poverty and poor diet. Thus housing improvements, initiated mainly by local government, outlawing the construction of 'back-to-back' housing, limiting the habitation of cellars and demolishing the worse slums, made an important contribution to lowering the fearsome mortality toll in nineteenth-century cities.

Changing internal character of infectious diseases
A contribution to mortality decline, entirely independent of economic development, may have been made by the ever-evolving relationship between micro-organisms and man. Thus a reduction in the virulence of an infective organism would bring about a spontaneous or autonomous mortality decline, and at least one historian (Chambers, 1972) assigns a critical demographic role to this random biological influence. Firm evidence is hard to come by, although there seems little doubt that scarlet fever behaved in this fashion in nineteenth-century Europe (McKeown, Brown and Record, 1972).

It has often been hinted that the disappearance from western Europe in the eighteenth century of perhaps the most dreaded epidemic killer, bubonic plague, was also of this form, although retrospective evidence is particularly difficult to evaluate because of the critical role of fleas and rats in the transmission process. Shrewsbury (1970) argues that plague disappeared from Europe essentially because the development of a direct oceanic trade route between Asia and western Europe destroyed the traditional caravan route to the Levant which had acted as a 'rodent pipeline' for the spread of plague from its Asiatic homeland. Others have cited the increasing ability of states to enforce quarantines and *cordons sanitaires* (Kunitz, 1986) or the displacement of the dominant rat species in Europe by one which was a less effective transmitter of plague to humans (Cartwright, 1972).

There is general agreement that the continued mortality decline from the eighteenth century in north-western Europe is due to a combination of the five processes discussed above, although the particular combination in terms of weighting varies temporally and spatially. There is little doubt that overall mortality decline has been achieved essentially by a reduction in the death toll of infectious diseases, both epidemic and endemic. This can be illustrated (Table 3.6) by cause-of-death data for England and Wales, although there are obvious problems of reliability and comparability of data arising from vagueness and inaccuracy of diagnosis, from changes in nomenclature and classification, and from exclusion of infanticide. Table 3.6 indicates that mortality from infectious diseases ('conditions attributable to micro-organisms') accounts for almost all of overall mortality decline in the second half of the nineteenth century and for the greater part of that during this century. The Scottish experience has been almost identical (Flinn, 1977), but in other developed countries the proportion of overall mortality decline attributable to the infectious diseases group has been less, although still substantial (Preston and Nelson, 1974).

Table 3.6 Standardized death rates (per 1,000) by cause, England and Wales*

	1848–54	1901	1971
Conditions attributable to micro-organisms:			
1. Airborne diseases	7.3	5.1	0.6
Respiratory tuberculosis	2.9	1.3	0.1
Bronchitis, pneumonia, influenza	2.2	2.7	0.6
Scarlet fever, diphtheria	1.0	0.4	0.0
Smallpox	0.3	0.0	0.0
2. Water- and food-borne diseases	3.6	1.9	0.0
Cholera, diarrhoea, dysentery	1.8	1.2	0.0
3. Other conditions	2.1	1.4	0.1
	12.9	8.5	0.7
Conditions not attributable to micro-organisms	8.9	8.5	4.7
All diseases	21.9	17.0	5.4

* Standardized to 1901 age and sex distribution
Source: McKeown (1976), tables 3.1 and 3.2.

The temporal pattern of mortality reduction by age groups is shown in Table 3.7 for England and Wales. The dramatic reduction in infant and child mortality, contrasting with very modest reductions in the old age groups, clearly reflects the conquest over death from infectious dieseases, to which young children, not having developed a natural immunity, have always been particularly subject. Indeed, the extent of mortality reduction in the under-five age group is underestimated by Table 3.7 since it is likely that some infant deaths, especially from infanticide, were still not being recorded in the early years of civil registration. Nevertheless, the infant mortality rate in England and Wales was remarkably resistant to change until this century, maintaining a level of about 155 per 1,000 throughout the second half of the nineteenth century. In contrast, Sweden, with few of the environmental and welfare problems associated with urban–industrial concentration, reduced its rate in the same period from about 150 to 100 per 1,000.

Table 3.7 Age-specific death rates (per 1,000), England and Wales

	1841–50	1901–10	1985
0–4	66.0	46.0	2.3
5–9	9.0	3.6	0.2
10–14	5.3	2.1	0.2
15–19	7.5	3.0	0.5
20–24	9.3	3.8	0.6
25–34	10.3	5.1	0.7
35–44	13.0	8.3	1.4
45–54	17.0	14.0	4.3
55–64	30.0	28.0	13.0
65–74	64.0	59.0	33.0
75–84	142.0	127.0	78.0
85+	301.0	261.0	186.0

Table 3.8 Percentage distribution of deaths by age group, England and Wales

	1838–44	1987
0–14	47	2
15–64	34	19
65+	19	79

Further striking illustrations of the changing age pattern of mortality are given in Table 3.8 and Figure 3.9. So dramatic have been the reductions in mortality in young age groups that survivorship curves for humans are now strikingly different from those for animals and plants (Figure 3.10).

North America

The pattern of mortality decline in relation to development which has been described for north-western Europe is likely to have been replicated, broadly, by North American experience, although nutritional improvements are unlikely to have played such a significant role, bearing in mind the always more favourable ratio between population numbers and agricultural resources in the New World. This is reflected in the rather lower estimated mortality rates for agricultural communities in colonial New England (Vinovskis, 1979) compared with those in Europe, although the unimportance of plague in America must have contributed.

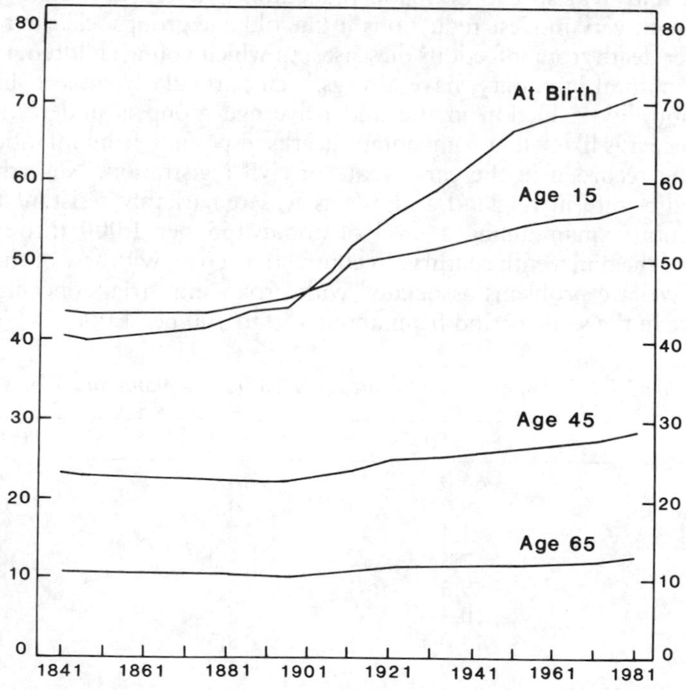

Figure 3.9
Expectation of remaining years of life at birth and at selected ages for males, England and Wales, 1841–1981
Source: Swerdlow (1987), figure 2. Crown Copyright.

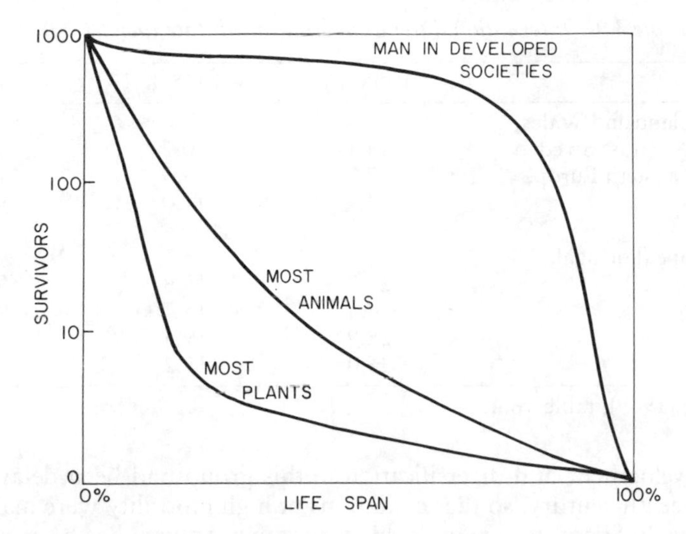

Figure 3.10
Number of survivors by age from a birth cohort of 1,000. Age is measured in relative units of mean life-span, which is the maximum length of life reasonably to be expected for a normally endowed organism under optimal environmental conditions
Sources: Deevey (1950), p. 59; Thomlinson (1965), figure 2. By permission of Random House Inc.

There is a dearth of reliable mortality data for much of North America before the twentieth century. Even in 1900–5 a minority of the population of the United States resided in states where registration systems were adjudged sufficiently accurate for them to belong to the national death registration area; and it was not until 1933 that a 100 per cent coverage was achieved by admitting Texas. Nevertheless, the somewhat sporadic data do bear out north-western European experience (Omran, 1977; Kunitz, 1984).

Life expectancies at birth for Massachusetts in 1860 of 46 for males and 47 for females (Vinovskis, 1979) and for the United States death registration area in 1900 of 46 and 48 respectively (US Bureau of Census, 1960) are comparable to those recorded for the same years in England and Wales (Figure 3.9). The levels and trend of infant mortality in Massachusetts are also very similar to those in England, with no significant reduction to below 150 per 1,000 being achieved until after 1900. Another close comparison with European experience comes from the finding of Condran and Crimmins-Gardner (1978) that almost 80 per cent of overall mortality decline in 28 cities of the United States between 1890 and 1900 is accounted for by deaths in four infectious disease categories: typhoid, tuberculosis, diphtheria and diarrhoeal diseases. One must beware, however, of inferring continental trends from the reasonably favourable mortality experience of the north-eastern United States.

Southern and eastern Europe, the Soviet Union, Japan

In the onset and pace of general mortality decline, this group has an intermediate position between the economically advanced countries of the West and the less developed countries of Africa, Asia and Latin America. Progress towards modern

Table 3.9 International trends in crude death rate (per 1,000)

	1906–10	1935–8	1955–8
Denmark, England and Wales, Netherlands, Norway, Sweden	14.1	10.7	9.3
Remainder of western Europe	17.7	13.3	11.3
USA	14.9	11.0	9.4
Southern Europe (Portugal, Spain, Italy, Greece)	21.7	15.6	9.4
Eastern Europe	24.5	15.7	10.0
USSR	28.9	19.5	7.7
Japan	25.0	17.2	7.8

Source: Bogue (1969), table 16.4.

economic development and diversification in this group had been delayed until the late nineteenth century, so that conditions of high mortality were maintained until that time. In Spain, for example, life expectancy at birth has been estimated at a mere 29 years in 1877–87 and 35 in 1900 (Livi-Bacci, 1972).

However, the rapid pace of mortality decline in these countries in the twentieth century is demonstrated in Table 3.9. Crude death rates were substantially above those in north-western Europe and the United States at the beginning of the century but had been reduced to equivalent levels by the 1950s.

In this intermediate group of countries it is generally agreed that specific advances in the sciences and technologies of public health and medicine, derived externally, have played as important a role in mortality reduction as the general forces of socio-economic development. In Italy, Cipolla (1965, p. 582) maintains that 'gains in infant mortality seem to be imputable more to improved medical and hygienic practices than to changes in economic conditions' and Taeuber (1958, p. 284) regards Japan as giving 'the first major demonstration that reduced and even low mortality could be achieved in somewhat unfavourable environmental conditions'. The remarkable mortality reduction in Japan in the decade following the Second World War is widely attributed to the introduction at that time, under the initial stimulus of the American occupation, of antibiotic wonder drugs, mass programmes of inoculation and vaccination, and the establishment of a dense network of community health centres.

While southern European countries and Japan have continued their mortality reduction (to the extent that throughout the 1980s Japan boasted the world's lowest infant mortality rate), there has been a remarkable reversal of mortality trends in the USSR and eastern Europe. From the mid-1960s to the mid-1980s life expectancy at birth stagnated, and actually declined by 1–2 years in Hungary and the USSR, where mortality increases were particularly concentrated among males aged 30–64 years. Thus in the mid-1980s life expectancy at birth averaged 71 years in eastern Europe and 69 in the USSR, but 74 in southern Europe and 75 in western Europe and northern Europe.

This demographic reversal, which has led Compton (1985) to suggest the emergence of a distinctive East European mortality pattern, has been a considerable embarrassment to the region's political regimes and was reflected in the disappearance for several years of detailed mortality data from official Soviet publications. Standard explanations in the West have centred on bureaucratic inefficiencies in health services and in food production and distribution, on industrial pollution and on increased alcohol consumption (Feshbach, 1982). But an additional or, indeed, alternative reason that is wholly demographic has been suggested perceptively by Dinkel (1985). He argues that the young adult group in the Second World War suffered major deprivation and was depleted by military deaths of its healthier male members through their greater selection for active service. Thus, when this cohort had aged significantly by the mid-1960s, it was peculiarly vulnerable to mortality. He also argues that the increase in infant mortality in the USSR (which is not found elsewhere in the region) is essentially a function of the changing spatial distribution of births, in that a considerably higher proportion of births now comes from the central Asian areas where cultural factors and poorer standards of health care and sanitation maintain relatively high infant mortality. The most recent analysis (Blum and Monnier, 1989) suggests a resumption in overall mortality reduction in the USSR from the mid-1980s, with an important contributory factor being the reduction in accidental or violent deaths consequent upon Gorbachev's campaign against alcohol abuse.

Less developed countries

At the beginning of this century, life expectancy at birth throughout Africa, Asia and Latin America was a mere 25–28 years (Arriaga and Davis, 1969). By 1985 the average figure for the three continents had more than doubled to 63 years, but was still some 12 years below the European and North American average.

Within the global set of less developed countries there is certainly a relationship between development and mortality. Thus Figure 3.2 shows, broadly, a life expectancy gradation from Latin America through Asia to Africa that must be functionally related to the comparable gradation in development and modernization levels (Table 3.10). Figure 3.2 reveals similar relationships at a more detailed scale: for example, the life expectancy differences between the North African littoral and the Sahel countries, and between Malaysia/Thailand and Cambodia/Laos. But the extent and rapidity of mortality decline – in some cases in countries showing only modest economic growth and diversification – suggest that factors other than development have contributed extensively to the concentrated mortality reduction. Through regression analysis of national scale data, Preston (1980) estimates that rather less than half of life expectancy gain in less developed countries between 1940 and 1970 may be attributed to development variables. The 'other' factors are essentially the medical and public health programmes imposed on often passive populations by colonial rules and by independent governments supported by the World Health Organization (WHO) and external aid agreements.

Table 3.10 Development indicators, c. *1985*

	Europe	Latin America	Asia	Africa
GNP per capita ($)	8,170	1,720	1,020	620
% population urban	75	68	30	30
Life expectancy at birth	74	66	61	52

Source: Population Reference Bureau, *World Population Data Sheet* 1988.

Programmed disease control

Many important control measures for tropical diseases were initiated by the needs of Allied troops during the Second World War, when, for example, the incidence of malaria among American troops at Guadalcanal in the Solomon Islands in 1942 reached a staggering 72 per cent (Madeley, 1988). So successful were the subsequent programmes based on vector control, vaccination and antibiotics that levels of infectious disease specifically and mortality generally fell spectacularly in the 1950s and early 1960s, despite the continuance of largely unfavourable economic, social and environmental conditions – an experience quite contrary to that of Western developed countries in the eighteenth and nineteenth centuries when medical developments played little part in mortality reduction.

A spectacular example of concentrated mortality decline (and the associated problems of population growth, congestion and pressure on resources) can be seen in the Indian Ocean island of Mauritius, where good-quality vital statistics have been available from the late nineteenth century Figure 3.11 shows that for much of this century there was little natural increase. There were appreciable fluctuations in the death rate, with peaks associated with the influenza epidemic of 1919, the depression and low wages in the all-dominant sugar industry in the early 1930s, and a devastating hurricane in 1945. But between the mid-1940s and the mid-1950s the death rate fell from about 30 per 1,000 to 12 per 1,000 (and life expectancy rose from 35 to 55 years), achieving in some ten years what had taken over 150 years in western Europe. The most important contribution was the malaria eradication programme, based on controlling the mosquito vector of the malaria parasites through insecticide spraying. Similar successful programmes, with equivalent impacts on mortality, were mounted at that time in Sri Lanka (Gray, 1974), Guyana (Mandle, 1970) and several other countries, although little progress was made in Africa, the continent with most malaria cases and deaths.

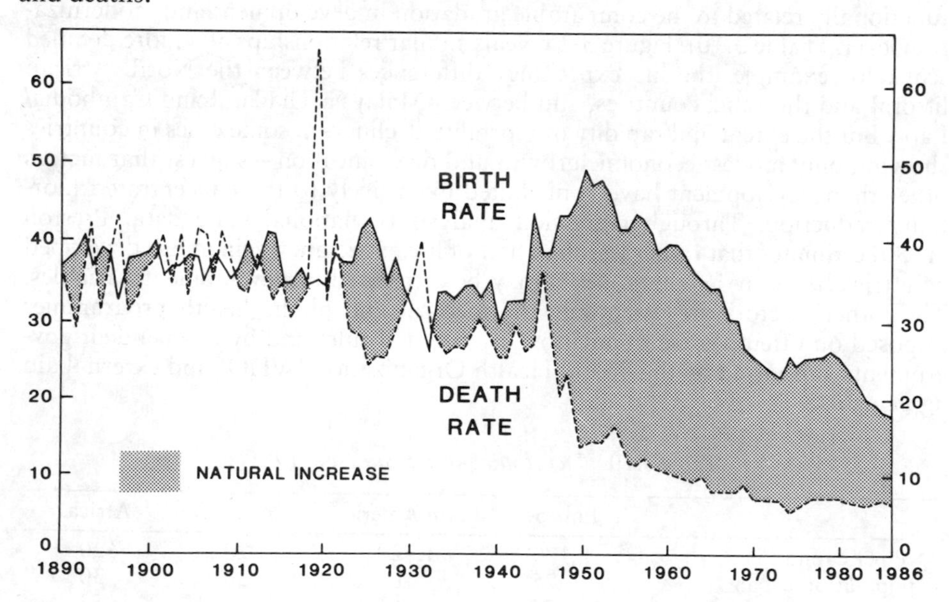

Figure 3.11
Crude birth rates and death rates, per 1,000, Mauritius, 1890–1986

The eradication of smallpox provides a second demonstration of the potential of externally derived disease control. One of the great scourges of mankind, this disease was endemic until recently in much of the inter-tropical world. But an intensified eradication programme by WHO from 1967 eliminated the disease worldwide in 1978, when the last victim was traced in Somalia and successfully treated. Subsequently, the only risk of infection has come from stocks of virus kept by research laboratories. This remarkable success can be attritubed to the assiduous detection and isolation of cases, to the development of a potent vaccine and to the international co-ordination of vaccination programmes. But, in addition, the nature of the disease facilitated management in that victims were immediately recognizable by characteristic pock marks, the disease is transmitted relatively slowly from acutely infected persons and there are no animal reservoirs or alternative hosts. Hence the disease could be tackled by a strategy of surveillance-containment involving case recognition, case isolation and vaccination of everyone who could have had contact with the case. Expensive and difficult-to-organize *mass* vaccination programmes were not required.

A third illustration of disease control, at a more intimate scale, has been provided by Orubuloye and Caldwell (1975). They select two villages in Nigeria closely matched in culture, socio-economic conditions, water supply and waste disposal, but contrasting in medical services provision. One village had a rural hospital and trained doctor for more than ten years, while the other did not even possess a dispensary or chemist's shop within 10 kilometres. Partly on the basis of mortality rates being twice as high in the latter village, the authors argue that mortality decline in rural Nigeria during the previous half-century may be almost wholly explained by the introduction of modern health services. Support for this general view is provided by a study (Faulkingham and Thorbahn, 1975) of a village in southern Niger at the heart of the Sahel zone. Drought and food shortages exacted heavy tolls throughout this area in several years following 1968. Yet in this particular village mortality rates continued to fall, the authors attributing this to the population's ready access to a nearby hospital and dispensary.

The technical-fix, intervention model of Third World mortality reduction has, therefore, been widely canvassed. Stolnitz (1975) estimated that disease control programmes, operating under existing medical technology and realistic funding, would alone raise life expectancy at birth to 50–55 years throughout the Third World. He argued (p. 233) that 'the evidence appears overwhelming that levels of living and lifestyle can be greatly offset or even dominated by what might be called programmed disease control'. Similarly, Arriaga (1970, p. 135) concluded his review of mortality trends in Latin America between 1930 and 1960 by stating that 'Public health programmes are no longer dependent on the country's economy but rather to a large degree, on the technology and concern of the most advanced countries.' We now know better.

Deceleration of mortality decline

Against all expectations, the earlier rapid tempo of mortality decline in many parts of the Third World slowed considerably in the 1970s and in several countries actually stagnated (Gwatkin, 1980; Arriaga, 1981; Sivamurthy, 1981). The implication is that the highly medicalized, single-disease campaign approach to health can reduce mortality levels so far, but no further.

It is the persistence of high infant and child mortality that prevents the great majority of Third World countries from approaching the life expectancy levels of

Table 3.11 Average age at death for males and females at birth, age one and age five, based on age-specific death rates (1984)

		At birth	Aged one	Aged five
Bangladesh	Male	55	63	67
	Female	55	61	66
United States	Male	71	72	72
	Female	78	79	79

Source: UN *Demographic Yearbook* 1986.

developed countries. While deaths of children under five years of age comprise 1–2 per cent of all deaths in Europe and North America, equivalent figures from some of the poorer Third World countries in the early 1980s are 48 per cent in Bangladesh, 39 per cent in Egypt and 32 per cent in the Philippines (1986 *Demographic Yearbook*). If children in less developed countries can survive the first few years of life, their life expectancy can approach that of developed countries, even in the comparative case (Table 3.11) of a particularly disadvantaged Third World country (Bangladesh) and a particularly advantaged developed country (United States).

Several factors can now be discussed as having been responsible for the deceleration of Third World mortality decline and for the persistence of unacceptably high infant and child mortality. More than anything, they indicate a resurrection of the determining role of socio-economic and cultural conditions.

Malnutrition and poverty

Although global food production per capita increased by a seemingly healthy 0.5 per cent per annum between the mid-1960s and the mid-1980s, most of the Indian sub-continent recorded little or no increase while in Africa there was a 1.0 per cent per annum decline (UN *Statistical Yearbooks*). The majority of African countries experienced even greater declines, the major contributory factors in this continent of crisis being widespread, prolonged drought in the Sahel and major wars in Uganda, Ethiopia, Mozambique, Somalia, Angola, Zimbabwe and Chad (Glantz, 1987).

Even where food production per capita has increased partly through increased irrigation and the adoption of Green Revolution high-yielding cereals, there is no assurance that the bulk of the population has benefited nutritionally. It has been widely alleged (e.g. Glaeser, 1987) that the Green Revolution has benefited a small, politically powerful élite of large, rich farmers at the expense of the majority of peasants who are unable to afford, and sometimes to understand, the package of fertilizers, pesticides and water input necessary for the successful adoption of new crop varieties. Particularly in Latin America, technological transformation of agriculture has led to an increasing concentration of land ownership and maldistribution of agricultural income that in turn produces increasing pauperization and malnutrition in the majority of the rural population. Nutritional surveys carried out in six Central American countries (Guatemala, El Salvador, Honduras, Nicaragua, Costa Rica and Panama) show that the prevalence of child malnutrition (measured by the relationship between a child's body weight and the desirable standard for his or her age and sex) increased substantially between the mid-1960s and the mid-1970s in all countries other than Costa Rica (Teller, 1981).

Although malnutrition itself is rarely a killer, it lowers resistance to often fatal infectious diseases. Particularly at risk are the growing numbers of landless rural population, the urban poor, infants and women – in other words, the groups disadvantaged in national and household 'share-outs'.

Defective water supply and waste disposal

The major killer of children in the Third World is the group of water-borne intestinal diseases which includes cholera, dysentery and, above all, diarrhoea. The most important cause is a water supply contaminated by nearby disposal of body wastes. WHO estimates that only 36 per cent of the rural population of less developed countries had access to safe water in 1985, the major problem being the financial cost of provision. Diarrhoeal diseases, even when they do not kill, prevent children from eating normally, and thus promote malnutrition and vulnerability to infection even when food supply is plentiful.

Urban bias

Modern medical facilities and public health services have been modelled on European and North American systems originally designed for urbanized societies, so that the spatial concentration they have developed in the Third World is inevitable, regardless of further bias given by governing urban élites in the spatial allocation of scarce resources.

> The most important class conflict in the poor countries of the world today is not between labour and capital. Nor is it between foreign and national interests. It is between the rural classes and the urban classes. The rural sector contains most of the poverty, and most of the low-cost sources of potential advance; but the urban sector contains most of the articulateness, organization and power. Scarce investment, instead of going into water pumps to grow rice, is wasted on urban motorways. Scarce human skills design and administer, not clean village wells and agricultural extension services, but world boxing championships in showpiece stadia.
>
> (Lipton, 1977, p. 13)

Effective medical programmes, therefore, have often failed to break out of their metropolitan beach-heads. In North Yemen, for example, the three largest settlements contain only 7 per cent of the country's population but just over half of the country's doctors and almost two-thirds of its trained nurses (Melrose, 1981).

Complacency and resistance

There were demonstrations in the 1970s of the sometimes fragile nature of externally imposed disease control when environmental conditions remain poor. A spectacular case has been the resurgence of malaria in many countries from which it had almost been eliminated. In India the number of known active cases rose from 100,000 in 1965 to 10 million in 1977, in Sri Lanka from less than 100 in 1963 to more than 1 million in 1969, and in the Solomon Islands from 3,500 in 1975 to 60,000 in 1981 (Madeley, 1988). The reasons seem to lie in the unacceptable costs of eradication campaigns (invariably a third or more of national health budgets), in the complacency with which programmes were conducted when elimination seemed just around the corner, and also, more disturbingly, in the way in which resistance has developed in some strains of mosquito to residual insecticides like DDT and in some forms of malaria to chloroquine, the most effective and widely used anti-malaria drug.

Another example is rinderpest, a cattle disease which affects human health through man's nutritional and economic well-being. This fatal cattle disease was brought under control in Africa in the 1960s, but the 1980s have seen major outbreaks in all cattle-rearing African countries. The major cause is lack of foreign exchange for adequate vaccine and veterinary services. These examples provide a salutary antidote to a blind belief in the technical-fix approach to human improvement.

West not always best

Non-governmental organizations, particularly charities like Oxfam and War on Want, have led the way in exposing the adverse health consequences of the often unscrupulous marketing in the Third World by Western companies of baby-foods, pharmaceutical drugs and cigarettes.

The benefits of breastfeeding in terms of balanced infant diet and child survival, as well as the subsidiary contraceptive effect through delayed resumption of ovulation, are well known, and there is no doubt that bottle-fed infants have a higher mortality wherever, as in much of the Third World, there is a lack of clean water, of facilities for sterilizing bottles and of refrigeration for the formula. It is scandalous, therefore, that bottle-feeding has been promoted as a modern advance in Third World countries at a time when breastfeeding is on the increase in the West among better-educated women. It was not until 1980 that WHO and UNICEF were able to persuade the major companies to agree to a Code of Marketing of Breast-Milk Substitutes that would, for example, ban advertising in clinics.

After tragedies like thalidomide, patients in the West are no longer uncritical consumers of medically prescribed drugs. In the Third World, however, the success of early medical campaigns led to a 'pill for every ill, magic of the needle' climate which has enabled Western drug companies to promote successfully products which are often only marginally useful, sometimes totally irrelevant and occasionally downright harmful; some of these drugs are actually banned in their countries of origin. A devastating attack on the profit-maximizing role of Western drug companies, which essentially regard health as a commodity rather than a service, has been provided by Muller (1982). Several countries, notably Bangladesh, Sri Lanka, Tanzania and Mozambique, have tried to restrict their drug purchases to a short list of essential products specified by WHO, but few have been able to maintain restrictions in face of the outcry from powerful pharmaceutical companies often backed by their governments' threats to cut aid payments.

The harmful promotional activities of the major tobacco companies are now widely regarded as notorious. Faced with increasing public opposition and decreasing numbers of male smokers in the West, the industry has increasingly targeted women, youth and the Third World. In Mauritius, for example, 60 per cent of adult males are smokers (*World Health*, August–September 1988), compared with 36 per cent in Britain (*General Household Survey*, 1984).

Current health policies: a rethink

The setbacks to continued mortality reduction in the 1970s so exposed the naïvety of the Western intervention model to health improvement that more culturally and environmentally sensitive policies are now increasingly advocated, based particularly on bridging the disciplinary gap between biomedical and social scientists (Moseley, 1984). What, then, have been the major policy responses?

Rural development and primary health care

There is little doubt that sheer poverty is the major cause of poor health in the rural Third World, so that there is a clear need for health administrators to participate actively in rural development in its widest sense (Abel-Smith, 1978). The whole emphasis of Third World development strategy has been changing:

> Some of the most fundamental development dogmas of the whole period since World War II are being challenged. These include the prior place of industrial growth, the desirability of shifting people out of agriculture into the 'modern sector', and the very concept of a 'modern sector' closely linked by technology and finance to the internationalist capitalist system as the principal engine of progress. An emergent doctrine of 'self-reliant development' is increasingly being advocated . . . Given that the poor remain poor, and that most of them are rural, there is new priority given to rural development.
>
> (Brookfield, 1978, pp. 124–5)

Such development is an integral part of the basic needs strategy that has become the dominant policy paradigm in Third World development.

WHO has become increasingly concerned about health-care deprivation among the great bulk of Third World rural populations who have no contact at all with modern health services, except for the occasional visits of mobile teams from the mass disease campaigns or the services provided by a few missionary hospitals. Moreover:

> The few trained workers who do work in rural areas – against their will – discover that they are completely unprepared to deal with the specific problems of the communities they serve: their training syllabuses were copied from European models and they received their training in establishments situated in urban areas. They work under bad conditions and are often poorly managed and supervised.
>
> (WHO, 1976, p. 14)

The major WHO policy initiative to counter the acute spatial and social mismatch of health-care needs and health-care resources has been the promotion of primary health-care systems (WHO, 1978) which attempt to provide the rural populations of less developed countries with at least a bare minimum of accessible health services. Such services do not necessarily require highly trained, scarce manpower and expensive facilities, as has been demonstrated by the successful 'barefoot doctor' programme in China and by pilot studies of villages where a local inhabitant, reasonably well educated and selected by the local community, has been trained successfully in a few months to provide basic health advice and some primary treatment. Similar successes have been achieved in Thailand by mobilizing for that role the Buddhist monks (the 'bare-head' doctors). There is also much potential in training traditional birth attendants in basic hygiene and methods of safe delivery, when one considers, for example, that 26 per cent of all infant deaths in a part of Bangladesh in the mid-1970s were caused by neo-natal tetanus (Chen, Rahman and Sardar, 1980).

Local, community-based programmes are now widely regarded as the key strategy in making health services accessible, affordable and socially acceptable

(Moseley, 1984). In complete contrast to the medicalized, single-disease, campaign approach of the 1950s and 1960s, it is recognized that local communities must become more actively involved in low-cost disease prevention. In the Solomon Islands, for example, villagers are encouraged to recognize mosquito larvae, to drain likely breeding sites, to clear undergrowth and to destroy water containers like old cans and inverted coconut shells (Madeley, 1988). Similarly, in China diarrhoeal mortality among infants has been controlled, not by any massive investment in wells, pipelines and latrines, but by local campaigns to persuade mothers to boil drinking water.

Social development

There is growing recognition that supply-orientated strategies are conditioned by social constraints on demand for, and effective use of, health services. Three components of social development may be regarded as of particular discriminating importance: education, female autonomy, and egalitarian policies.

The close inverse relationship between mortality and education has been demonstrated in Table 3.4 and Figure 3.1. Similarly, Caldwell's (1986) examination of mortality finds a closer statistical relationship with education than with any other hypothesized causal factor in a data set of 99 Third World countries. Of particular policy importance is a group of low mortality–low income–high education countries, notably China, Sri Lanka, Cuba, Costa Rica, Jamaica and Thailand (Halstead, Walsh and Warren, 1985). In southern India, Caldwell, Reddy and Caldwell (1983) have shown that, with increased schooling, parents are much more likely to identify with modern health services, to seek out such services for their sick children and to follow properly prescribed treatments. In Latin America, Behm (1979) and Palloni (1981) have shown that this inverse relationship between child mortality and maternal education is particularly strong in very poor countries. Caldwell (1986) has also emphasized the importance of being able to recruit in every village educated young women to train and work as health auxiliaries or midwives.

The role of education, then, is clear, but consideration should also be given to the reasons for differing levels of education provision. One revealing contrast is between cultures or religions like Buddhism that emphasize the pursuit of enlightenment and those societies in large parts of Latin America and the Middle East where entrenched class and gender élites feel threatened by mass education.

The health implications of female status and autonomy have been demonstrated clearly in India by Dyson and Moore (1983), Caldwell (1986) and Basu (1989). They show how the better position of women in southern India, especially Kerala, encourages them to take independent responsibility in seeking health care for themselves and their children. In northern India, on the other hand, the low status and often isolation of the incoming wife in the husband's family home reduce her capacity to manipulate the personal environment of herself and her children. Dyson and Moore also consider the broader geographical setting, suggesting that the north Indian kinship system and its demographic characteristics can be viewed as part of a wider West Asian system, whereas the southern Indian model is related to the South and East Asian kinship constellation. In other words, India is at the transition of two major cultural zones with the fundamental fashioning influences possibly lying in differences of religion and agrarian ecology.

It is also clear that at any level of economic development the greatest progress in mortality reduction has been made by those countries with distributional

Table 3.12 Population statistics of India and Kerala State, c. 1980

	India	Kerala
Crude death rate (per 1,000)	14	7
Infant mortality rate (per 1,000)	130	55
Crude birth rate (per 1,000)	35	26
Males per 1,000 females	1,069	967
% adult males literate	47	74
% adult females literate	25	65
GNP per capita ($)	230	190

Source: Visaria and Visaria (1981), Tables 3 and 6.

policies that explicitly emphasize improvement of social welfare of the population at large through reasonable equity of access to land, employment, education and health care (Repetto, 1979; Flegg, 1982; Ruzicka and Hansluwka, 1982). The Communist countries of China, Cuba and in the 1980s, Nicaragua, provide clear illustrations, but other examples include Sri Lanka (at least until the growing instability and violence of the 1980s) and the Indian state of Kerala.

Table 3.12 indicates that although the per capita income of Kerala is well below that of the country as a whole, its mortality (and fertility) rates are also much lower; moreover, its sex ratio is one of the few in India not to show the male excess that normally indicates severe neglect of female children. This desirable demographic position is widely attributed to a long tradition of political activism, radicalism and leftist governments, expressed in land reform, subsidized food and dense networks of schools and health centres. All this demonstrates that mortality reduction is enhanced by mass, rather than élite, participation in the development process.

Refined medical intervention programmes

One must beware of 'throwing the baby out with the bath water' in that Western medical skills and technology still have an important role to play if they can be directed to providing health care at low cost by health workers rather than scarce doctors. UNICEF has been particularly active in this field with a programme for poor countries involving:

1. promotion of the benefits of breastfeeding;
2. mass production and distribution of packets of salts for the treatment of diarrhoeal dehydration, the world's major child killer. Each packet, containing glucose, sodium chloride, sodium bicarbonate and potassium chloride, costs less than 20 cents, yet such is the efficacy of the salts mixture that a *Lancet* editorial has described it as 'potentially the most important medical advance this century';
3. use of growth charts to monitor the weight and therefore the nutritional status of infants;
4. mass immunization programmes, taking advantage of recent advances in vaccine technology, against the six common childhood diseases that are preventable by vaccination – diphtheria, tetanus, whooping cough, measles, polio and tuberculosis. This extended programme of immunization (EPI) has had to face major logistical problems in persuading, accessing and servicing widely

distributed populations, but imaginative promotion in countries like Brazil (national immunization days propagandized at football matches) has achieved remarkable results. In 1988 WHO estimated that half of the infants born in less developed countries received the full three doses of DPT and oral polio vaccines.

It does seem, then, that further progress in mortality reduction in most Third – World countries will probably depend on fundamental societal change, particularly in improved access to land, employment, education, food and primary health systems. This is a socio-political rather than medical or economic matter. There is, however one great unknown – the progression of a modern epidemic disease which as yet has no cure and no easy prevention.

AIDS

This is a wholly exceptional infectious disease in that during the 1980s it has been increasing its incidence and mortality toll in both developed and less developed countries. Its greatest severity is probably in sub-Saharan Africa, where estimates from WHO data (Bongaarts, 1988) suggest that less than 10 per cent of the world's population there contains perhaps 30 per cent of all people infected with the HIV virus, the causative factor of AIDS. HIV prevalence among the region's adults was perhaps one per cent in the late 1980s, reaching 10 per cent in some urban areas in East and Central Africa. On the basis of several assumptions based on late 1980s conditions and on no successful medical interventions, Bongaarts estimates that HIV prevalence in a seriously affected Central African population could well reach 20 per cent of adults by the year 2000, when the crude death rate would be 26 per 1000 (instead of a likely 13 per 1000 in the absence of AIDS).

The essentially behavioural reasons for the particularly high concentrations in sub-Saharan Africa have been discussed by Caldwell et al. (1989). They emphasize that in traditional African social systems lineage bonds are stronger than conjugal ones, encouraging widespread polygyny (multiple wives) and relatively relaxed attitudes towards pre-marital and extra-marital sexual relations. Clearly this encourages the spread of all sexually transmitted diseases. They conclude that the future level of AIDS is more likely to be decided by lifestyle trends than by medical developments – something, at least, that does conform to the more sensitive pre-AIDS views on the future path of Third World mortality.

4

MORTALITY: VARIATIONS WITHIN COUNTRIES

In the recognition, description and interpretation of cross-sectional mortality variations within countries, there are two, often distinct, academic approaches. First, there is the traditional interest of demographers and medical sociologists in the field known as differential mortality – the way in which mortality varies in relation to social categories of class, marital status, ethnicity, community of residence, etc. Second, there is the central concern of medical ecologists, epidemiologists and medical geographers with spatial patterns of disease and associative environmental factors. It is unfortunate that demographers often seem unaware of the spatial implications of differential mortality, notably the way in which social categories are spatially concentrated. Likewise, some medical geographers fail to appreciate that disease distributions can be understood perhaps as much through the spatial distribution of social categories as through basic environmental influences like water supply and atmospheric pollution. Consequently, a major theme of this chapter will be an attempted integration of the two approaches.

Attention will be focused initially on what arguably are the two most important independent variables in differential mortality: social class and community of residence. Many other variables conventionally recognized as influencing mortality are often themselves controlled by these two primary variables.

Social class

Social class is a multi-dimensional concept of social stratification based on the broad relationships which exist between specific components of socio-economic status like education, occupation and income. 'Social class is a structural reality throughout the world and has to be used as one of the principal, scientific categories explaining social behaviour' (Townsend, Phillimore and Beattie, 1988, p. 149).

The British evidence

The best-known and most authoritative evidence for the association between social class and mortality is derived from mortality data by occupation assembled at ten-year intervals from as early as 1851 by the Registrar General in England

Table 4.1 Standardized mortality ratios by social class, males 15 (or 20)–64 years, England and Wales

<table>
<tr><th></th><th colspan="7">Social class</th></tr>
<tr><th></th><th>I</th><th>II</th><th>IIIN</th><th>IIIM</th><th>IV</th><th>V</th><th>All</th></tr>
<tr><td>1930–1932</td><td>90</td><td>94</td><td colspan="2">97</td><td>102</td><td>111</td><td>100</td></tr>
<tr><td>1949–1953</td><td>86</td><td>92</td><td colspan="2">101</td><td>104</td><td>118</td><td>100</td></tr>
<tr><td>1959–1963</td><td>76</td><td>81</td><td colspan="2">100</td><td>103</td><td>143</td><td>100</td></tr>
<tr><td>1970–1972</td><td>77</td><td>81</td><td>99</td><td>106</td><td>114</td><td>137</td><td>100</td></tr>
<tr><td>1979–1983*</td><td>66</td><td>76</td><td>94</td><td>106</td><td>116</td><td>165</td><td>100</td></tr>
</table>

*Great Britain
Source: Registrar General (England and Wales), *Decennial Supplements.*

and Wales. Occupational mortality rates are derived from two data sources: the death registration system provides the last full-time occupation of the deceased, and the decennial census of population records occupation. Accordingly, for any particular occupational group one can relate national deaths in a period of years straddling census year (numerator) to the census population (denominator) to provide an occupational mortality rate. It is difficult to define occupation meaningfully and consistently for retired older persons, so that occupational mortality rates are typically limited to the 15–64 age group. A three- or five-year period is used to provide sufficiently large numbers of deaths to ensure statistically reliable results in all but the smaller occupational groups and the minor causes of death.

Since 1921 occupations have been grouped into five social classes, with the implication that occupation is a meaningful indicator of living standard and life-style, embracing income, education, housing, leisure, diet, etc. (Leete and Fox, 1977).

I Professional (e.g. doctor, lawyer).
II Intermediate (e.g. manager, teacher).
IIIN Skilled non-manual (e.g. clerk, shop assistant).
IIIM Skilled manual (e.g. miner, bricklayer).
IV Partly skilled (e.g. postman, bus conductor).
V Unskilled (e.g. cleaner, labourer).

Social class is not assigned to the unemployed, disabled, armed forces, full-time students and inadequately described occupations.

The appreciable differences in mortality by social class revealed by the England and Wales data for the last half-century are shown in Table 4.1 by standardized mortality ratio (SMR), the index commonly used to standardize for differences between categories like class in age composition.

The SMR is the ratio of observed to expected deaths for each category of the total population being considered:

$$\text{SMR} = \frac{\text{observed deaths}}{\text{expected deaths}} \times 100$$

The expected deaths are those that would occur if the age-specific death rates for the total population were to apply in the specific category being studied. To provide a worked example, the mortality data from Table 4.2 can be used to calculate the 1970–1972 SMR for social class I (males 15–64) as follows:

Table 4.2 Selected mortality data, males 15–64 years, England and Wales, 1970–1972

	All social classes			Social class I	
	Population 1971	Deaths 1970–1972	3-year death rate (per person)	Population 1971	Deaths 1970–1972
15–24	3,584,320	9,935	0.0028	95,190	193
25–34	3,065,100	9,238	0.0030	214,680	431
35–44	2,876,170	19,911	0.0069	171,060	854
45–54	2,965,880	64,045	0.0216	137,080	2,079
55–64	2,756,510	170,000	0.0617	100,000	5,029
Total	15,247,980	273,129	0.0179	718,010	8,586

Source: Office of Population Censuses and Surveys (1978b), table 2.3. Crown Copyright.

$$
\begin{aligned}
\text{Observed deaths in social class I} &= 8{,}586\\
\text{Expected deaths in social class I} &= 95{,}190 \times 0.0028\\
&+ 214{,}680 \times 0.0030\\
&+ 171{,}060 \times 0.0069\\
&+ 137{,}080 \times 0.0216\\
&+ 100{,}000 \times 0.0617\\
&= 11{,}222
\end{aligned}
$$

$$\text{SMR} = \frac{8{,}586}{11{,}223} \times 100 = 77$$

SMR of 100 indicates that the age-standardized mortality level in the category being studied is the same as in the overall or standard population (observed deaths = expected deaths), so that SMR of 77 in social class I indicates a mortality level well below that of the overall population.

From the very beginning of occupational mortality study in Britain by William Farr, the first compiler of statistical abstracts at the General Register Office, nineteenth-century social statisticians were well aware of occupational and class variations in mortality which favoured the well-to-do (McDowall, 1983; Woods and Woodward, 1984). Table 4.1 demonstrates strikingly that these variations have not only continued into the twentieth century, but may well have increased in intensity. Meaningful comparisons can be made only within each row of the table, where there is a progressive increase in adult male mortality from class I to class V at every cross-section in time. Far from there being a theoretically plausible narrowing in mortality differentials associated with the modern welfare state, there seems to have been a widening of the class gulf in survival. These appreciable class gradients are confirmed by mortality among single women classified by own occupation and among infants, children and married women classified by occupation of father or husband (Office of Population Censuses and Surveys (OPCS), 1986, 1988).

Criticisms of data

So politically contentious have been the implications of such findings that careful examination is required of data quality. There have been several criticisms of the compilation, analysis and interpretation of the occupational mortality data (e.g. Illsley, 1986).

Because of difficulties in assigning occupation and therefore class to retired persons, the published data are confined to deaths of persons below 65 years of age. Inevitably this restricts analysis to a minority of all deaths, especially in more recent years. There have also been changes in class composition which complicate comparisons over time. These are due partly to the classification changes by successive Registrars General in occupation allocation and partly to structural shifts in the pattern of employment which have led to more non-manual and fewer manual jobs. In particular, the size of class V has been eroded by upward mobility. A very different form of mobility is when some people, because of poor health, are forced to move down the occupational scale and find themselves in jobs prior to death that are below the status of their normal occupations. The causation here is from health status to class, and not vice versa.

Finally, there is the problem known as numerator-denominator bias, which is thought to inflate artificially the class mortality gradient, especially in recent censuses. The occupations recorded on death certificates by relatives (the numerator) are much less precise than the self-recorded occupations at census (the denominator), so that a significant proportion of deaths has to be assigned to residual categories in the occupational classification. This objection, however, is invalid for analyses of infant and child mortality since, in the case of infant deaths, both numerator and denominator data are collected through birth and death registration processes; and for child deaths the information is invariably given by the same person, the parent, at census and at death registration.

Criticisms confronted

That the criticisms above cannot effectively question the presence of an appreciable mortality gradient by class in contemporary Britain has been established by two recent forms of enquiry, one which reworks the standard official data and one which collects and analyses completely new longitudinal data.

Many of the analytical problems disappear when the data are aggregated into a smaller number of occupational and class groups like the manual and non-manual categorization of Marmot and McDowall (1986). Their aggregation of the 1979–83 data (Table 4.3) confirms the appreciable mortality difference between the two groups, not only for overall mortality but also for each of the major causes of death, even for coronary heart disease which has sometimes been associated with stressful executive-type occupations. They also show that the class mortality gap has genuinely widened in recent decades, a finding confirmed by the analysis of Townsend, Phillimore and Beattie (1988) on aggregated data for classes I–II and classes IV–V and by Pamuk's (1985) recalculation of SMRs over time based on standardized occupation units.

Table 4.3 Standardized mortality ratios for selected causes of death in Great Britain, 1979–83, among males age 20–64*

	Non-manual occupations	Manual occupations
All causes	80	116
Lung cancer	65	129
Coronary heart disease	87	114
Cerebro-vascular disease	76	120

* For each cause the SMR for all males aged 20–64 is 100.
Source: Marmot and McDowall (1986). By permission of The Lancet Ltd.

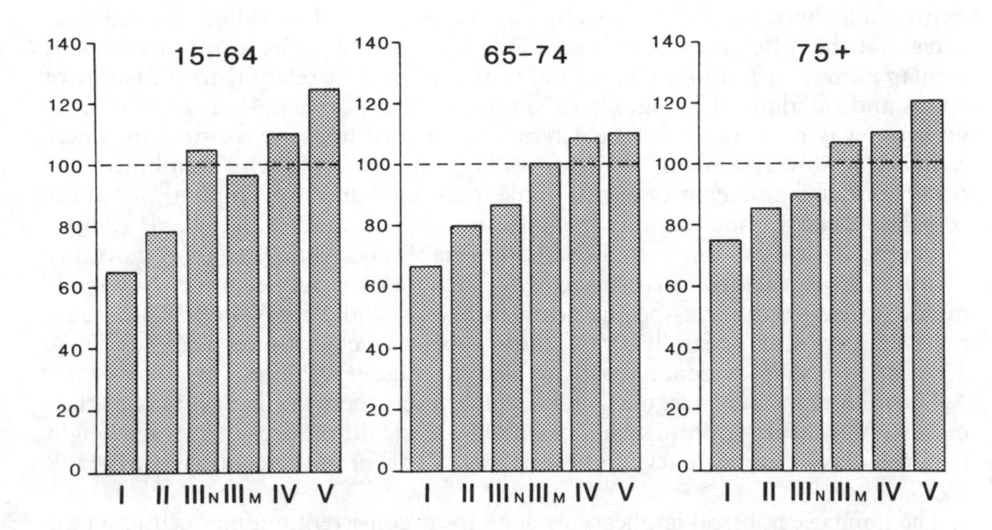

Figure 4.1
Standardised mortality ratios, 1976–81, for males in England and Wales by social class and age group
Source: Fox, Goldblatt and Jones (1985), figure 3. By permission of A. J. Fox.

Invaluable, unbiased data on mortality are now being routinely provided by the OPCS Longitudinal Study (Fox and Goldblatt, 1982). A 1 per cent sample was drawn from the 1971 census of England and Wales and has been supplemented subsequently by a small proportion of births and of immigrants, so that the sample continues to represent 1 per cent of the total population. Vital events, notably registrations of births, deaths and cancers, affecting members of the sample are incorporated into the data set. In the subsequently compiled mortality data, there are no numerator-denominator biases; the problem of downward mobility in status because of poor health is much reduced; and the assigning of a meaningful occupational class group to the retired now becomes feasible.

Table 4.4 Standardized mortality ratios, 1971–81, for males aged 15–64 and females aged 15–59 in Great Britain by housing tenure and access to cars*

	Males	Females
Housing tenure:		
Owner-occupied	84	83
Privately rented	109	106
Local authority	115	117
Access to cars:		
One or more	85	83
None	121	135

* For all members of each sex the SMR is 100.
Source: Whitehead (1988), Table 2, using unpublished data from the OPCS Longitudinal Study.

The evidence from the Longitudinal Study (Figure 4.1) confirms the presence of appreciable mortality differentials by class among men of working age, and indicates that the differentials continue, moderated only slightly, into retirement. A similar pattern is found when mortality is analysed in relation to measures of status and standard of living other than occupation (Table 4.4). These measures go some way to recognizing the determining role of household situation. There remains, however, a major need for a diagnostic variable which embraces the occupational characteristics of the whole household and not simply those of the so-called 'head' of household with all its out-moded connotations (Pahl, 1984).

There can be little doubt, then, that appreciable class differentials in mortality continue to characterize British society in the late twentieth century. New-born males subject to the age-specific death rates of social class V in 1970–2 could expect to live seven years less than those subject to class I rates (Fox, 1977, p. 13). There is also evidence from the Health and Lifestyle Survey (Cox *et al.*, 1987) for a remarkably regular, and in some cases steep, gradient by social class on several measures of fitness and morbidity, while differences in average height between classes are as marked in the 1980s as earlier in the century (Carr-Hill, 1988).

The immense political implications of all these consistent findings centre on the extent to which such inequality can be regarded as injustice. Indeed, two powerful official reports which investigated health and survival inequalities (reported by Townsend and Davidson, 1982; Whitehead, 1988) were given a churlish reception and restricted distribution by a Conservative government committed to the values of private enterprise and the market-place and hostile to the interventionist, collective-responsibility stance of the reports' recommendations.

Explaining health inequalities

There are four major determinants.

Occupational risks

Manual workers, in general, are more subject to specific industrial hazards and accidents than non-manual workers (e.g. male SMRs in 1979–83 of 168 for dockers, 180 for steel erectors and scaffolders, but 61 for schoolteachers and 74 for lawyers). Higher proportions of the less skilled than of the professional classes work out of doors or in noisy factories, and have poorer access to good wages, sick pay, pensions, holidays, job security and job satisfaction – all of which can be expected to affect health.

Access to health care

Some geographers have demonstrated that the spatial patterning of health-care delivery systems often compounds the disadvantages of poor people living in poor neighbourhoods. This occurs particularly under a free-enterprise health-care system as in the United States (Shannon and Dever, 1974), but even within supposedly egalitarian welfare-state systems like Britain, inequalities of access persist. Knox (1978) has demonstrated the marked under-doctoring of peripheral public-housing estates in British cities, Haynes (1987, ch. 7) has described the reduced quality, if not the density, of general practitioner services in inner cities, while Hart (1971, p. 412) eloquently formulated the inverse care law that 'the availability of good medical care tends to vary inversely with the need of the population served'.

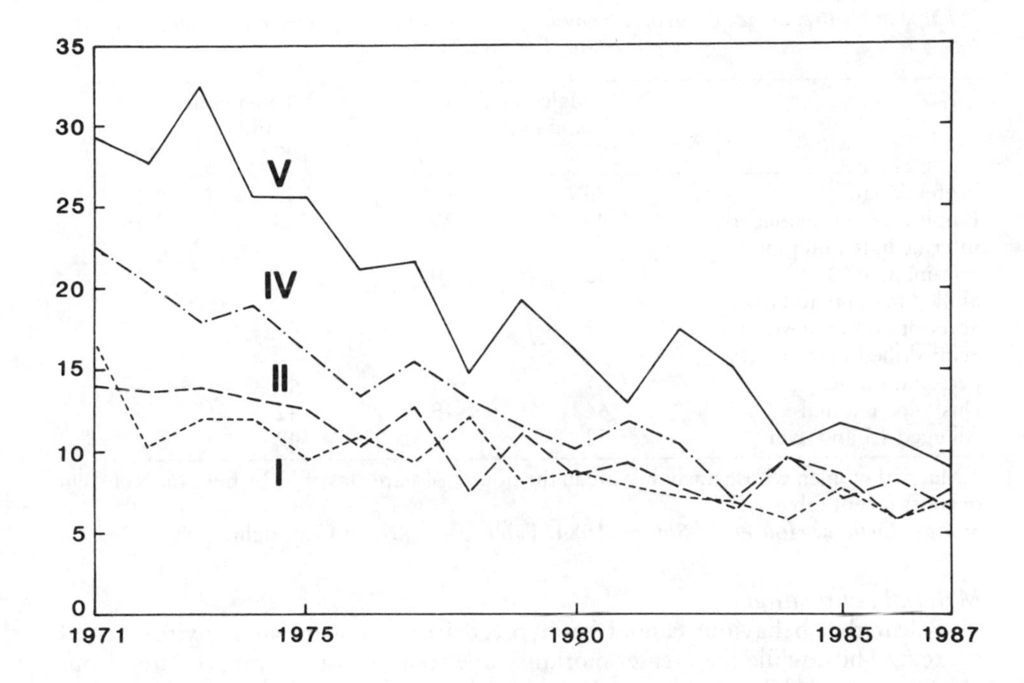

Figure 4.2
Infant mortality rates (per 1,000) in social classes I, II, IV and V, Scotland 1971–87
Source: Registrar General Scotland, *Annual Report 1987*, HMSO, Edinburgh, p. 92.

When adequate medical care is provided equitably, or with some discrimination in favour of disadvantaged groups, it can certainly reduce health inequalities. Thus Figure 4.2 charts the considerable narrowing of class differentials in infant mortality in Scotland from 1971 in response to the provision of community-based services in the antenatal, postnatal and child health fields. In particular, health visitors have played an important role in the identification and treatment of infants at risk (Walker, 1985).

Behavioural influences
These are the ways in which members of different social groups *choose* to lead their lives. It can be argued that members of lower status groups follow more harmful life-styles than the higher social groups, exhibiting a 'culture of poverty' (Lewis, 1959) embracing attitudes of fatalism, apathy and lack of long-range planning. There is abundant evidence to indicate that lower status groups have the highest proportions of smokers (Table 4.5) and of heavy drinkers; that they have been less responsive to recent nutritional advice emphasizing the need to eat more fibre, fresh fruit and vegetables and less sugar, salt and animal fat; that they participate less in exercise-type leisure activities; and that they are more infrequent users of preventive services like antenatal facilities, cervical screening, dental checks and child immunization. Much more debatable is the extent to which this class-based behaviour is voluntary, since, at its crudest, this argument smacks of 'blaming the victim'.

Table 4.5 Prevalence of cigarette smoking by sex and socio-economic group, Great Britain, 1972 and 1984*

	Males aged 15 and over (%)		Females aged 15 and over (%)	
	1972	1984	1972	1984
Professional	33	17	33	15
Employers and managers	44	29	38	29
Intermediate and junior non-manual	45	30	38	28
Skilled manual and own account non-professional	57	40	47	37
Semi-skilled manual and personal service	57	45	42	37
Unskilled manual	64	49	42	36
All aged 15 and over	52	36	42	32

* Married women whose husbands are in the household are classified by husband's current or most recent job.
Source: *General Household Survey*, 1984, Table 10.5. Crown Copyright.

Material deprivation
It is clear that behaviour cannot be divorced from its social and environmental context. Thus, while the greater mortality due to accidents among children from classes IV and V households may sometimes be attributed to feckless and reckless parental behaviour, a more sustainable explanation focuses on factors like high-rise housing, lack of private gardens, unsafe play areas and insufficient income for good child-minding services. Indeed, a Marxist explanation would focus on the structural role of class as part of the central mechanism of the capitalist mode of production.

There is no doubt that the combination of low income and poor housing is harmful to health (Byrne *et al.*, 1986; Townsend, Phillimore and Beattie, 1988). Yet, after some decades of gradual reduction in the wealth gap between classes in Britain, government policies have widened that gap again in the 1980s.

The international evidence

Analyses of death registration and census data, using similar methods and providing broadly similar results to the British analyses, have been conducted in the United States (Guralnick, 1963; Kitagawa and Hauser, 1973). The European evidence (Fox, 1988; Whitehead, 1988, ch. 5) also confirms the presence of important class differentials in morbidity and mortality, although it is significant

Table 4.6 Infant mortality rate (per 1,000 live births) by maternal educational level, Peru (1981–6), Colombia (1976–86) and Dominican Republic (1976–86)

	Peru	Colombia	Dominican Republic
No education	124	60	102
Primary	85	41	76
Secondary	42	28	56
Higher	22		34

Source: Studies in Family Planning (1988), Vol. 19, pp. 125, 194, 308.

that these are least evident in Denmark, Sweden and Norway, where there has been a more sustained government commitment than elsewhere to egalitarian policies.

In less developed countries it is a socio-spatial group, the urban élite, which enjoys considerable health and survival advantages through its superior access to education, wealth and medical services (Hobcraft, McDonald and Rutstein, 1984). The particularly important role of maternal education in child health has been discussed in the previous chapter and can be illustrated further by infant mortality data for selected countries (Table 4.6).

Community of residence

A spectacular differential in mortality associated with economic development and urbanization in nineteenth-century Europe and North America is the rural–urban gradient. The insanitary, congested living conditions which spawned epidemics of often calamitous proportions in the major cities of that era have been vividly described by social historians and novelists. The particularly high mortality rates of Paris, Lyon and Marseille at that time have already been noted (Figure 3.8), while in Britain male life expectancy at birth in 1841 was a mere 24 years in Manchester, 25 in Liverpool and 35 in London, compared with 40 for England and Wales as a whole and 44 for the then rural county of Surrey (Glass 1963–1964, p. 265).

Although urban–rural differentials have narrowed considerably in this century, they are still recognizable in most, although not all, developed countries (Federici, 1976). A major interpretative problem is to identify those dimensions of urban life which promote higher mortality. These might be grouped into either class-related factors such as occupation, income, diet and smoking, or factors associated more specifically with the physical nature of urbanism, notably housing, residential and working densities, recreational opportunities and atmospheric pollution.

But the crudity of imposing a simple urban–rural categorization on the complexity of settlement patterns is now increasingly recognized. Accordingly, British mortality data from the OPCS Longitudinal Study have been profitably analysed by 34 ecological clusters derived by Webber and Craig (1976) from grouping 5,500 wards on the basis of 40 census variables. The results (Table 4.7) reveal a clear pattern of low mortality in high-status clusters and high mortality in low-status clusters, but secondary analysis (Fox, Jones and Goldblatt, 1984, p. 313; Fox, Jones and Moser, 1985, p. 13) establishes that there is an important ecological or area contribution to the pattern that is independent of determination by social class composition.

All four rural clusters (13, 14, 15, 31) in Table 4.7 have low mortality rates, but rural populations as an aggregate group pose a particular problem of interpretation since they span the spectrum from affluent big-city commuters to the physically isolated rural poor. Thus in the United States some of the lowest age-standardized mortality rates are found in the wealthy 'rururban' fringes of the north-eastern cities, while very high rates are found among the black rural poor of the south. A smaller-scale example is found in eastern England, where the generally low mortality rates of rural Norfolk obscure appreciable differences between very low rates in prosperous, growing areas close to the main towns and relatively high rates in the remoter, more sparsely peopled areas (Bentham and Haynes,

Table 4.7 Standardized mortality ratios of males, 1971–81, in Great Britain by ecological cluster of residence

Areas of young and growing population		
1.	New towns	99
2.	Planned developments, smaller towns	103
3.	Very new council housing	110
4.	Modern, low-cost, owner-occupier housing	108
5.	Not-owner-occupied housing in areas of growth	97
6.	Modern high-status housing, young families	87
7.	Military bases	73
Areas of older settlement		
8.	Edwardian development	104
9.	Older industrial settlements with low stress	105
10.	Market towns	100
11.	Inner areas with low-quality older housing	130
12.	Poor-quality housing in areas of economic decline	114
Rural areas		
13.	Villages with some non-agricultural employment	92
14.	Rural areas with large land holdings	90
15.	Rural areas with small land holdings	84
Urban council estates		
16.	Overspill estates	126
17.	Local authority housing in Scotland and North East	115
18.	Urban local authority estates with good job opportunities	114
19.	Mining areas	115
20.	Inter-war local authority housing	97
21.	Inner-city council estates	123
22.	Areas of local authority housing with single people	110
Areas of multi-occupancy, students and immigrants		
23.	Inner London	105
24.	Multi-occupied inner London	104
25.	Multi-occupied and immigrant areas	112
26.	Student areas and high-status Central London	94
27.	High-status rooming-house areas	90
Areas of extablished high status and resorts		
28.	Modern high-status areas	86
29.	Mock-Tudor areas	92
30.	Established high-status areas	87
31.	Rural established high-status areas	83
32.	Very-high-status areas	85
33.	Residential retirement	87
34.	Seaside and retirement	92

Source: Fox, Jones and Goldblatt (1984) Table 1. By permission of A. J. Fox.

1985). Indeed, it is thought that this pattern applies in most of rural Britain (Haynes, 1987, ch. 6), with the highest mortality being found in areas experiencing population decline (Bentham, 1984), suggesting the health-selective role of out-migration.

Remoter rural areas, then, may no longer be retaining their privileged mortality position in developed countries. Indeed, there has been growing concern about the acute disadvantage suffered by some rural dwellers in terms of access to health-care personnel and facilities (Moseley, 1979; Shaw, 1979). In rural Britain the fundamental problem is that rising car ownership has undermined both the traditional village services as well as the public transport system that is needed by the vulnerable car-less poor to access centralized services. In the United States, Shannon and Dever (1974) have shown that during the course of this century physicians, particularly younger ones, have been leaving and avoiding rural areas due to the pull of urban client wealth, the urban home background of an increasing majority of medical students, the attraction of urban cultural facilities, and, perhaps most important of all, the growing dependence of physicians on the equipment and auxiliary personnel of city hospitals. They also demonstrate that there is a clear distance-decay function in the use of medical facilities and that people living at greater distances tend to make their visits for curative rather than preventive purposes. These findings are confirmed by a study of rural health care in the American South (Davis and Marshall, 1979). Nevertheless, it is possible to exaggerate the contribution of medical care to health provision in developed countries, since rural populations in general are still not experiencing higher mortality than urban populations.

In less developed countries the pattern is very different. It is widely recognized that urban mortality rates are generally lower than those in rural areas (Federici, 1976; Gilbert and Gugler, 1982). Certainly the reliability of spatially disaggregated mortality data is questionable, but it is likely that such data *understate* urban–rural differentials, since data on infant and child mortality are normally derived from retrospective surveys. Given extensive urbanward migration, this means that some deaths reported as urban actually occurred in rural areas. The lower urban mortality reflects the better, although still inadequate, provision of clean water, sanitation, education and modern medical facilities in the urban areas. This is, however, an over-simplification, because of the major contrasts within the larger urban settlements between the élite sectors and the poorer quarters, which suffer as much poverty and probably more environmental deprivation than the rural areas. Many Third World countries still exhibit a colonial, enclave mode of health care originally developed to protect the army and the European civil population and then maintained to benefit the spatially segregated urban élites of the new independent states (Ramasubban, 1984). Such disparities are only recently beginning to be redressed by basic needs strategies and primary health-care policies that are socially and spatially more equitable.

Other differentials

Demographers have recognized mortality variations within societies by social categories based on several other variables, including marital status, parity and ethnicity. The last variable has been reviewed by Curson (1986) in relation to the elevated mortality of Aboriginals in Australia, blacks and Hispanics in the United States, Asians and West Indians in Britain, and Maoris and immigrant

Polynesians in New Zealand. But it is often difficult to specify the independent role of such variables, since they are often correlated with the two primary variables discussed above.

Consider the long-standing, although slowly narrowing, racial mortality differential in the United States. In 1985 life expectancy at birth was 75.3 years for whites and 69.5 years for blacks, and their respective infant mortality rates were 9.3 and 18.2 per 1,000. Biological differences in survival potential are unlikely to have made a significant direct contribution. Rather the answer lies in the disadvantaged position of blacks in housing, income, education, diet and health care. In other words, 'the race differentials, in general, are consistent with the inverse relationship between mortality and socio-economic status' (Kitagawa and Hauser, 1973, p. 102). This is not to deny that there are sometimes behavioural factors specific to some ethnic groups that prejudice their health, like the much higher smoking rates among Maoris and immigrant Polynesian men than European-origin population in New Zealand (Heenan, 1983).

Spatial patterns

The spatial distribution of disease mortality and morbidity within countries is the central concern of medical geography, although recent work in this branch of geography has focused rather more on studies of the spatial provision of medical services grounded in location theory and concerned with optimizing access, efficiency or equity (Mayer, 1982). Whether traditional medical geography should be regarded more appropriately as a component of population geography, for which a strong theoretical case can be made, or as an independent sub-discipline, which is its current status, is a fairly insignificant matter of academic organization. It is more important to appreciate the three overlapping dimensions which characterize the geographer's approach to studies of disease.

Ecological

Ecology is the study of relationships between living organisms and their surroundings. Since geography developed primarily as an ecological discipline, it can be argued that 'the geographer's contribution to medical knowledge can properly be expected to lie chiefly in the field of environmental studies and the relation of disease distribution to other geographic variables' (McGlashan 1967, p. 333).

The roots of medical geography are often traced to the environmental awareness of Hippocrates, notably his appreciation of the links between climate and disease in *De aere, aquis et locis* (*On Airs, Waters and Places*). But for practical purposes the foundations of medical geography were laid in the 1950s. In the Soviet Union there was a concentration by Pavlovsky and others on 'landscape epidemiology' which 'delimited the foci of infectious, zoonotic diseases by analysing the associations of vegetation, animal and insect life, soil type and acidity, precipitation regime, and other elements of the natural landscape. General landscape patterns could be used to assess the likely presence of the infectious agents of diseases' (Meade, 1977, p. 380). Soviet medical geography has always been closely associated with biogeography and with the planning of newly developed territories.

In the West, the critical influence was that of May. In his prolific writing (e.g. May, 1961) he guided medical geography towards understanding the processes of medical ecology. Since his interest was largely in tropical vector-borne diseases, he paid particular attention, as in the Soviet studies, to natural environmental risk factors, in which he distinguished between inorganic influences (climate, hydrology, geology, etc.) and organic influences (especially the cycles of interdependence between infective micro-organisms, plants, animals and man). His working links were with ecologists, entomologists, medical scientists and biogeographers rather than social scientists. But humanity has always been a critical element in the ecosystem, as May and his followers came to recognize in their increasing appreciation of socio-cultural influences. As societies become modernized and more complex it is increasingly difficult to attribute disease to the natural environment, as the previous studies of differential mortality based on socio-economic variables have demonstrated. Even in less developed societies it is the human being who often creates the necessary habitat conditions for foci of disease, as subsequent case studies of malaria in Sri Lanka and schistosomiasis in Africa will demonstrate.

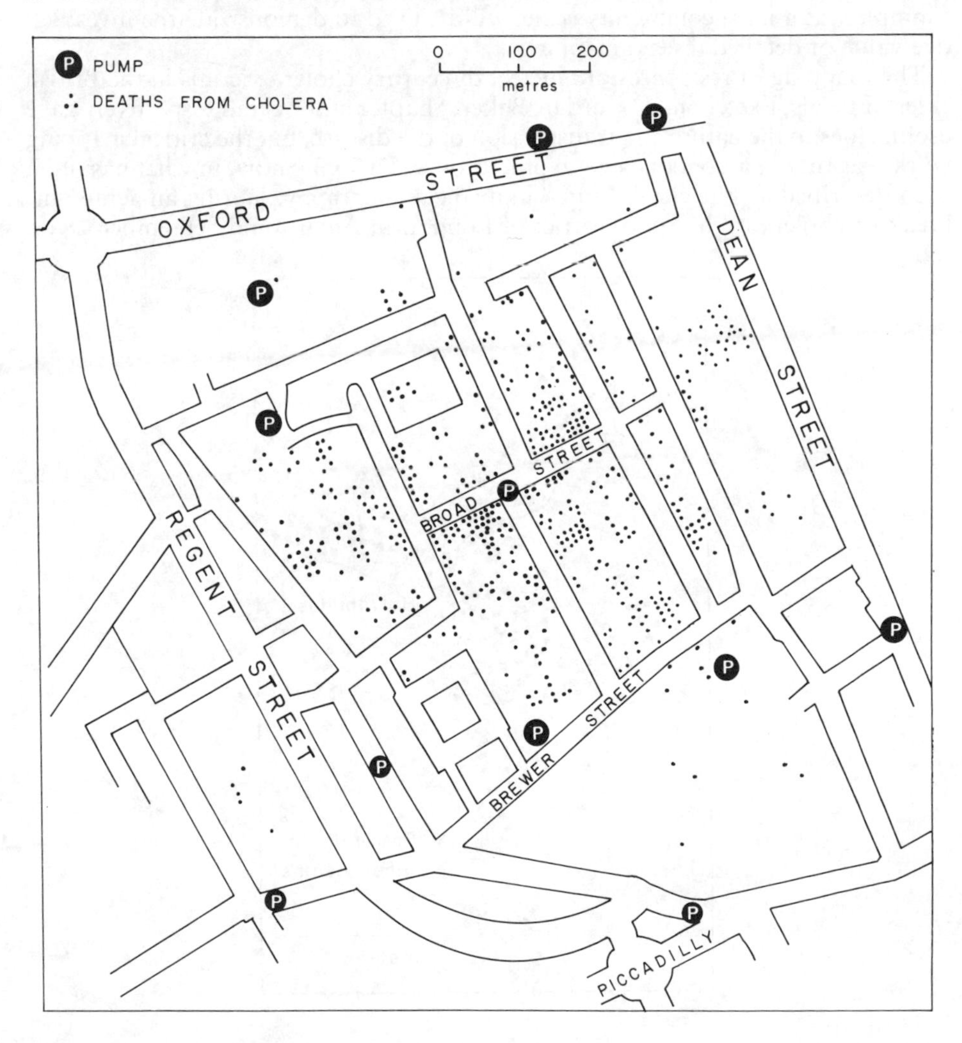

Figure 4.3
Snow's map of cholera deaths in the Soho district of London, August–September 1848
Source: Stamp (1964), figure 9.

Cartographic

The spatial occurrence of disease has been neglected until fairly recently in most health fields outside epidemiology, largely because of traditional medical preoccupation in understanding disease through clinical, laboratory-based, experimental work. This has stimulated some geographers to make what they feel is a positive contribution to the understanding of disease aetiology through the mapping of disease incidence and of possibly causal associative factors. An early and influential expression of this view was by Stamp (1964, pp. 66–7), who was 'convinced that no more important work is waiting to be done than the very detailed mapping of the incidence of different forms of cancer and the search for possible correlation with causative factors. With all the many millions being spent on cancer research, this approach is being seriously neglected.' Three British examples, at a local community scale, will be used to demonstrate the investigative value of detailed disease mapping.

The mapping of residences of nineteenth-century cholera victims in the British cities of Leeds, Exeter and Oxford by Baker, Shapter and Acland respectively gave useful clues to the nature and transmission of the disease, but the critical mapping work was that of a London general practitioner, Dr John Snow, in what has often been described as 'the classic study' in medical geography. During an acute outbreak of cholera in the Soho district of London in August and September 1848,

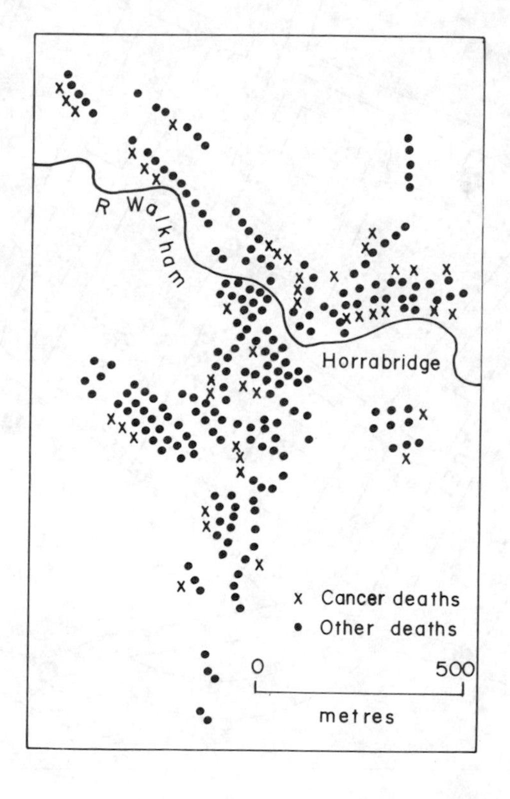

Figure 4.4
Distribution of deaths from cancer and from all other causes in Horrabridge, UK, 1939–58
Source: Simplified from Allen-Price (1960), figure 2.

Snow marked on a large-scale map the residences of the 500 victims (Figure 4.3). He was obviously testing a hypothesis, since he also marked the stand-pumps which supplied drinking water to the population. As Figure 4.3 indicates, there was a heavy clustering of deaths around the Broad Street pump, and when the pump handle was removed at Snow's insistence the cholera outbreak subsided. Sceptics argue that the outbreak was self-limiting and that the dismantling of the pump was essentially symbolic, yet it was Snow's map that clearly established the critical role in cholera transmission of contaminated drinking water. It was not until 1883 that the infective micro-organism, *Vibrio cholerae*, was identified by the German bacteriologist Robert Koch.

Links between cancer mortality and water supply have been suggested strongly by the mapping of over 5,000 deaths, classified as either cancer deaths or deaths from other causes, in West Devon over a 20 year period (Allen-Price, 1960). An extract from the maps is shown in Figure 4.4 for the village and environs of Horrabridge.

> Here there is a homogeneous group, following the same occupations, eating the same food, and in an identical environment, and merely separated one from another by the natural boundaries of the river Walkham, which is crossed by a bridge. Here, for generations, the people have intermarried freely, and their social activities have been confined, yet each artificial section of the community has a widely different cancer mortality. As far as can be assessed, the only difference that could account for this is their water supply.
>
> (Allen-Price, 1960, p. 1238)

Allen-Price was able to show that for West Devon as a whole the highest incidence of cancer was in populations deriving their water from highly mineralized rocks, and subsequent micro-chemical analysis of these water supplies revealed the presence of significant amounts of radon.

A third example is the mapping by Young (1974) of cases of illness diagnosed as food poisoning in the English town of Consett. He was able to demonstrate a pattern dominated by locations downwind from an iron and steel plant. A strong supposition arose that the majority of food poisoning notifications were, in fact, derived from iron oxide dust pollution.

The cartographic contributions of medical geographers towards establishing patterns of disease have perhaps been most evident at regional and national scales. Murray (1967) was the first to attempt a portrayal of areal variations of mortality in the United States on any kind of detailed basis. His depiction of age-standardized death rates by county indicated a strong spatial association between high mortality and areas of traditionally poor socio-economic circumstances – the south, northern inner cities, Indian reservations and Spanish-American districts. In contrast, low mortality characterized much of the agricultural interior, California and the Pacific Northwest. There has been comparable work in Australasia (Heenan, 1975; McGlashan, 1977), while in Britain Howe (1970, 1986) has provided a series of contributions to mapping the national pattern of disease-specific mortality, particularly notable for the development of cartograms or the so-called demographic base map (Figures 4.7 and 4.8) as a less visually biased alternative to the traditional choropleth map. This illustrates that geographers have skills in the conceptual and technical aspects of cartographic presentation and analysis that should not be undervalued, particularly when maps are being used as serious tools of analysis and not simply as illustrations.

Building on these pioneering geographical contributions and benefiting greatly from computerized data analysis and automated cartography, there has been a flood of national and cross-national disease atlases in recent years, including England and Wales (Gardner, Winter and Barker, 1984), Scotland (Lloyd *et al.*, 1987), the United States (Mason *et al.*, 1975), Japan (Japan Health Promotion Foundation, 1981) and the European Community (Holland, 1988). The major analytical, as opposed to educational, role of such atlases is to allow the identification of areal clusters with particularly high rates of specific disease incidence that warrant detailed analytical examination by case-control, cohort or environmental studies. For this purpose, 'geographical studies are relatively quick and economical in terms of both money and manpower because they utilize routinely collected and readily available data' (Davies and Chilvers, 1980, p. 87). This is fair comment on mortality statistics, but it is much more difficult to acquire morbidity data of reasonably uniform accuracy throughout a country. Portrayals of mortality, then, often reveal only the tip of the disease iceberg.

There should also be concern about what may be becoming an excessive concentration on, and replication of, simple descriptive patterns. For example, one recent disease atlas openly describes its contribution in remarkably limited terms:

> The interpretation of the relationship between health service inputs, indicators of social conditions, and avoidable deaths is complex and requires further research; at this stage we are just making relevant data available to enable appropriate questions to be posed and to make people aware of the problems.
>
> (Holland, 1988, p. 2)

Statistical

Medical geography has benefited from the development of more objective and statistically rigorous methods in three key areas.

Probability tests

Tests are now applied routinely to distributions to assess whether or not patterns are likely to have occurred by chance. A simple example is the use in Table 4.8 of the chi-squared test on mortality data from Devon represented in Figure 4.4. There is a rejection of the null hypothesis, which postulates that there is no significant difference between north and south of the river in the frequency of cancer deaths; this null hypothesis would be confirmed 'by chance' less than five times in a hundred.

Several atlases of disease mortality have also addressed the problem that chance factors might well be affecting the distribution patterns of diseases with small numbers of deaths in less populous areas. One response in two recent atlases

Table 4.8 Number of cancer deaths north and south of the river shown by Figure 4.4

	North of river	South of river
All deaths	98	148
Observed cancer deaths	25	18
'Expected'* cancer deaths	17.2	25.9

$\chi^2 = 5.62$ $p < 0.05$

* On the assumption that the proportion of cancer deaths in each of the two areas would equal the proportion in the whole area.

(Gardner, Winter and Barker, 1984; Lloyd *et al.*, 1987) has been the adoption of a scale of values which recognizes both the level of an area's SMR and the statistical significance of its deviation from the national figure. Other responses to the same problem have adopted the Poisson probability formula (Lovett and Gatrell, 1988) or the signed chi-square statistic (Visvalingam, 1978).

Multivariate analysis

Several methods in this field have been used to measure the degree and form of correspondence between the spatial incidence of disease and of postulated associative factors. Classic cartographic correlation by visual inspection of patterns is still important, particularly at an early stage of investigation, but its value is limited by subjectivity, particularly in assessing the multivariate relationships of most diseases which occur only when several factors coincide in time and space.

Spatial data can normally be arranged in a matrix – by variable and by unit of observation – and are thus suitable for correlation and regression analysis at bivariate and multivariate levels. The basic functional multivariate model of disease is:

$$Y = f(X_1, X_2 \ldots X_n)$$

where Y = disease incidence by area
$X_1 - X_n$ = environmental and socio-economic characteristics by area.

Such models have been operationalized extensively at the intra-urban level in, for example, Exeter (Griffiths, 1971), Bristol (Townsend, Simpson and Tibbs, 1985), Sheffield (Thunhurst, 1985), the Northern region of England (Townsend, Phillimore and Beattie, 1988), Sydney and Melbourne (Burnley, 1982), Wollongong (Wilson, 1979) and Chicago (Pyle and Rees, 1971). Invariably such studies demonstrate the positive relationship between indices of mortality and of deprivation, with particularly high values in industrial zones, the inner city and working-class estates.

Factor analysis has sometimes been adopted to examine the fundamental structure of the spatial relationships involving ill-health and associative variables (e.g. Thomas and Phillips, 1978), and trend surface analysis has been used for urban disease mapping by Pyle and Rees (1971). For 271 districts of Chicago they calculate an overall relationship between death rate (dependent variable) and district spatial co-ordinates (independent variables). When mapped (Figure 4.5), these relationships provide generalized or smoothed patterns of disease mortality which override non-systematic, local and chance variations.

There are interpretative problems that analysts should be aware of in any form of multivariate spatial analysis (King, 1979). The most important are the modifiable areal unit problem (Openshaw and Taylor, 1981) and the ecological fallacy. The first relates to the way in which measures of association can vary according to the scale of aggregation of the raw spatial data; correlation measures will generally increase as the size of the areal units is enlarged. The ecological fallacy refers to the dangers of using ecological data to test and establish hypotheses about individuals. In this respect there can be no substitute for behavioural and, if at all possible, longitudinal studies of individuals, but the operational difficulties, considerable time-scale and costs involved in such work mean that ecological, cross-sectional studies, when responsibly used, remain an important analytical method in medical research.

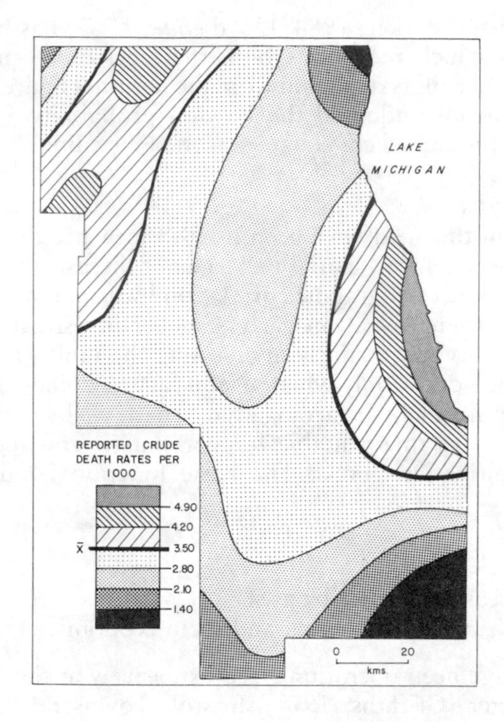

Figure 4.5
Total heart disease mortality, Chicago, 1967
Source: Pyle (1971), figure 14.

Diffusion studies
The term 'diffusion' implies a spreading in time and space from an innovation centre, so that such studies are particularly relevant in the analysis of infectious diseases. There is certainly scope for descriptive cartographic reconstruction of actual disease spreads, as exemplified by Curson's (1985) portrayal of disease patterns in colonial Sydney. There has also been some mathematical modelling of epidemic waves spreading from an innovative centre in relation to postulated barrier and enhancement effects (Brownlea, 1972; Kwofie, 1976; Haggett, 1976; Cliff *et al.*, 1981). Such studies, clearly inspired by Hagerstrand's (1967) seminal work on spatial diffusion, are particularly important for the possible predictive utility of their models (Cliff and Haggett, 1989).

Geographical Information Systems (GIS)

GIS comprise one of the most important and fastest-growing applications of information technology in the social and medical sciences (Department of the Environment, 1987; Openshaw, 1988). Such systems are based on computerized geographic base files where recording of individual events can be by street section, city block, unit postcode or National Grid reference. The areal matching of health files with housing, planning, social work, crime, census and other geocoded files is thus permitted, as well as the ready aggregation of data to any prescribed area and, given the availability of explicit spatial co-ordinates, the automated drawing of maps.

The considerable potential applications of GIS to health research have been outlined for the United States by Pyle (1979, ch. 8) and for Britain by Gatrell (1988). The NHS is the custodian of probably the largest repository of massively under-used, virtually locked-up databases in Britain (Wrigley, Morgan and Martin, 1988), including births, deaths, stillbirths, perinatal deaths, neonatal deaths, cancer registrations, staff records for over a million employees, and admissions, discharges and diagnoses of hospital patients. Following the strong recommendations of the Korner Committee (1980) on information handling in the NHS, many health authorities have started to apply GIS techniques, particularly in the postcoding of records and the combining of such records with other databases, especially census small area statistics.

There is danger, however, that some fundamental analytical problems may be overlooked in the rush to produce increasingly detailed statistical and cartographic output. Several of these problems relate to small areas or small populations (Carstairs, 1981; Alderson, 1987): their small number of medical events may be subject to appreciable chance variation; local migration can cause problems in a search for possible environmental causes of disease, given lengthy incubation periods; and the presence of hospitals and homes for the elderly can inflate local numbers of deaths, even although deaths within six months of admission in Britain are routinely attributed to former residence. Many of these problems can, of course, be addressed by aggregation of data over time and/or space, and, given the flexibility of GIS, such aggregation can be to user's specification and not constrained by administrative boundaries.

Consideration can now be given to an interpretation of selected spatial patterns of disease mortality and morbidity.

Western Europe

There are difficulties in cross-national comparison of mortality from specific diseases because of variations between countries in the diagnosis, certification and coding of deaths (Kelson and Heller, 1983), so that cross-national patterns are best studied at the level of overall mortality. Figure 4.6 shows that high mortality is concentrated in two types of area. The first comprises older industrial areas (especially involving mining, metallurgical and chemical industries) suffering from environmental problems as well as employment contraction. Notable cases are the Glasgow area, South Wales, North East England, Luxembourg, southern Belgium, Nord, Pas-de-Calais and Alsace-Lorraine in France, Saarland and the Ruhr in West Germany, and Aosta, Bergamo, Brescia, Cremona, and Udine in northern Italy.

The second focus of high mortality is in some rural areas which have long histories of economic difficulties and restricted opportunities, notably Brittany, northern Scotland, western Wales and much of Ireland. One might perhaps have expected to see southern Italy in this category, but any excess mortality there, outside Naples, is confined to infants.

Great Britain

Figure 4.6 shows a substantial mortality gradient from the older industrial areas of the North and West to the more affluent and economically dynamic regions of the South and East – a pattern shown by the Registrar General's Decennial

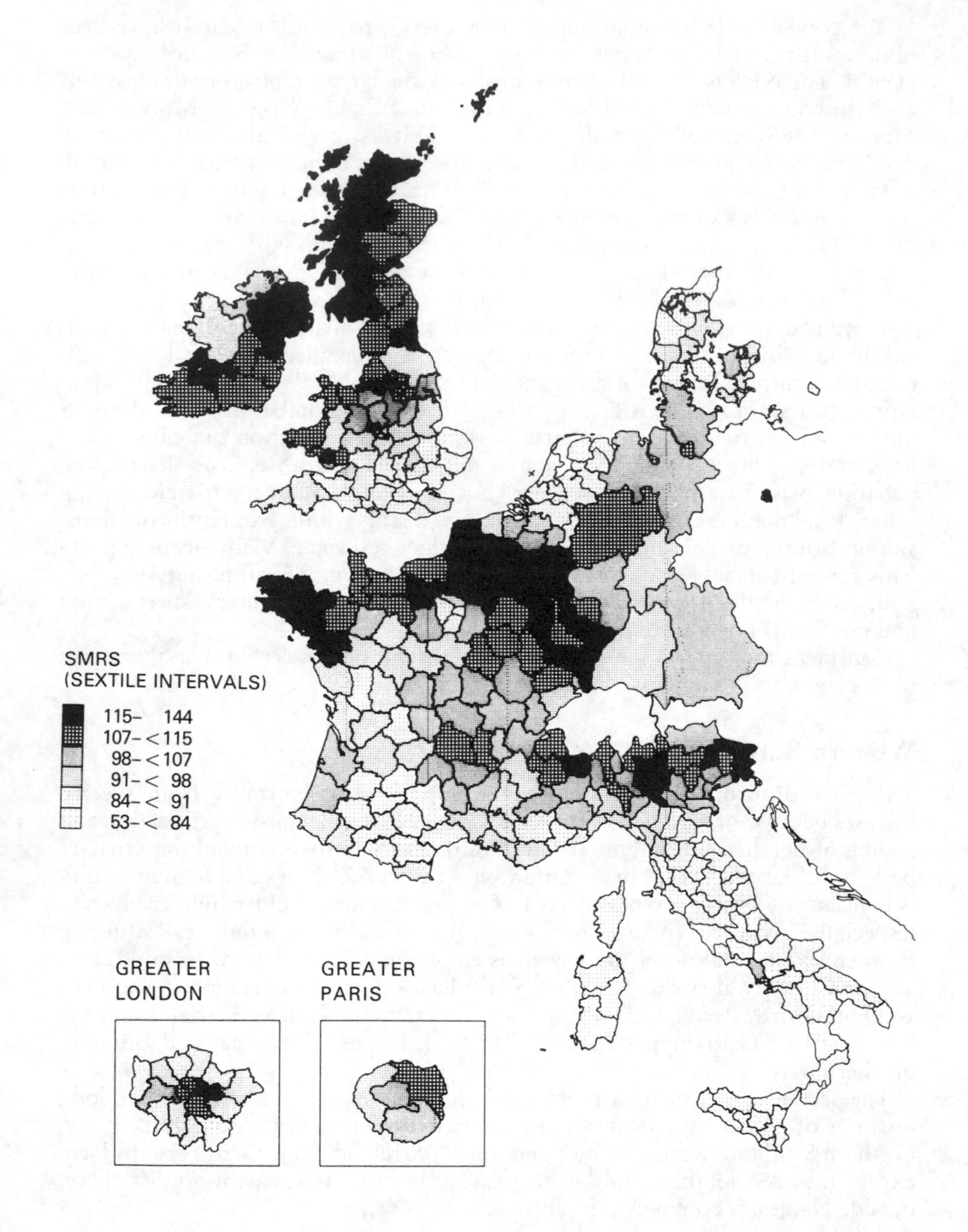

Figure 4.6
Standardized mortality ratios for all causes of death, 1974–8, for the total population aged 5–64
Source: Holland (1988), p. 74.

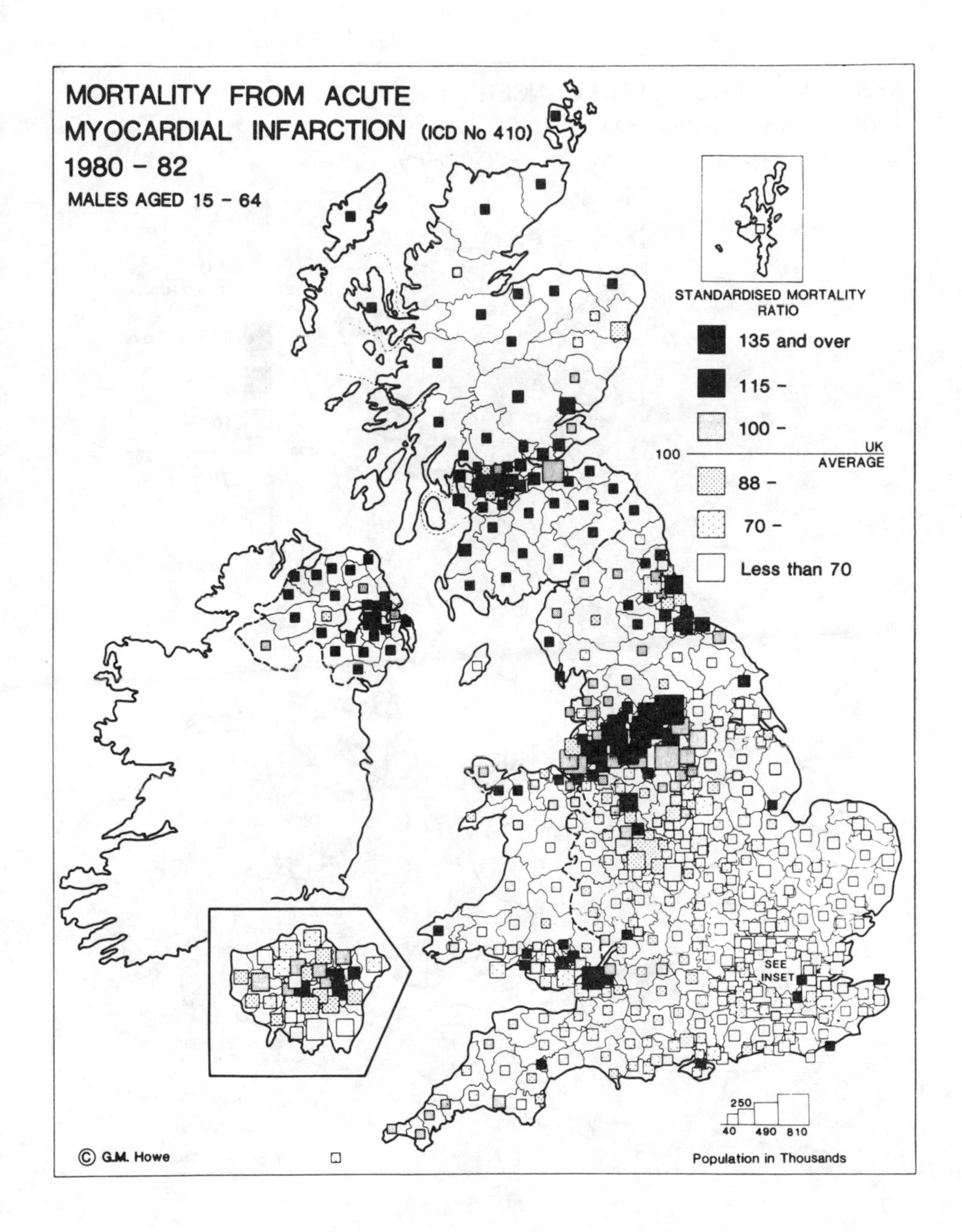

Figure 4.7
Mortality from coronary heart disease, 1980–2, UK males aged 15–64
Source: Howe (1986), figure 4. By permission of G. M. Howe.

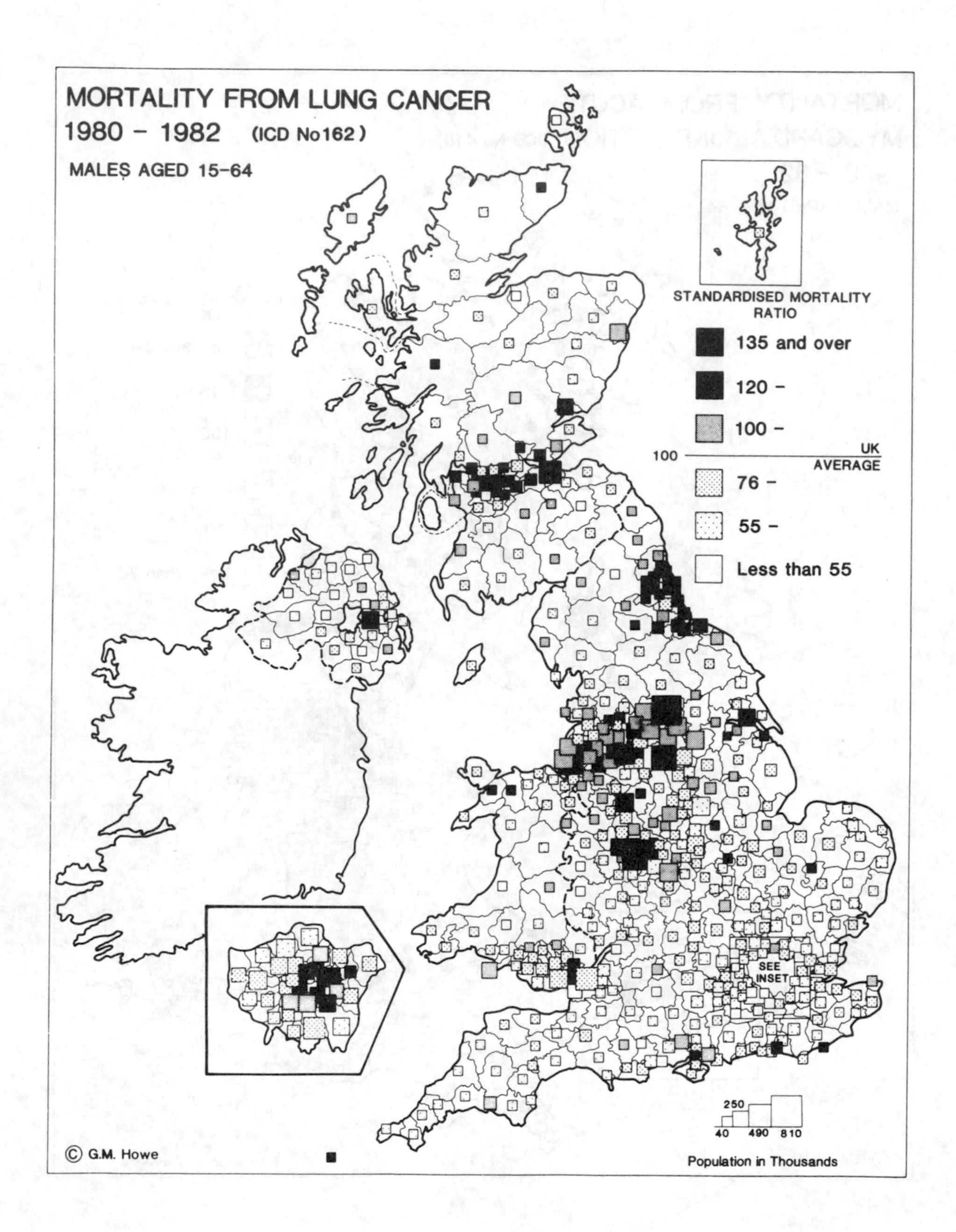

Figure 4.8
Mortality from lung cancer, 1980–2, UK males aged 15–64
Source: Howe (1986), figure 9. By permission of G. M. Howe.

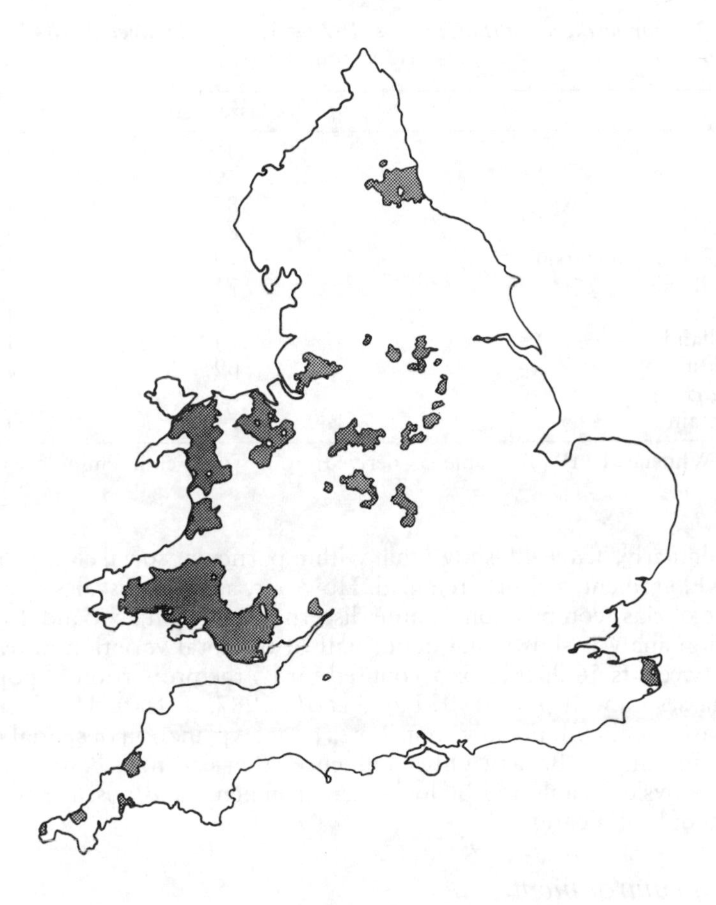

Figure 4.9
Male mortality from pneumoconioses, England and Wales, 1968–78; local authority areas where the standardized mortality ratio is significantly above the national rate at the 95 per cent level of statistical significance
Source: Gardner, Winter and Barker (1984), p. 32. By permission of John Wiley and Sons Ltd.

Supplements to have been remarkably stable throughout this century. Maps of specific disease mortality (Figures 4.7, 4.8, 4.9) at a much more detailed district scale also confirm the regional gradient, although obviously there are differences of detail between the three individual patterns that reflect rather different sets of causes.

In interpreting the regional pattern, it is useful to bear in mind the appreciable variations in mortality by social class which have already been identified. Since the regional distribution of social classes is not uniform, it seems plausible that class distribution might control the regional mortality pattern. In other words, the poorer survival prospects in Wales, Scotland and northern England might simply reflect a relatively high proportion in their populations of the disadvantaged classes IV and V. Yet when regional mortality data are standardized for class as well as age, there is only a modest narrowing of the regional differential (Fox,

Table 4.9 Standardized mortality ratios, 1979–83, for males aged 20–64 by region and class grouping

	Classes I–II	Classes IV–V
North	81	152
Wales	79	144
Scotland	87	157
North West	83	146
Yorkshire and Humberside	79	134
West Midlands	75	127
South East	67	112
East Midlands	74	122
South West	69	108
East Anglia	65	93
Great Britain	74	129

Source: Whitehead (1988), Table 5, derived from Townsend, Phillimore and Beattie (1988).

1977). Similarly, Table 4.9 shows that within particular social class groupings the N.W.–S.E. gradient is clearly retained. However, at smaller scales of analysis the influence of class composition is more discernible. Within Scotland, for example, correlation analysis shows that about half of the total variation in overall mortality between its 56 districts is accounted for by the proportion of population in social classes I and II ($r = -0.71$; Lloyd *et al.*, 1987, p. 160). The distribution of social class, then, undoubtedly contributes to an explanation of spatial patterns of disease mortality in Britain. Other influences that can now be discussed derive from the physical, built and biological environments and possibly from the distribution of health care.

Physical environment

Britain exhibits considerable variety in physical environmental conditions. The relationship of weather and climate to disease has been reviewed by Tromp (1973), but more important are thought to be the geological influences operating on health through background radiation, trace elements in the soil and water quality. The sedimentary rocks of lowland Britain (the South and East) have a lower content of radioactive elements and thus less gamma-ray background than the igneous rocks of highland Britain (the North and West), although in terms of human exposure it is type of building material, which may not be derived locally, that seems to be the decisive factor. Nevertheless, there was considerable concern in 1988 about dangerously high levels of Radon-222 (a natural radioactive gas derived from the decay of subterranean traces of uranium) found in many properties in Cornwall.

Trace elements are present in the atmosphere and especially in the soil, where they act as important micro-nutrients. Soils derive their trace elements from the underlying geology and transmit them to the vegetation and thereby to human diets. Both deficiencies and excesses of certain trace elements are known to be harmful to health (Cannon and Hopps, 1972; Warren, 1989). Excesses of nickel, cadmium, mercury and lead are notorious hazards, and British readers may recall the formal warning given in 1979 by the Department of the Environment to villagers at Shipham, Somerset, not to grow vegetables in their gardens because of

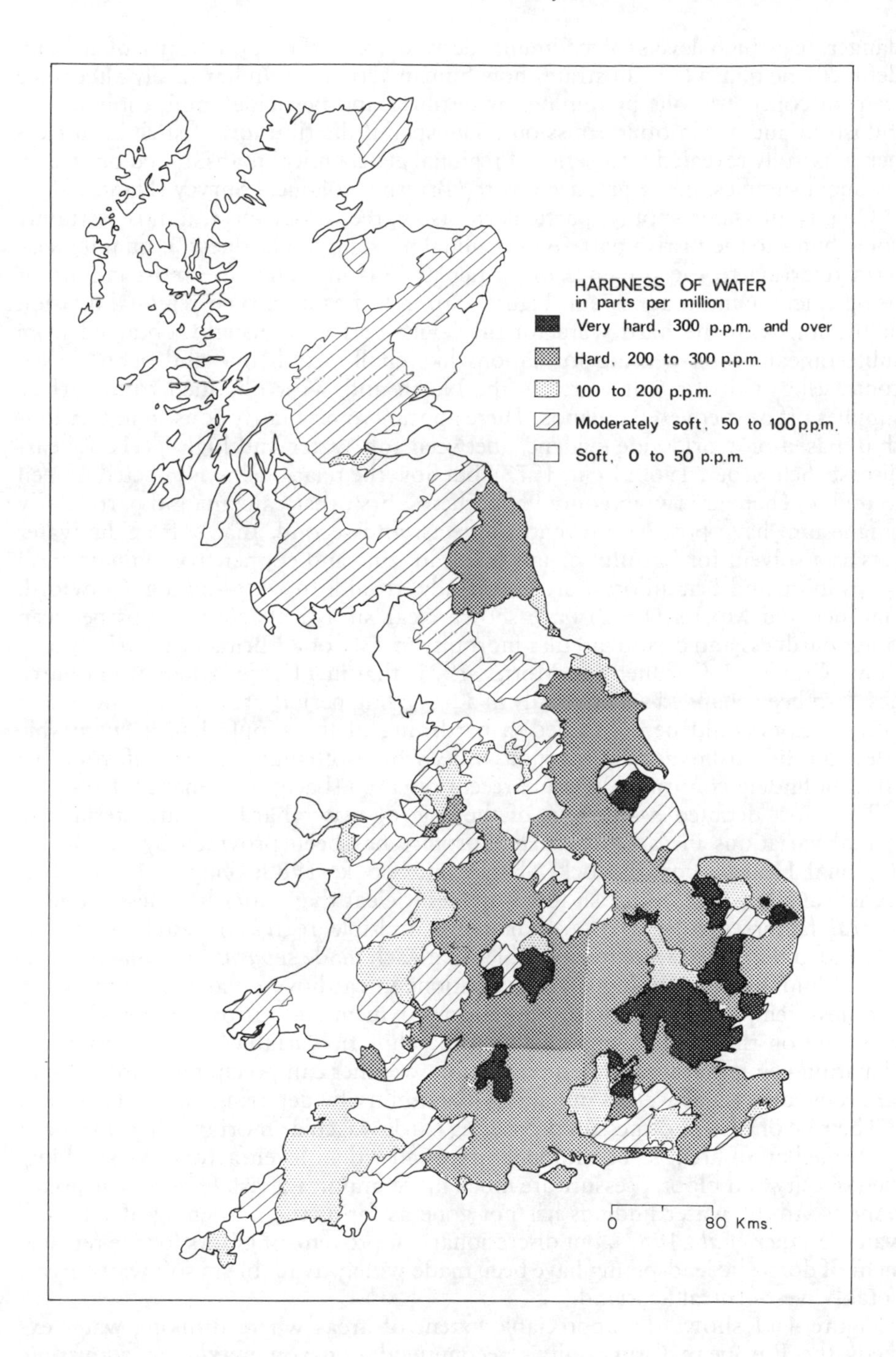

Figure 4.10
Degree of hardness of water for direct supply water undertakings in Great Britain
Source: Howe (1972), figure 20, based on data from *Water Engineer's Handbook 1968*.

dangerously high levels of cadmium derived from the spoil heaps of a long-defunct zinc mine. This illustrates how human activities can harmfully alter trace element concentrations by mining, by fertilizer and pesticide application and by industrial and automobile emissions. The spatial distribution of such hazards is being usefully revealed by a series of regional geochemical maps, based on stream sediment samples, being prepared by the British Geological Survey (Plant, 1987).

Quality of water supply, particularly its degree of hardness, almost certainly contributes to the British pattern of regional mortality. 'Hardness' is an imprecise term referring to the presence of particular ions in water, the most important being calcium and magnesium. Figure 4.10 indicates a marked regional division in Britain, with the hard water of the South and East, usually obtained from subterranean water-bearing formations like chalk, sandstone and pebble beds, contrasting with the soft water in the North and West obtained from surface supplies off peat-covered uplands. There appears to be a fairly consistent relationship, based on worldwide evidence, between soft water and high levels of heart disease (Schroeder, 1960; Reid, 1973), but how the relationship is effected is open to doubt. There are two favoured hypotheses: first, that calcium and particularly magnesium have positive preventive effects; and second, that soft acidic water acts as a solvent for harmful metals like lead, zinc and copper from mining spoil heaps in upland Britain or, more universally, from lead water-piping. Crawford, Gardner and Morris (1968) have shown clear statistical relationships between water hardness and cardiovascular mortality in a set of 83 British cities. They also show (Crawford, Gardner and Morris, 1971) that in 11 cities where water hardness had been changed substantially in a preceding period of 30 years (and where other factors could be controlled) a hardening of the supply had a favourable effect on the cardiovascular mortality rate, while softening had an unfavourable effect, a finding confirmed by more recent research (Lacey and Shaper, 1984).

The most detailed assessment of the role of water hardness in determining spatial variations in cardiovascular mortality has been provided by the British Regional Heart Study (Pocock, Shaper and Cook, 1980; Shaper, 1984). This statistical study of 253 British towns makes it clear that water hardness is one of *several* factors involved. After exhaustive multiple regression analysis, it was concluded that there were five variables that each made *separate* and *independent* contributions to explaining spatial variations in cardiovascular mortality: water hardness, temperature, rainfall, manual employment and car ownership. One can speculate on the mechanisms involving the climatic variables, the preferred explanation being that wet and especially cold weather can precipitate cardiovascular events directly as well as inducing vulnerability by deterring physical exercise.

The role of water softness in promoting cardiovascular mortality has thus been confirmed at an areal level, although at an individual level factors like smoking, diet, obesity and blood pressure are much more important risk factors. The policy response to the water findings has not gone as far as the hardening of very soft water (Shaper *et al.*, 1983), but discretionary improvement grants for the replacement of domestic lead-piping have been made widely available in soft water areas, notably west-central Scotland.

Figure 4.11 shows the appreciable extent of areas where drinking water exceeds the European Community's recommendations on maximum admissible concentration for specific toxins. Excessive concentrations of lead and aluminium in the North and West contrast with the excessive levels of pesticides and nitrates in the South and East, where modern arable farming is implicated. But since many

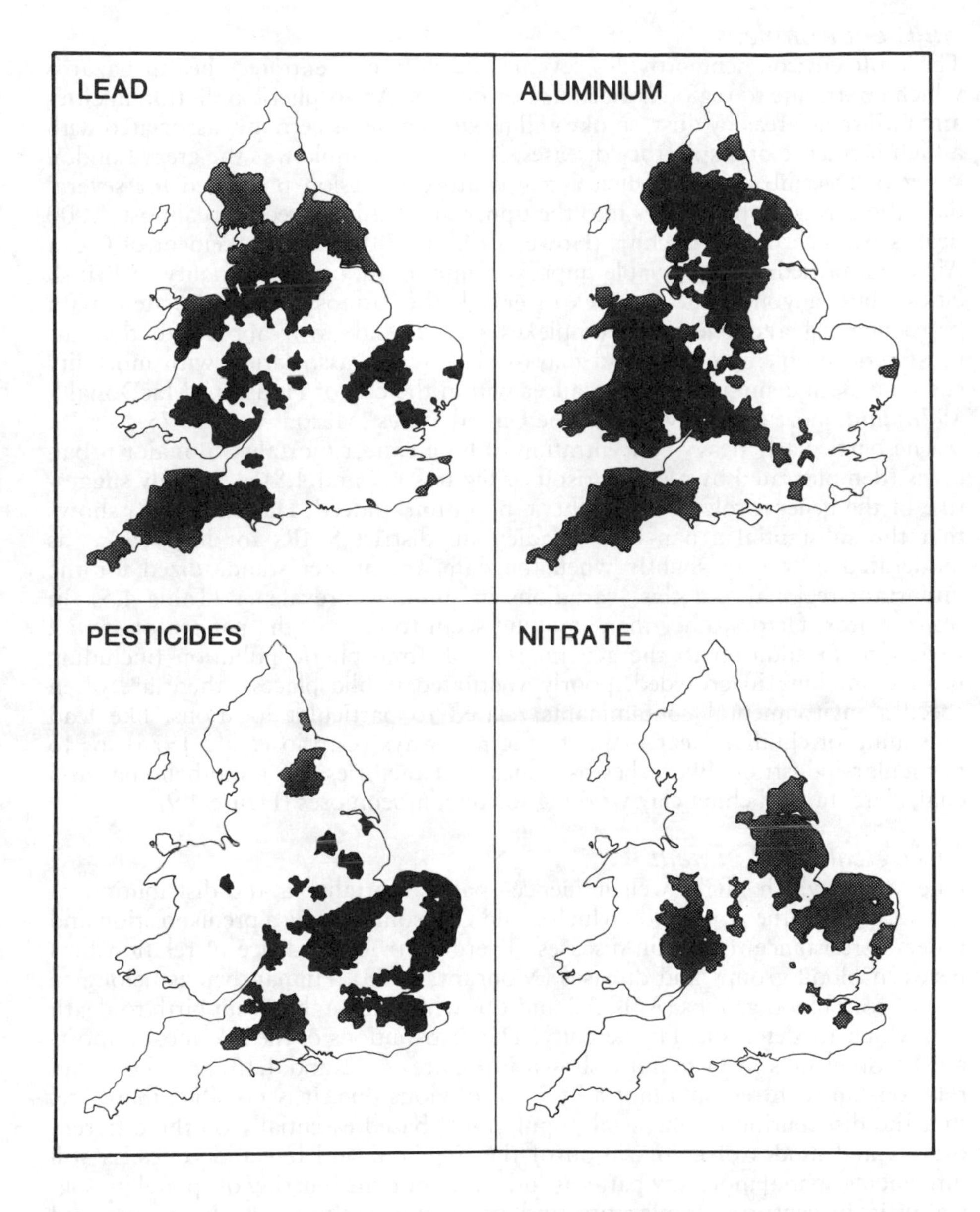

Figure 4.11
Areas of England and Wales where tap water exceeds the European Community's maximum admissible concentration for specified types of chemical pollution, 1989. Data supplied to Friends of the Earth by water authorities and companies
Source: *The Observer* Magazine, 6 August, 1989. By permission of *The Observer*.

diseases, especially cancers, have a long incubation and/or latency period, and since the application of pesticides and chemical fertilizers has increased greatly in recent decades, there is unlikely to be any spatial correlation between current levels of, say, stomach cancer mortality and of nitrates.

Built environment

The built environment provides several, spatially concentrated, health hazards which contribute to regional mortality variations. Atmospheric pollution in cities and industrial areas by dust, smoke and noxious gases is certainly associated with a high incidence of respiratory diseases. A classic example was the great London smog of December 1952, when a temperature inversion prevented for several days the escape of impurities into the upper air and led directly to almost 4,000 deaths from chronic bronchitis (Howe, 1972, p. 78). The enforcement of Clean Air Acts has led to appreciable improvements in atmospheric quality in British cities, but anyone who has experienced the atmosphere associated with petrochemical and chemical complexes like Teeside will appreciate that atmospheric problems persist and may well show an association with mortality such has been demonstrated for cancer within the city of Houston (MacDonald, 1976) and, more generally, within the United States (Mason *et al.*, 1975).

The particularly heavy concentration of lung cancer mortality in major urban areas (demonstrated by a comparison of Figures 4.7 and 4.8) is strongly suggestive of the general role of atmospheric pollution. Indeed, Haynes (1988) shows that the substantial urban–rural gradient in district SMRs for lung cancer is moderated only very slightly when the data are further standardized for the important regional and class variations in smoking prevalence (Table 4.5). In other words, factors other than smoking seem to control the pattern of spatial variation. In addition to the general role of atmospheric pollution (including passive smoking in crowded, poorly ventilated public places), there are often specific environmental contaminants related to particular locations, like lead poisoning of children near major traffic highways (Caprio *et al.*, 1975), or to particular industries, like asbestos mines and factories for mesothelioma, and coal, slate, tin and china clay working for pneumoconioses (Figure 4.9).

Biological environment

One factor which might well influence spatial mortality is the distribution of genetic traits in the population which could give some people a predisposition and others a resistance to certain diseases. There is strong evidence of relationships between blood groups and diseases (Mourant, 1978). Human beings belong to one of four blood groups (A, B, AB and O), which are stable from birth to death and which are determined by heredity. The distributions of the two most important blood groups in Britain are shown in Figures 4.12 and 4.13, and the broad relationship with regional mortality is an obvious one. It is possible, therefore, that the distribution of ancestral populations, based essentially on the different origins and modes of colonization of the highland and lowland zones, is still influencing spatial mortality patterns today, despite the blurring of spatial biological traits by centuries of migration and intermarriage. It has also been suggested that selective migration of healthy individuals from north to south in twentieth-century Britain may have contributed to regional mortality differences.

Distribution of health care

Before the establishment of the NHS in 1946 the regional distribution of health-care facilities clearly favoured the South and East, because of the unfettered attraction of services and facilities to client wealth. These regional imbalances (in England and Wales at least, since health services in Scotland and Northern Ireland are administered separately) continued well into the 1970s because of the

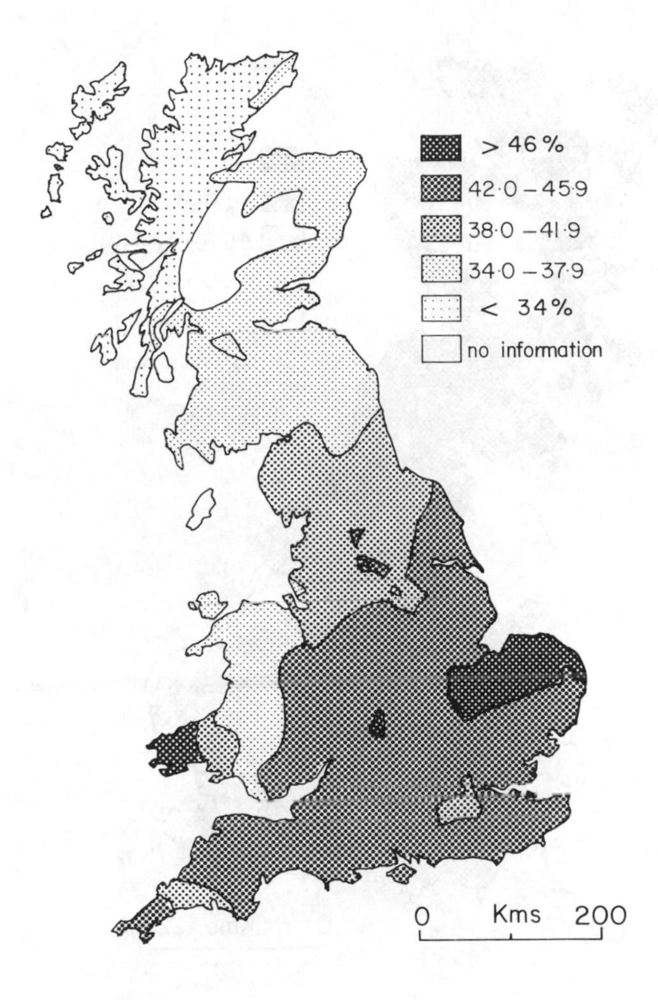

Figure 4.12
Percentage frequencies of the blood group A gene in Great Britain, based on blood donor evidence
Source: Simplified from Howe (1972), figure 4.

need of the NHS to maintain its inherited stock. But from 1976 regional resources have been allocated much more equitably on the basis of a formula measuring regional need by population size, composition and mortality level (Haynes, 1987, ch. 2). On the other hand, the recent expansion of private medical facilities has been heavily concentrated in the South (Mohan, 1988), another example of the widening in the 1980s of the 'Two Nations' gulf or the 'North–South Divide'. Clearly, the long-standing mortality and morbidity regional gradient in Britain, which shows no sign of easing, has its roots in the country's political economy (R. Martin, 1988; Lewis and Townsend, 1989).

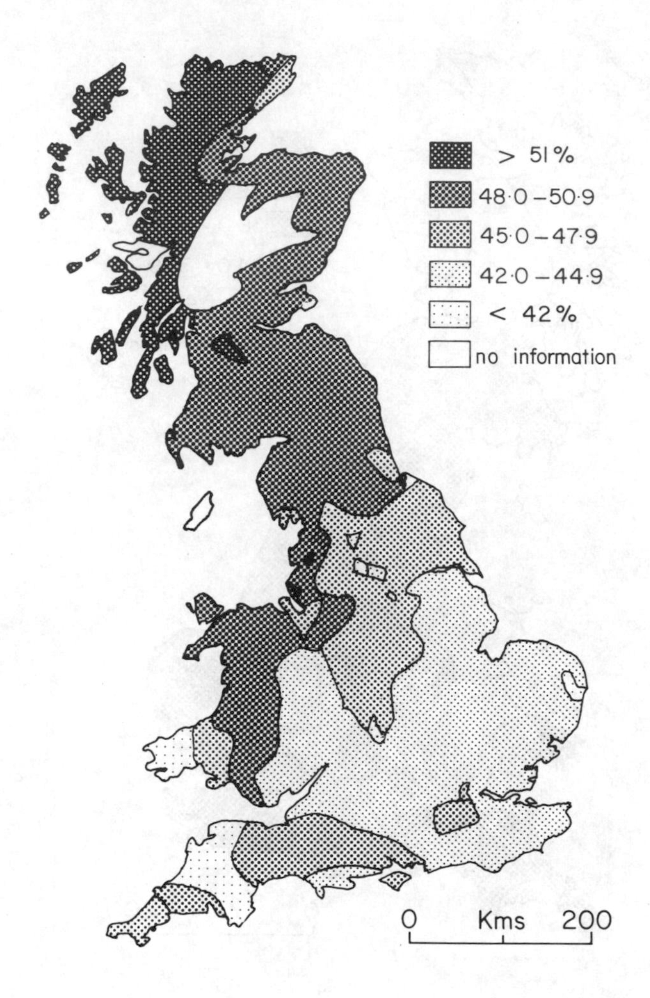

Figure 4.13
Percentage frequencies of the blood group O gene in Great Britain, based on blood donor evidence
Source: Simplified from Howe (1972), figure 5.

Malaria in Sri Lanka

The important role in mortality reduction of the post-Second World War programme of malaria eradication has been referred to in Chapter 3, but the spatial pattern of malaria incidence at that time can now be considered. It provides a clear demonstration of how ecological forces can fashion spatial patterns of disease in less developed societies.

Malaria is a three-factor disease, embracing the causative organism, malaria parasites of the genus *Plasmodium*; the host, humans; and the vector (strictly

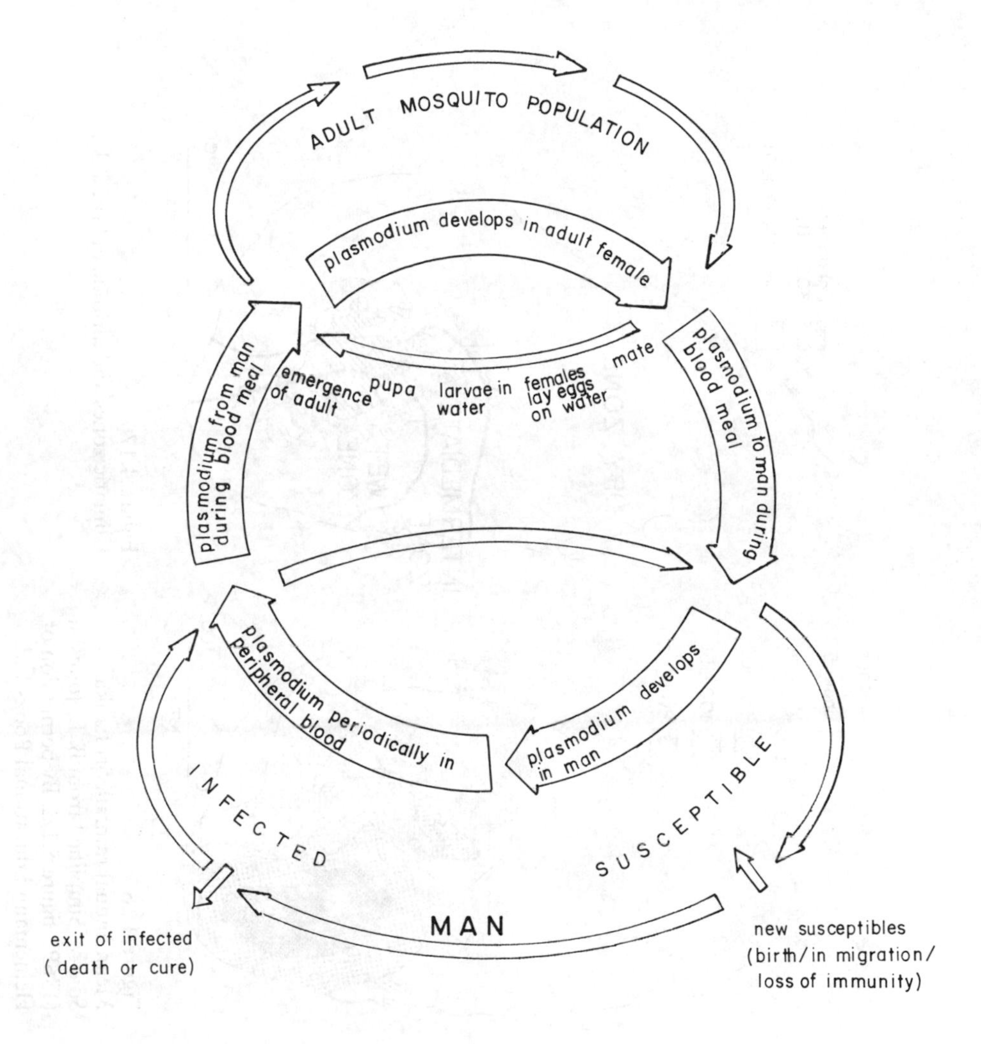

Figure 4.14
The malaria cycle viewed as the interlocking relations of parasite, vector and man
Source: Learmonth (1977), figure 16, and *The Population Explosion, an Interdisciplinary Approach*, Understanding Society Course, The Open University Press, 1971, figure 33.6.

speaking an alternative host), female mosquitoes of the genus *Anopheles*. There are thus three separate ecologies in that each individual factor interacts with its total environment, which includes the other two factors. Several of these interactions are summarized in a particularly useful representation (Figure 4.14) of the malaria cycle provided by Learmonth (1977).

Figure 4.15 depicts the typical distribution of malaria in Sri Lanka in the pre-control era. The index of malaria prevalance is the spleen rate, the percentage of a sampled population with the splenic enlargement associated with malaria. The distribution pattern is closely related to the island's physical and human geography, particularly to the fundamental and universally recognized regional division between Wet and Dry Zones (Figures 4.16 and 4.17).

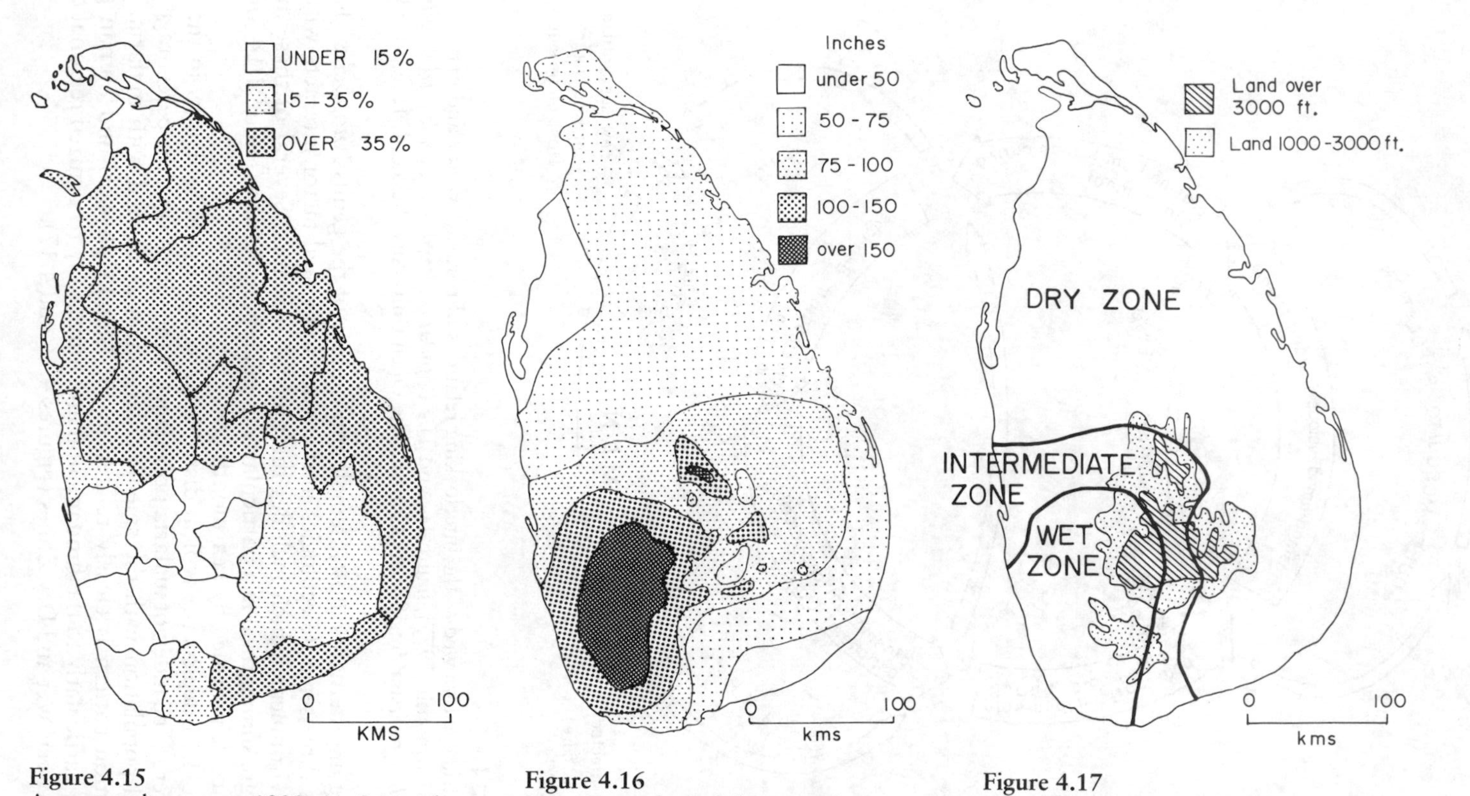

Figure 4.15
Average spleen rates, 1938–41, Sri Lanka
Source: Data from Gray (1974), Table 1.

Figure 4.16
Mean annual rainfall, Sri Lanka
Source: Simplified from B. L. Johnson (1969), figure 5.1.2. By permission of Heinemann Educational Books Ltd.

Figure 4.17
Climatic zones and altitude in Sri Lanka

The Wet Zone receives the bulk of its appreciable rainfall from onshore winds during the south-west monsoon from May to July, but its southerly latitude ensures the adjacency of the Inter-Tropical Convergence Zone throughout the year, so that there is normally no rainless season. The Dry Zone, on the other hand, in the north and east of the island, has its more limited and essentially seasonal rainfall during the relatively weak and unreliable north-east monsoon from October to February, since the uplands in the south-centre of the island impose a rain-shadow effect on the Dry Zone in the south-west monsoon period.

The epidemiological significance of these contrasting climatic conditions lies in the breeding habits of the vector mosquito *Anopheles culcifacies*, which requires fresh and still water sites like river pools, irrigation ditches and rain puddles for egg laying and the development of larvae. Water flow in the Wet Zone is normally continuous and fast enough to keep the stream beds flushed and clear of larvae, and the limited amount of irrigation practised is not of the tank-storage type. However, when the south-west monsoon occasionally fails, stagnant water does

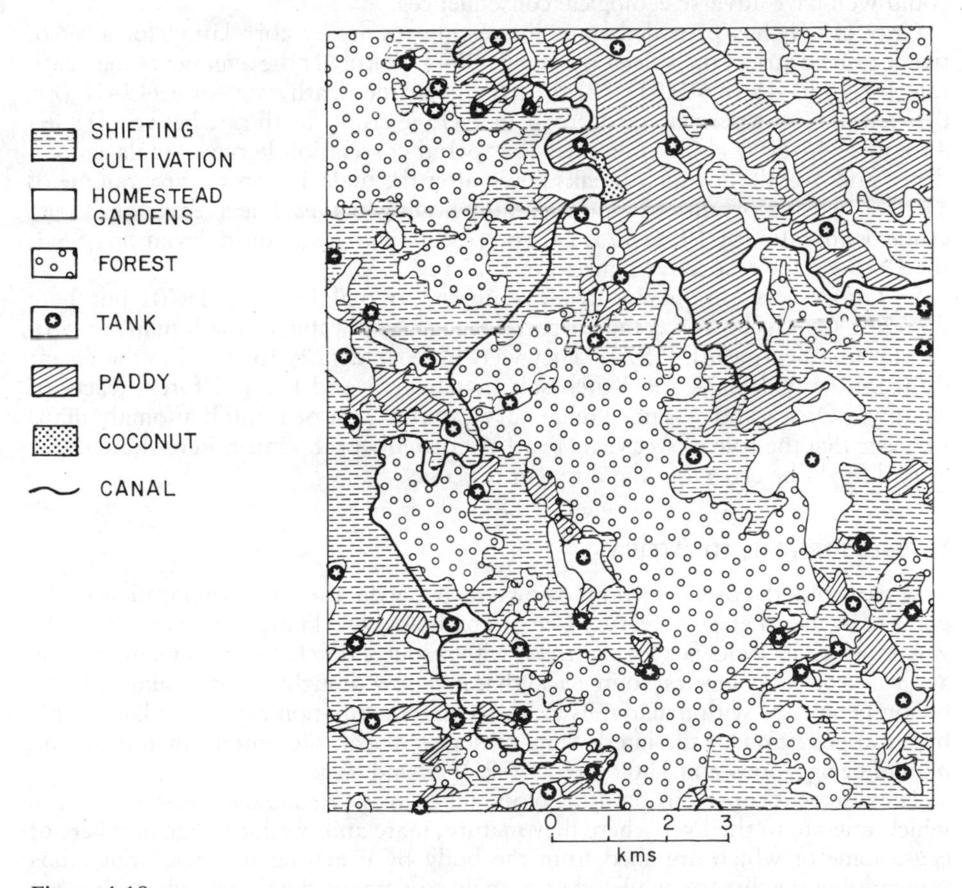

Figure 4.18
Land-use in the Dry Zone of Sri Lanka: a representative area
Source: B. L. Johnson (1969), figure 5.2.5. By permission of Heinemann Educational Books Ltd.

collect in river-bed pools, leading to the rapid multiplication of mosquitoes and epidemic outbreaks of malaria. Such conditions often result in high mortality, as in the devastating outbreak of 1935, since the Wet Zone population, not being continually exposed to malaria, has little natural resistance to the disease.

In the Dry Zone, optimum breeding grounds for larvae are provided by the reduced river flow of the dry season and by irrigation tanks, canals and ditches, particularly when silted up and abandoned. Figure 4.18 shows the abundance of tanks and canals in a typical part of the Dry Zone. A few large tanks are formed by barrages on fair-sized rivers, while others derive their water by diversion weirs from rivers; but the great majority of tanks are small reservoirs tapping localized catchments. There is only one exception to the Dry Zone pattern of endemic malaria, as Figure 4.15 indicates: the Jafna district in the extreme north. Here limestone prevents surface rivers, and well irrigation is the mainstay of agriculture. There are no tanks, and the industrious Tamils have refurbished the ancient irrigation channels, allowing constant water flow and consequent avoidance of malarial hazards, although recent political strife and military action in this area could well have adverse ecological consequences.

Gray (1974) has provided some interesting results by correlating for a set of twenty-one districts the spleen rate for 1938–41 with: (1) the average crude death rate for 1936–45 ($r = 0.91$), (2) the average crude death rate for 1949–52 ($r = 0.17$), and (3) the percentage decrease in average crude death rate between 1936–45 and 1946–60 ($r = 0.91$). Clearly the spatial correlation between malaria incidence and overall mortality breaks down after the malaria control programme of 1946–8, when mortality becomes spatially homogeneous. The average of district crude death rates for 1936–45 is 24.5 per 1,000, with a standard deviation of 6.3; for 1946–60 the respective figures are 11.7 and 1.3.

Sri Lanka shared in the worldwide resurgence of malaria in the 1970s, but there does not seem to be any detailed spatial study comparable to the Indian study of Akhtar and Learmonth (1977). They were able to identify four foci in the recent diffusion of malaria – the Kutch salt marsh area and the hill forest tracts of Madhya, Orissa and Assam. Preliminary examination of rainfall anomaly maps suggests that the role of wet years in areas normally arid, semi-arid or moderately humid might be crucial in the different process.

Schistosomiasis in Africa

This particular disease, sometimes known as bilharzia after the identification of the parasite by the nineteenth-century German pathologist, Bilhartz, is remarkable in the way in which its incidence has *increased* with resource development in many tropical and sub-tropical countries, so much so that it is now thought to afflict almost half a billion people. As with malaria, its incidence and distribution can be explained only by an appreciation of the interlocking ecologies of parasite, intermediate host and human being (Kloos and Thompson, 1979; Weil and Kvale, 1985).

The disease is caused by parasitic blood flukes or worms known as schistosomes which migrate to the liver where they mature, mate and produce huge numbers of eggs, some of which are shed from the body by urinating or defaecating, thus transmitting the disease, while others remain to damage the tissues where they are lodged. The severity of symptoms in the host depends on the number of worms and eggs, but invariably the disease has a debilitating effect which seriously prejudices both work productivity and resistance to other infections.

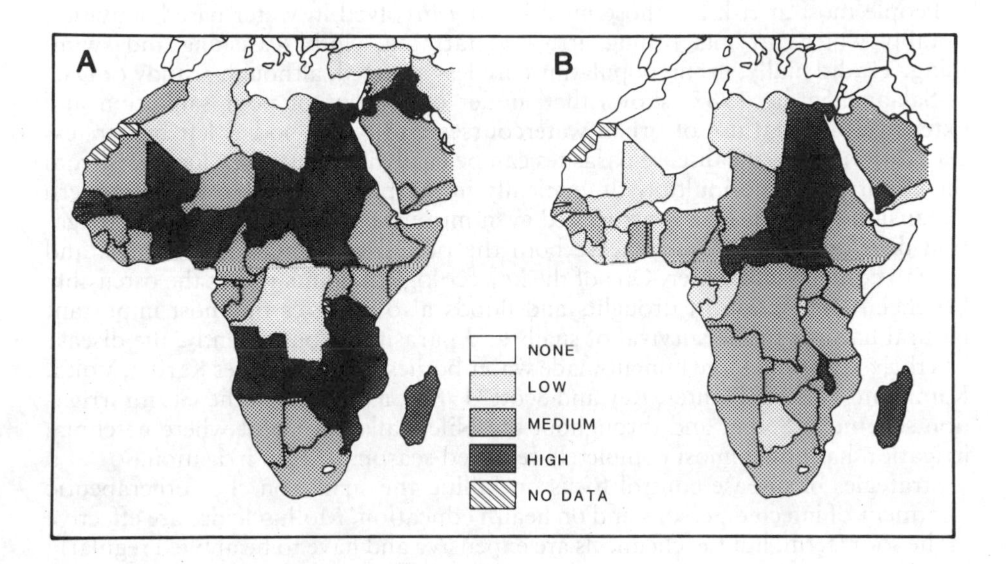

Figure 4.19
Level of endemicity in Africa of (A) schistosoma haematobium, and (B) schistosoma mansoni
Source: Weil and Kvale (1985), figures 2 and 3.

The transmission cycle of the disease can now be considered. Eggs that are shed from an infected person need warm, fresh water to hatch and release free-swimming larvae known as miracidia, which, to survive, must find and penetrate within about one day the bodies of certain types of aquatic snails. Within the snail the schistosomes multiply asexually, each miracidium producing several hundred larvae known as cercariae, which emerge from the snail after several weeks'incubation. These then have a two-day survival period in the water in which to find and penetrate a human host to continue their life-cycle (there are no significant animal hosts). Penetration can be by drinking or by burrowing through the skin.

This understanding of the transmission cycle provides the two explanatory bases of the distribution pattern: that there must be suitable habitat conditions for both parasite and snail, and that humans must have ready contact with the larvae shed from the snails. Figure 4.19 depicts the broad distribution of the two dominant forms of schistosomiasis within the most afflicted continent, Africa. Although the spatial scale is too coarse to capture the impact of some important environmental factors, the maps do reflect the lower rates of incidence in deserts and in equatorial rain-forest caused by the aversion of snails to drought, to fast-flowing water and to deep shade. At cooler water temperatures the shedding of cercariae by snails is retarded, so that the disease is rarely found above 2,000 metres, while in the extreme north and south of the continent transmission and infection are restricted to the warm season. It is shallow, fairly stagnant, fresh water with surface vegetation for food that provides the optimum environment for snails, but man's critical contact with snail- and schistosome-infested water is subject to a host of economic and behavioural influences.

People most at risk are those most heavily involved in water-based activities, notably water-carrying, fishing, irrigated farming, clothes-washing and swimming. Traditionally, urban populations are less affected, although a study of Dar-es-Salaam (Sarda, 1987) shows that, under conditions of poor sanitation and extensive informal use of urban watercourses and water bodies left by careless excavation work, the disease parasites can be easily introduced by incomers from the countryside and multiply dramatically in the population. However, modern expansion of the disease is associated even more with water resource and irrigation developments which increase both the preferred habitat of the snails and human contact with water. One of the key ecological points is that the ostensibly beneficial elimination of droughts and floods also removes the most important natural hazards to the survival of snails and parasites. Consequently, the disease has thrived around large human-made water bodies, notably Lakes Kariba, Volta, Kainji and Nasser (Hunter, Rey and Scott, 1983), as well as in the Gezira irrigation scheme in Sudan and throughout the Nile Valley in Egypt where perennial irrigation has now almost completely replaced seasonal, basin irrigation.

Strategies of disease control focus on killing the snails, on chemotherapeutic treatment of infected persons and on health education. Molluscicides are effective in the short term, but the chemicals are expensive and have to be applied regularly because of rapid repopulation by snails. There is growing concern about wider toxic effects, but the possible replacement of molluscicides by biological predators of snails is only in its infancy. The use of modern drugs has been successful in relieving disease symptoms, but again the financial cost is high and little progress has been made in the development of a vaccine to provide protection. Health education attempts to change behavioural patterns by restricting both the number of eggs that reach water and the human contact with snail- and parasite-infested water. Policy emphasis has shifted from almost exclusive reliance on molluscicides to integrated programmes that now incorporate the provision of drugs, clean domestic water and latrines, snail control and health education.

5

FERTILITY PATTERNS: DEVELOPED COUNTRIES

Geography and fertility

In multi-disciplinary fields of study it is customary for writers to emphasize the importance of their own discipline's contribution. But in the case of population geography there can be no avoiding the very modest nature of its contribution to fertility studies. There has been no sustained equivalent to the work by medical geographers on mortality, morbidity and access to health care, presumably because the behavioural influences to which fertility responds are often regarded, rightly or wrongly, as less intrinsically geographical than those which influence health.

Until very recently, most of the scattered work on fertility by population geographers, whether based on cartographic or statistical methods, has been essentially descriptive. The prime interest was in recognizing, depicting and elucidating spatial variations in fertility at all levels of scale from neighbourhood to nation. Unfortunately, such objectives were invariably regarded as ends in themselves, so that little was contributed to theory in a field dominated by demographers, sociologists, economists and social historians.

More perceptive modern work has begun to recognize that spatial patterns of fertility should more properly be regarded as means to ends – being able, potentially, to throw light on some important, unresolved issues in population studies. Three particular applications of the geographer's spatial perspectives and skills are capable of making important contributions to fertility studies:

1. Spatial diffusion theory can be used to assess the spread from innovative centres of fertility knowledge, attitudes and behaviour, while distance–decay mechanisms are often relevant to understanding the variable use of centralized services like family planning clinics.
2. An appreciation of regional cultural identities, grounded in the best traditions of classical French regional geography, could be particularly helpful in understanding patterns of leads and lags and of convergence and divergence in the development of demographic regimes across space.
3. Territorially disaggregated data are being used increasingly to supplement, or substitute for, temporal data in the analysis of evolving relationships between

development and fertility. Modern geographical expertise is invaluable in countering some of the pitfalls of such cross-sectional or 'ecological' multivariate analysis, by:

(a) appreciating the limitations of classical inferential procedures stemming from their assumption of independence of observations when, in fact, geographical data invariably exhibit systematic ordering in space (spatial auto-correlation);
(b) understanding that inferences about individual behaviour cannot routinely be made from group or areal observation ('the ecological fallacy'); and
(c) being aware of analytical problems caused by scale generally and, in particular, by the so-called modifiable areal unit problem in which different areal arrangements of the same data often produce different results.

Measures of fertility

Before proceeding to analyse fertility patterns, it is obviously necessary to appreciate the ways in which fertility is commonly measured. The principles and procedures are very similar to those of mortality measurement presented in Chapter 3.

Crude birth rate

This is the simplest and commonest measure, being defined as the number of births in a year per 1,000 of the mid-year population.

$$\frac{B}{P} \times 1{,}000$$

It has the advantage of showing the exact rate at which additions are made to a population through births, but for temporal or spatial comparisons it has the disadvantage of including in its denominator a large mass of males and of young girls and older women not 'exposed' to any possibility of child-bearing.

It is not, however, such a 'crude' demographic measure as the crude death rate, because the proportion of the population 'at risk' to child-bearing (women aged 15–49 years) does not vary so much as a proportion of the total population as does the population most at risk to dying (the elderly). In the mid-1980s, women aged 15–49 comprised virtually the same proportion of the total population in many countries at very different stages of development (e.g. Paraguay 23 per cent, Tunisia 24 per cent, Japan 25 per cent, Sweden 24 per cent).

Nevertheless, there are more refined and analytically useful measures of fertility that do standardize for variable age and sex compositions, so that births are related specifically to the female population 'at risk'.

General fertility rate

This measures the number of births in a year per 1,000 women of normal reproductive age:

$$\frac{B}{P_{f15\text{–}44}} \times 1{,}000$$

Sometimes the somewhat larger 15–49 age group is used, but since women of 45–49 contribute relatively few births, it is preferable to avoid dilution of the denominator by this low-risk group. Even so, there is still only a partial standardization for age, since populations differ in age composition within the 15–44 age group. This is important for fertility measurement since the rate of child-bearing is normally appreciably higher in the 20–29 group than in the 15–19 and 30–44 groups. Fertility is inhibited in the 15–19 group by the cultural factor of delayed marriage and in the post-30 groups by the biological factor of reduced fecundity.

Child:woman ratio

This is the ratio of children under five years of age to women of child-bearing age:

$$\frac{P_{0-4}}{P_{f15-49}} \times 1{,}000$$

In this case the 15–49 rather than the 15–44 group is conventionally used, since children under five may have been born up to five years prior to the census, when the mothers were up to five years younger. Its great operational advantage is that the basic data are available from census age-tables. This is important in less developed societies with poor systems of birth registration, while even in developed countries the child:woman ratio is often the only means of measuring fertility in small areas for which births data are not tabulated. The child:woman ratio is thus used widely in population geography for micro-scale spatial studies of fertility.

Nevertheless, the ratio has several deficiencies. It is only part standardized for age, not taking account of age distribution within the 15–49 range; the number and age of young children may be misreported to census enumerators; the ratio is essentially a measure of survivors of births, so that it is affected by differences between populations in child mortality; and there are further complications caused by migration, as in the case of women in some less developed countries moving to urban areas but having their children brought up by grandparents in the countryside.

Age-specific birth rate

Such a rate measures the number of births in a year to women of a given age group (B_{fa}) per 1,000 women in that age group (P_{fa}):

$$\frac{B_{fa}}{P_{fa}} \times 1{,}000$$

Sets of age-specific birth rates allow detailed but cumbersome comparisons between populations. For more effective comparisons, demographers have evolved single-figure indices from sets of age-specific birth rates.

Standardized birth rate

This can be calculated from a set of age-specific birth rates in the manner demonstrated in Table 5.1. It indicates what the crude birth rate would have been in a population if the age and sex composition of that population was the same as in a population selected as the standard. In the so-called direct method of calculation used in Table 5.1 the age-specific birth rates of the population being studied (England and Wales, 1988) are applied to the age structure of the standard population (Great Britain, 1981).

Table 5.1 Age-specific birth rates, standardized birth rate and total fertility rate, England and Wales, 1988

Age	Age-specific birth rate per 1,000 women	No. of females in standard population* (thousands)	Expected births in standard population
15–19†	32.4	2,185.2	70,800
20–24	94.8	1,954.1	185,249
25–29	123.6	1,797.1	222,122
30–34	82.6	1,999.8	165,183
35–39	27.9	1,688.8	47,118
40–44	5.1	1,533.9	7,823
Total	366.4		698,295

Standardized birth rate (per 1,000 population)

$$\frac{698,295}{53,556,991} \times 1,000 = 13.04$$

Total fertility rate (per 1,000 women):

$$366.4 \times 5 = 1,832$$

* Great Britain, 1981 census; total population (male and female) 53,566,911.
† The small numbers of births to mothers of under 15 and over 44 years are attributed to the 15–19 and 40–44 age groups respectively.

Total fertility rate

This is the measure of fertility that is most widely used by demographers. It indicates the number of children that would be born to 1,000 women passing through the child-bearing ages assuming, first, that none of the women dies during this period and, second, that age-specific birth rates remain the same as for the year of calculation. The number is obtained by summing the age-specific birth rates for the relevant age intervals and multiplying by the number of years in the age interval. In the case of quinquennial groups:

$$5 \cdot \sum_{15\text{–}19}^{45\text{–}49} \left(\frac{B}{P_f} \times 1,000\right)$$

Table 5.1 demonstrates how a total fertility rate of 1,832 per 1,000, or 1.83 per woman, is calculated for England and Wales in 1988.

It is important to appreciate that the total fertility rate is an entirely hypothetical figure based on age-specific birth rates in one year only. It has sometimes been used to assess whether the population is replacing itself, particularly when it is modified to form the *gross reproduction rate* (taking account of female births only) and the *net reproduction rate* (taking account of female births only and also of female mortality between birth and normal completion of child-bearing). But calendar year reproduction rates in any shape or form are now recognized as being very imperfect measures of replacement. Only a persistent trend of annual rates substantially above or below unity might be an indication of a growing or declining population.

Rates adjusted for marital status and duration

It can be argued convincingly that the age-standardized fertility indices discussed above do not entirely restrict their denominators to persons truly exposed to the 'risk' of child-bearing, since they include unmarried as well as married females. Accordingly, demographers have often found it useful to compute measures like the general fetility rate, child:woman ratio and total fertility rate on the basis of the currently married or the ever-married female population. One can refine further by standardizing for duration of marriage.

It is always important to understand the marital customs of the society that one is analysing. Routine adjustments for legal marital status are inappropriate in societies where a substantial proportion of births occurs to unmarried women. Such societies particularly comprise those where there is no significant social disapproval of child-bearing cohabiting relationships, as in contemporary Scandinavia and former slave territories in the Caribbean.

Cohort measures

All the fertility indices so far presented are based on births occurring in a particular year; they are thus described as annual or period measures. But births in any one year occur to only a cross-section of women, and changes in period measures of fertility may reflect changes in the tempo of family formation as well as in completed family size. Thus a fall in the total fertility rate might simply reflect a lengthening – temporary or permanent – of intervals between births.

In order to avoid the interpretative problems associated with measures of one-point-in-time reproductive performance, demographers have adopted measures of cohort fertility – a cohort being regarded in these cases as a group of women born or married in a particular period. An extensively used measure is that of completed or lifetime fertility, which may be regarded to all intents and purposes as the average number of live births achieved in their lifetime by women of fifty years of age. The measure can be calculated for all such women, or restricted to the ever-married, the still-married or to women married at particular ages.

A particularly refined measure of cohort fertility has been adopted by a population geographer in a spatial study of fertility in Scotland. M. G. Wilson (1978a) derived for 31 areal units a cumulative marital fertility ratio standardized for age at, and duration of, marriage. The ratio is that between, on the one hand, the number of children ever born to once- and still-married women of under 45 years in the area and, on the other, the number expected there on the basis of Scottish national rates of fertility being applied to each area's particular structure of age at, and duration of, marriage.

Proximate determinants of fertility

Our understanding of fertility variations in time and space is greatly enhanced by an appreciation that there are two categories of fertility determinants: first, a series of cultural, socio-economic and environmental 'background' influences; and, second, a set of proximate or intermediate determinants comprising those biological and behavioural factors through which the background variables must operate to affect fertility (Davis and Blake, 1956; Bongaarts and Potter, 1983). Thus the background variables can influence fertility only indirectly, by modifying the proximate (i.e. nearest) determinants, as Figure 5.1 demonstrates schematically and as this chapter and the next will elaborate.

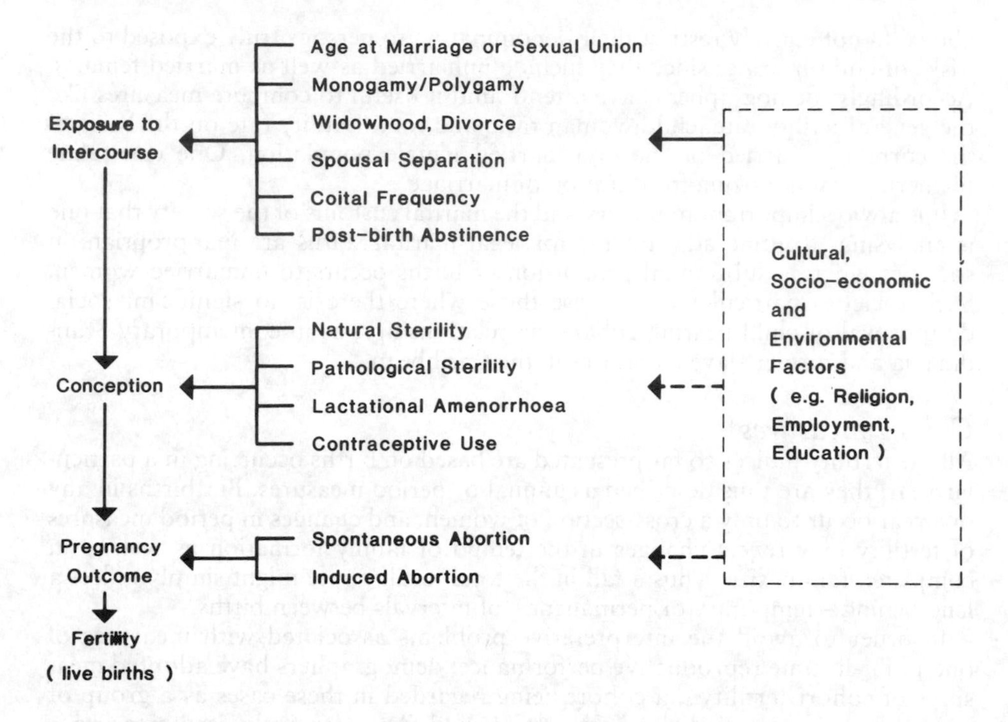

Figure 5.1
The determinants of fertility

Fertility decline theory

The classical demographic transition model, formulated in the second quarter of this century by demographers Warren Thompson and Frank Notestein, describes adequately the major, concentrated fall in fertility that occurs during the transition from pre-industrial societies to economically diversified modern societies. But the reasons proposed for this fertility transition were never entirely convincing, explanation being attributed rather vaguely to changes variously and often interchangeably identified as industrialization, bureaucratization, urbanization, commercialization, development and modernization.

It is clear that any single theory of fertility decline must consciously confront two particular problems. First, it has to provide a single framework capable of explaining two major fertility declines – that in the West from the late nineteenth century and that in the contemporary Third World. Such a theory must do justice, therefore, to some fundamental differences in historical and cultural specificity. Second, it must accommodate the very different threshold levels, as measured by conventional socio-economic indicators, at which fertility decline is initiated.

Theory of inter-generational wealth flows

This has been the most promising explanatory theory to emerge, although it is sufficiently open-ended to leave problems for individual researchers to resolve in particular settings. It has been presented in several papers in *Population and*

Development Review and *Population Studies* that are brought together in consolidated, and sometimes repetitive, form in Caldwell (1982). The theory can now be summarized.

Caldwell argues that, in general, fertility behaviour is rational in societies of every type and stage of development. There are two basic kinds of society – one where high fertility is economically advantageous to family leaders, and the other where it is not. The one society changes to the other fairly quickly and the period of transition is an unstable one, being associated with fundamental changes in the family's internal economic structure. The transition occurs specifically when there is a change in the net balance or net direction of the two inter-generational wealth flows: the one from children to parents and the other from parents to children. Wealth flows are regarded by Caldwell as *lifetime* flows of money, goods, labour services, protection and guarantees, but since economists use the term 'wealth' for a stock rather than a flow, a better term may have been 'income flows' or 'benefit flows'.

What is it that turns the net wealth flows from upwards from children to downwards from parents (or, more accurately in the case of extended families, from upwards from younger generations to downwards from older generations)? The fundamental factor is likely to be the transition from a familial mode of production, where the family is the basic organizational unit (as in subsistence farming), to a capitalist mode of production with labour markets external to the family. In the former, the older generation controls and profits from both production and reproduction, while in the latter the new contractual relationship between worker and employer severs the functional link within the family between reproduction and production as well as undermining the institutionalized obligations which flow traditionally from young to old.

Wealth flows and fertility in less developed countries

Demographic transition theory was formulated in relation to European demographic history before being applied prospectively to recent population change in the Third World. Caldwell's wealth flows theory has evolved very differently, emerging from detailed field studies in West Africa and being applied in its most convincing form to modern fertility reduction in the Third World. It is appropriate, therefore, to elaborate the theory initially in that context.

The earliest pre-capitalist modes of production were hunting and gathering, shifting cultivation and pastoral nomadism, which are still practised in remoter parts of the contemporary world by so-called primitive societies. While their basic modules of society are generally networks of relatives, decision-making is often by groupings larger than the family, and net wealth flows are directed less specifically from children to parents as from younger to older generations. There is no compelling link between production and reproduction, so that optimum fertility is one that balances the need for group security through numbers with the requirement to sustain basic resources.

The dominant and most recent pre-capitalist mode of production is the 'Asiatic' or 'communal' mode of Marx and Meillassoux (1972), involving peasant subsistence farming organized on tribal or communal land by sedentary, village-based, extended families. Such families are groups of close relatives larger than the nuclear family, living in close proximity and having mutual economic interests and obligations. It is essential to appreciate the nature of stable high fertility in such peasant societies so that one can understand the circumstances under which destabilization can occur. Caldwell argues that high fertility is advantageous to parents in these societies because of the following, largely economic, benefits conferred on them by their children through the internal relations of the family in which children act essentially as investment, rather than consumption, goods:

1. From very young ages, children provide a range of services that are regarded wholly or partly as children's work – weeding, caring for livestock, looking after younger children, sweeping, and the carrying of wood, water and messages.
2. Adult children, even when no longer living in the parental household, regularly provide labour on their parents' land, as well as supplying gifts.
3. In the absence of wider community or state social security, adult children care for their aged or infirm parents materially as well as emotionally.
4. They carry out ceremonies and rites for deceased parents, an important consideration in much of sub-Saharan Africa where traditional religious belief is that the dead survive as spirits, but only if their descendants regularly revere them and, indeed, try to contact them.
5. Power and prestige in the local community are invariably related to the number of adult, especially male, supporters that the head of the family, the patriarch, can call upon. This support can be marshalled by a judicious combination of high fertility and arranged marriages.

These benefits are maintained by a framework of family morality which ensures that children work hard, demand and consume little, and respect the authority of their elders. That authority derives from their greater knowledge (real or assumed), from their ownership or control of land, and from their direction of inheritance and marriage. Thus, 'uncontrolled fertility is . . . the central aspect of the cultural superstructure that maintains the relations of production' (Caldwell, 1981, p. 27).

This traditional economic, social and demographic system is being undermined in many parts of the Third World by the separate and combined impacts of mass schooling, capitalism and Westernization. The mass provision of primary schooling has occurred widely in recent decades at stages of economic development a good deal earlier than in the nineteenth-century West. Almost at a stroke, this removes children from household production, while adding substantially to the costs of children through fees, uniforms and school materials and, indirectly, through newly stimulated demands and expectations. The child is often alienated from traditional household chores, and, in the longer term, the knowledge and skills acquired at school are often incompatible with retention of traditional veneration for the superior wisdom and authority of the old.

As wage employment external to the family develops, younger family members often acquire sufficient income to support a separate, autonomous conjugal family, thus removing young couples from patriarchal domination. A capitalist production system also increasingly provides and propagandizes household consumption goods as a substitute for the labour-intensive services traditionally provided by children. Thus the market increasingly replaces the family as the provider of household needs, undermining the whole rationale for high fertility.

The Westernization of social systems in the Third World is obviously related to capitalism, but there need not always be a close symbiotic relationship. Caldwell argues that a good deal of modern fertility reduction in the Third World is dependent on the recent spread of the Western type of nuclear family, dictated not so much by the needs of emergent capitalism as by the family models propagandized by modern mass schooling and mass media. Schools, newspapers, magazines, radio, television and cinemas draw heavily on Western materials, with their normative family unit comprising a father, a mother of almost comparable decision-making status, and a small number of schoolchildren of the Janet and John type passively receiving considerable benefits from their doting parents. Such family models ignore traditional elements like polygyny, arranged marriages, child labour and veneration of elders. There is thus imported an alien European concept of family relationships and obligations that is grounded as much in feudalism and Protestantism as in capitalism. The impact of imported family models and therefore of fertility decline is, of course, mediated by the receptivity or opposition of the receiving culture. In the Islamic world, for example, there remains strong religious bolstering of traditional family structures.

Wealth flows and fertility in developed countries

The basic theory that the onset of major fertility decline is associated with a reversal in inter-generational net wealth flows caused by the breakdown of labour-intensive familial modes of production has been extended to the Australian (Caldwell and Ruzicka, 1978) and European experience (Caldwell, 1980; Seccombe, 1983; Lesthaeghe and Wilson, 1986; Levine, 1987) of late nineteenth-century fertility transitions.

The pre-decline social structures in the West differed from those described above for less developed countries in that the extended family of mutual obligations had in many areas (certainly in England) long been replaced as the fundamental organizational unit by the nuclear or conjugal family (Laslett and Wall, 1972); the roots of this change can be found in the Protestant Reformation, in feudalism and even in the Roman Empire. Nevertheless, production, which was dominantly subsistence-agricultural but also involved pre-capitalist artisan, proto-industrial and commercial activity, was essentially organized on a family basis, where the net wealth flow was clearly from children to parents.

What is intriguing, and at first sight perplexing, is that sustained fertility decline in much of Europe was not initiated until the *late* nineteenth century – well after the establishment of a dominantly capitalist economy and an appreciable level of industrialization and urbanization in Britain and elsewhere. The explanation is that the morality or cultural superstructure of the familial mode of production continued for some time under capitalism, in so far as children not only continued to consume relatively little and to assist their mothers in household chores, but often contributed through external earnings to the common family budget controlled by the father.

Caldwell argues that the wealth flow turned *decisively* downward, thus making high fertility no longer advantageous, only when universal compulsory schooling was initiated (generally in the last quarter of the nineteenth century in northern Europe, the United States, Canada, Australia and New Zealand), thereby withdrawing children from the internal and external labour force of the family and raising expectations for their upward social mobility as wider opportunities arose in a diversifying economy. At much the same time, the manufacturing system had evolved to a stage when it was able to provide and persuasively advertise alternatives in the market to commodities and services hitherto provided internally by family labour. The working-class family thus changed from a unit of reproduction and production to one of reproduction and consumption. Finally, the legitimization and moral approval by Christian teaching of the traditional family structure and its internal relationships was being undermined by growing secularization, social reformist movements and embryonic socialism (Lesthaeghe and Wilson, 1986).

A more explicit reliance on production relations is provided by Seccombe (1983) and Levine (1987) in their explanation of the delay in fertility reduction in urbanized and industrialized Britain. They argue that the decisive change in wealth flows within the family was a *second* phase of industrialization in which the factory finally achieved dominance over the cottage and workshop. The workplace was now decisively separated from the home, making it more difficult for mothers to combine working and child-rearing. Moreover, large-scale mechanization at last superseded the sorts of jobs which women and particularly children had customarily performed in the early proletarian period of the Industrial Revolution. According to this view, the decline of child labour and of its contribution to the family budget was as much the cause as the result of the 1872, 1876 and 1880 Acts introducing compulsory primary education and the Factory Acts curbing the worse excesses of child labour. The high-fertility family-income economy of the early proletarians had been replaced in the later decades of the nineteenth century by a pattern of compartmentalized roles within the nuclear family, which was now sustained by the 'living wage' of the male 'breadwinner'. Changes in reproduction are thus related centrally to the Marxist notion of changing social relations of production.

Fertility in pre-industrial Europe

Uncritical acceptance of the classical demographic transition model or, indeed, of any grand theory of fertility decline can sometimes foster an impression of uniformly high fertility in pre-industrial populations. The empirical evidence reveals a very different situation.

Through the assiduous work of historical demographers in recording and collating demographic material from ecclesiastical registers of baptisms, burials and marriages in centuries preceding the regular collection by the state of vital statistics, it is now well known that fertility levels in wide areas of north-western Europe between the sixteenth and eighteenth centuries were rather lower than those presumed to occur elsewhere in the world at that time and known to occur in less developed countries in the twentieth century. Crude birth rates of about 35–40 per 1,000 were typical, although in Scandinavia rates barely exceeding 30 per 1,000 were common in the eighteenth century (Figures 2.9 and 5.2).

This widespread, although by no means dramatic, lowered fertility in agrarian, pre-industrial societies – sometimes referred to as Europe's pre-modern fertility transition – is attributed largely to a pattern of late marriage and widespread celibacy which Hajnal (1965) identified as emerging from the sixteenth century in Europe to the west of a line drawn from Leningrad to Trieste. He demonstrated that 15–20 per cent of women abstained altogether from marriage and that at any particular time married women comprised only about 40–50 per cent of all women aged 15–44, compared with proportions of 60–70 per cent in eastern Europe, Africa and Asia. From many micro-level reconstitution studies (where continuous runs of parish registers permit the linkage of vital events over individual life-cycles and thus their aggregation to provide summary demographic measures) Flinn (1981) estimates that the average age at first marriage for females in most north-western European countries was 25–27 years. Since ex-nuptial births (a preferable

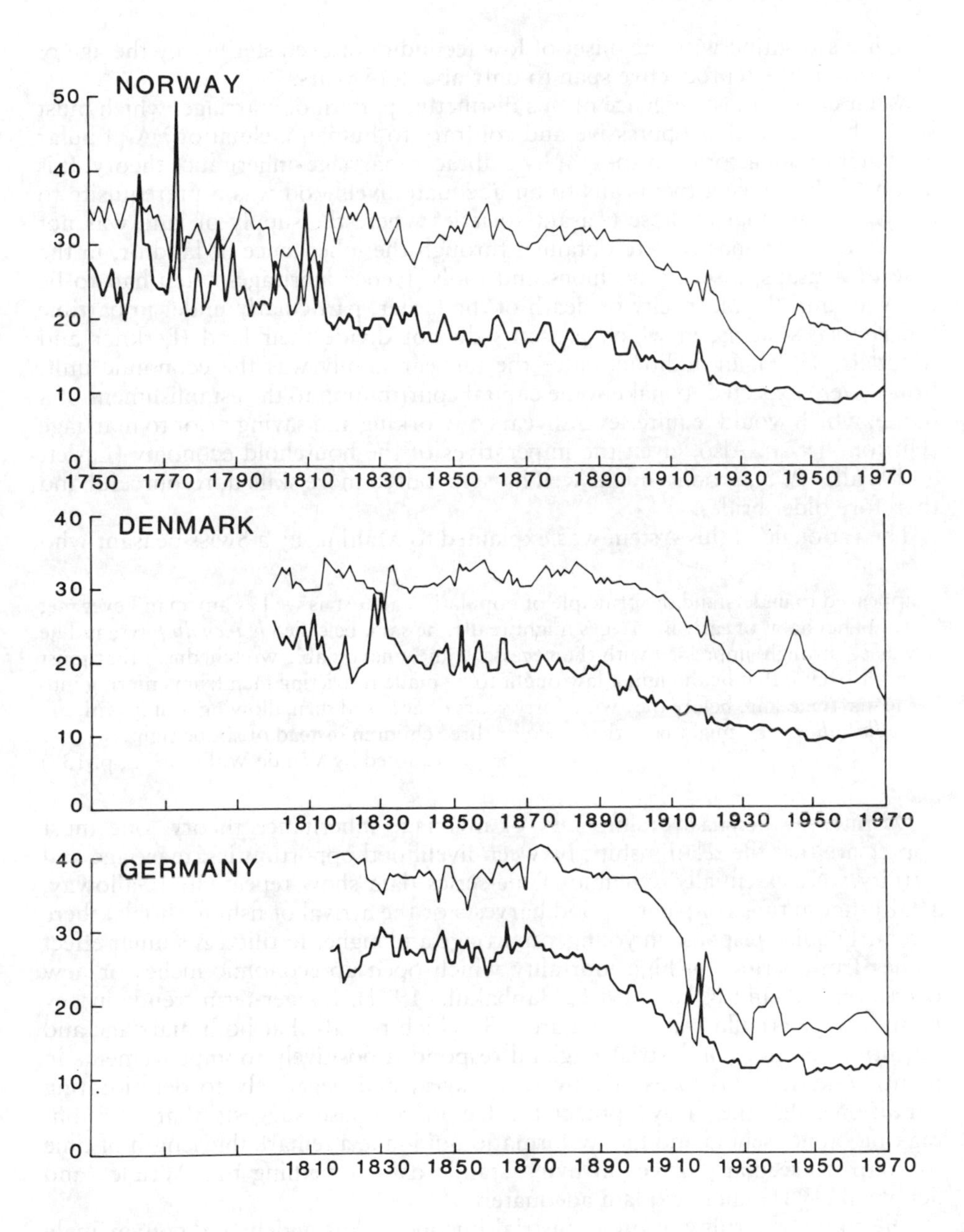

Figure 5.2
Crude birth rates (upper line) and crude death rates (lower line) (per 1,000), Norway, Denmark and Germany, 1750–1970
Source: Grigg (1982), p. 103. By permission of The Geographical Association.

term to illegitimate births) were heavily frowned upon, particularly by authorities responsible for poor-relief, they rarely comprised more than 3 per cent of all births (Shorter, 1976). It is clear then, that a delay in the age of marriage until the late

20s, in association with the onset of low fecundity or even sterility by the age of 40, restricts the reproductive span to only about 15 years.

What caused the emergence of this distinctive pattern of marriage, which must surely be regarded as oppressive and contrary to human inclination? A popular explanation in historical demography embraces marriage–inheritance theory. It is assumed that having the means to an adequate livelihood was a prerequisite to marriage, and that in those peasant societies where the supply of land was not expanding such means were obtained through the inheritance of land or, in the case of artisans, a set of machines and tools. Hence marriages often had to be delayed until the incapacity or death of the father, particularly under impartible inheritance systems, in which peasants do not divide their land (Berkner and Mendels, 1978). In addition, since the nuclear family was the economic unit, brides were expected to make some capital contribution to the establishment of a home, which would require several years of working and saving prior to marriage (Hufton, 1975). Also, given the imperatives of the household economy (Laslett and Wall, 1972), grooms might well have favoured more skilled, resourceful and therefore older brides.

The rationale of this system was explained to Malthus by a Swiss peasant who

> appeared to understand the principle of population almost as well as any man I ever met . . . [The] habit of early marriages might really, he said, be called *le vice du pays*; and he was so strongly impressed with the necessary and unavoidable wretchedness that must result from it that he thought a law ought to be made restricting men from entering into the marriage state before they were forty years of age, and then allowing it only with *des vieilles filles*, who might bear them two or three children instead of six or eight.
>
> (quoted by Van de Walle, 1972, p. 137)

Despite the persuasive simplicity of marriage–inheritance theory, one must appreciate that the relationships between livelihood opportunities, marriage and fertility were essentially dynamic. Time-series data show repeatedly (Galloway, 1988) that in times of plenty (good harvests or the arrival of fishing shoals) there is a fairly quick response in younger marriage and higher fertility. A similar effect comes from periods of high mortality which open up economic niches or new places on the land (Ohlin, 1961; Habbakuk, 1971). Longer-term trends in key relationships are illustrated in Figure 5.3, which reveals that both marriage and fertility rates in pre-industrial England responded positively to improvements in living conditions (as measured by real wages) and negatively to deteriorating conditions. But the fairly appreciable lag in response suggests that decision-making on household and family formation did take a remarkable length of time to adapt to secular changes in living standards – something that Wrigley and Schofield (1981) fail to explain adequately.

The subdued fertility in pre-industrial Europe is thus widely and convincingly attributed to a distinctive, socially regulated marriage pattern. Much more controversial is the extent to which, if at all, there was voluntary fertility control within marriage – by coitus interruptus, folk abortion, sheaths of linen or animal intestines, or whatever. Indirect demographic techniques can be used on family reconstitution data to detect inferentially the presence of such intervention; the shape of the curve of age-specific marital fertility is particularly revealing, as are birth intervals and age of mother at last birth. Such techniques have certainly revealed the practice of conscious fertility control within marriage among some

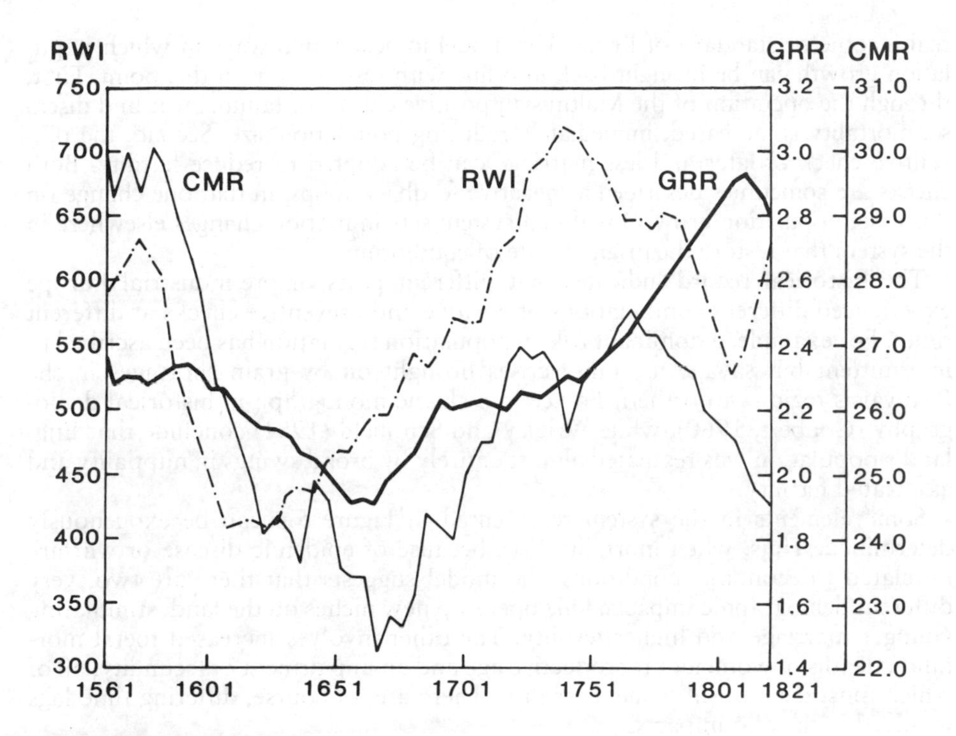

Figure 5.3
Crude marriage rates (marriages per 1,000 persons aged 15–34) and gross reproduction rates in England compared with a real-wage index. All three variables are represented by 25-year moving averages
Source: Wrigley and Schofield (1981), Figures 10.9 and 10.10.

high-status groups, notably the French, British and Italian aristocracy and the bourgeoisie of Geneva and Rouen (Henry, 1956; Henry and Lévy, 1960; Livi-Bacci, 1986). However, most reviews of the evidence from the great mass of rural populations (Van de Walle and Knodel, 1980; Flinn, 1981) conclude that, apart from parts of France and isolated examples elsewhere, marital fertility schedules are consistent with natural (i.e. uncontrolled) fertility. They argue from literary and other qualitative material that the very notion of fertility control within marriage was alien, indeed unthinkable, to pre-industrial peasants who simply accepted, although sometimes reluctantly, as many children 'as God sends'. Indeed, more important as a means of family limitation, even if adopted unconsciously, were child neglect, abandonment and infanticide. In this last respect, it is important to recognize, contrary to some entrenched views, that high infant mortality was more the *result* of high fertility than its cause.

To summarize the discussion up to this point, Figure 5.4 provides a simple model of the essentially self-regulating or homeostatic demographic regime that characterized much of pre-industrial Europe by roughly matching procreation in the community to the carrying capacity of the environment. The model is heavily based on Malthus, in that under circumstances of an agricultural economy with little virgin land available and few technological advances in productivity, any significant increase in population inevitably leads to higher food prices and therefore a lower real income or standard of living. The model indicates two ways in which population growth can be brought back into line with resources from this point. First, through the operation of the Malthusian positive checks of famine, war and disease, mortality is increased, immediately reducing population size. Second, the preventive check of later and less marriage can be adopted to reduce fertility. Both checks are sometimes described as negative feedback loops, in that one change (in this case, population growth) within a system sets in motion changes elsewhere in the system that restore the original state of equilibrium.

The historical record indicates that different parts of pre-industrial Europe experienced different combinations of positive and preventive checks at different times. For example, a dominant role in population regulation has been ascribed to intermittent but savage mortality crises brought on by grain shortages in the Beauvaisis region of northern France in a classic monograph in historical demography (Goubert, 1960), while Wrigley and Schofield (1981) conclude that England's population was regulated almost entirely by broad swings in nuptiality and associated fertility.

Some elements in the system represented in Figure 5.4 can be exogenously determined. Thus, when mortality rises because of epidemic disease or warfare unrelated to economic conditions, the model suggests that there are two, very different demographic impacts. One opens up new niches on the land, stimulating younger marriage and higher fertility. The other involves increased foetal mortality, deaths of women of reproductive age and an impairment of fecundity, all of which must depress subsequent fertility. There are, of course, differing time-lags involved in the two impacts.

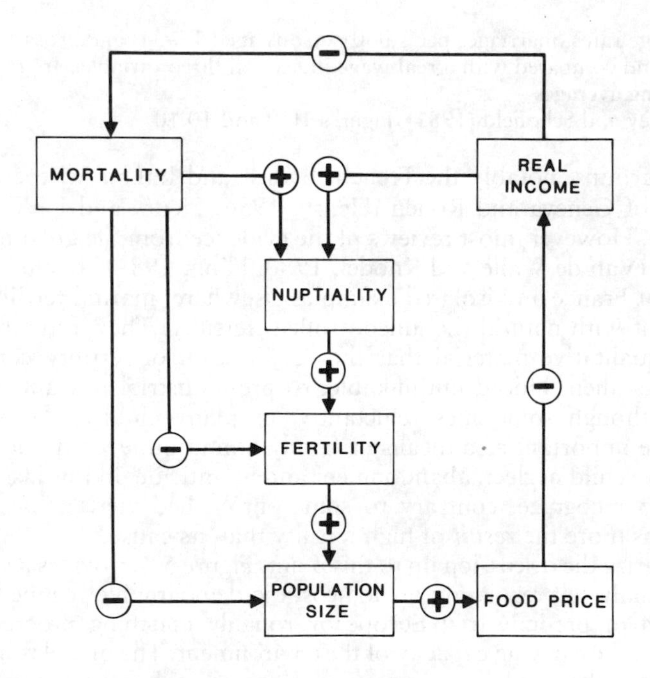

Figure 5.4
A model of population regulation by positive and preventive checks
Source: Amended from Wrigley and Schofield (1981), Figure 11.1.

The essentially homeostatic regime of Figure 5.4, and particularly its restrictive element of late marriage, are much less relevant when land was not fully settled. This was the case in eastern Europe, although Tilly (1978) suggests that an additional factor here was the early proletarianization of the rural population on great estates, so that peasant inheritance concern was not present. Likewise, in the Finnmark area of northern Norway early marriage and high fertility in the nineteenth century were essentially due to land availability in an area made habitable by the newly introduced potato. 'To it came those who wanted to marry, yet could not find the means to do so in the southern part of the country' (Drake, 1969, p. 87).

The relevance of marriage–inheritance theory declines even further whenever new employment opportunities, detached from land inheritance, become available to Europe's peasantry. Proto-industrialization emerged widely as a new mode of production in rural areas of eighteenth-century Europe during the lengthy transition to industrial capitalism (Mendels, 1972). Its basis was the domestic cottage industry of simple, small-scale commodity production organized by urban merchant capital initially putting to work rural labour surpluses at agriculturally slack periods of the year. It permitted the establishment of holdings on marginal land like the lower slopes of the Pennines as well as the subdivision of existing holdings, thereby facilitating new household formation and early marriage. Earnings, rather than property, became the prerequisite for family formation. Moreover, since production was for an external market by family labour, children were widely regarded as economic assets; they were parts of a household production team. In addition, the reduced breastfeeding practised by working mothers would have stimulated fertility.

Well-documented examples of enhanced fertility associated with proto-industrialization are villages in the East Midlands of England (Chambers, 1965; Levine, 1977) and the countryside around Zurich (Braun, 1978), but the importance, the wide geographical extent and the continuation deep into the nineteenth century of proto-industrialization and its associated demographic regime have only recently been recognized (Levine, 1984).

The modern fertility transition in Europe

This is the substantial and prolonged fertility decline schematized in the demographic transition model and addressed by fertility decline theory. It is represented in the form of falling birth rates from the late nineteenth century by Figure 5.2 for Norway, Denmark and Germany and by Figure 2.9 for Sweden.

Adequate explanation, at a high level of generality, encompasses the processes of economic and social transformation which induce the demise of the familial mode of production and the consequent reversal of inter-generational wealth flows. Yet it has proved impossible to determine anything like a threshold of development at which significant fertility decline commences. Van de Walle and Knodel (1980) attempted this task by considering indicators of development in fifteen European countries at the commencement of modern fertility decline – the year in which an index of marital fertility, during a sustained period of decline, fell to a level 10 per cent below its highest level. They were confounded by the huge variations revealed by the indicators: infant mortality, from 76 per 1,000 in Norway to over 200 in Austria, Germany and Hungary; proportion of male labour force in agriculture, from 15 per cent in England and Scotland to 70 per

Table 5.2 Gross reproduction rate (per woman) in three predominantly coalmining and heavy industrial counties, England and Wales, 1851–1911.

	1851	1861	1871	1881	1891	1911
Glamorgan	2.35	2.56	2.69	2.61	2.57	2.08
Monmouth	2.32	2.50	2.70	2.55	2.53	2.07
Durham	2.58	2.73	2.98	2.75	2.57	2.08
England and Wales	2.22	2.28	2.35	2.30	2.01	1.42

Source: Friedlander (1973–4), table 3, By permission of The University of Chicago Press, © 1973 by The University of Chicago.

cent in France, Bulgaria and Hungary; proportion of population in towns of over 20,000 population, from under 10 per cent in France, Bulgaria, Switzerland and Finland to over 40 per cent in England, Scotland and the Netherlands.

One of the prime confounding factors is that in several parts of Europe, fertility actually increased at the beginning of industrialization. We have noted this effect with the growth of cottage industries in the eighteenth century, and similarly in the early years of nineteenth-century industrialization the positive effect of new employment opportunities on fertility seems to have been mediated through a relaxation of inhibitive marital customs. A large proportion of those moving to the new industrial areas were young adults of marriageable age, moving away from the parental and cultural constraints of rural communities, so that 'it was changes in the age and frequency of marriage itself which played the pivotal role in inaugurating the new proletarian fertility regime' (Seccombe, 1983, p. 36).

Data from the 1861 census reveal that the five counties of England with the youngest mean age at first marriage of women born between 1826 and 1841 and married between 1841 and 1861 were the rapidly industrializing counties of Durham, Staffordshire, Northampton, Lancashire and Yorkshire West Riding, while the five counties at the other end of the marriage-age spectrum were essentially rural: Somerset, Berkshire, Shropshire, Hereford and Rutland (Crafts, 1978). Coalmining areas, in particular, showed this fertility increase. Wrigley (1961) identified several such areas in Europe, while British examples are provided by South Wales and Durham (Table 5.2). Friedlander (1973–4) and Haines (1977) demonstrate that the higher fertility of mining areas was due not only to early marriage but also to high marital fertility. This would be stimulated by the lack of female employment opportunities, particularly after 1842 legislation forbidding underground work by women, and by the nature and health risks of mining which curtailed the effective working life of miners, thereby creating a need for unmarried working sons to support the family.

Geographers should be particularly interested in the territorially disaggregated approach of the European Fertility Project organized over two decades by the Office of Population Research at Princeton University, under the directorship of Ansley Coale, to clarify the links between European fertility declines and development. A major aim has been to uncover more detailed spatial patterns than those available from national aggregate data, while at the same time achieving a more comprehensible overview than any based on a mass of detailed village data. Accordingly, indices of fertility and nuptiality have been calculated for more than 600 areas within Europe at ten-year intervals. The specialized indices need not be described fully here, except to say that they attempt a standardization for age in the absence of age-specific fertility data.[1] Some of the mapped patterns are shown

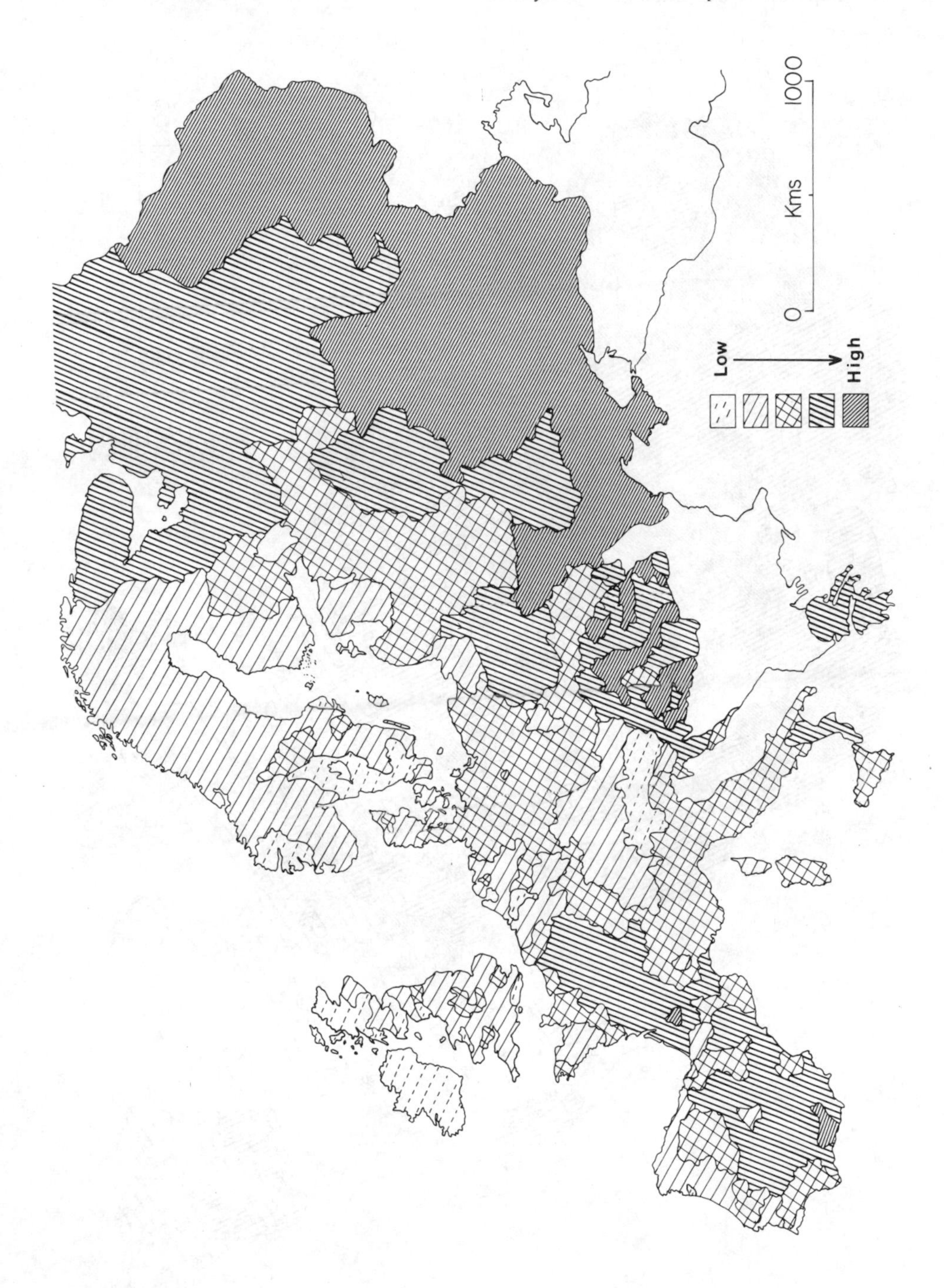

Figure 5.5
Index of proportion married among women of child-bearing age, 1900
Source: simplified from Coale (1969), map 1. By permission of The University of Michigan Press, © 1969 by The University of Michigan.

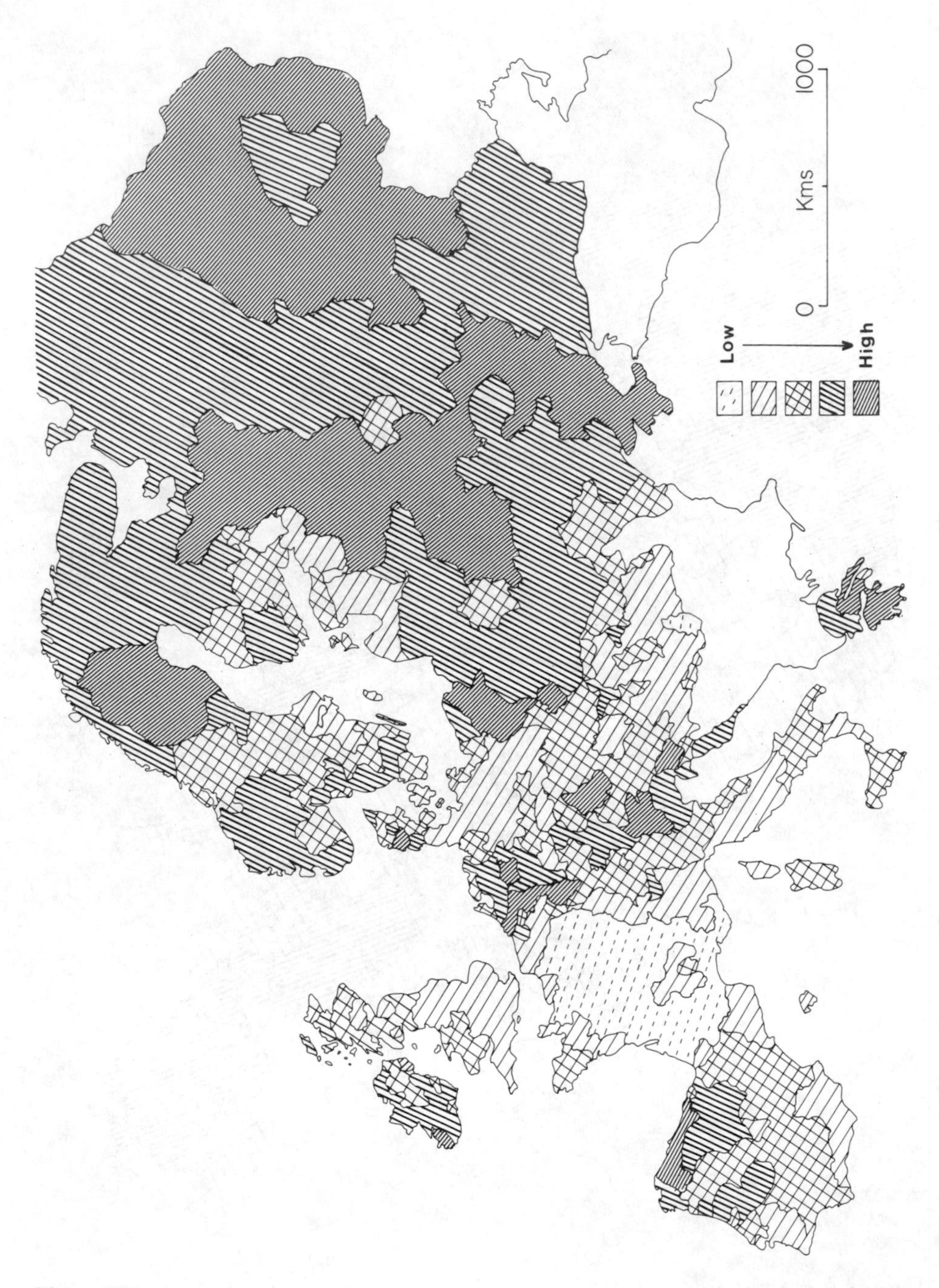

Figure 5.6
Index of marital fertility, 1900
Source: simplified from Coale (1969), map 2. By permission of The University of Michigan Press, © 1969 by The University of Michigan.

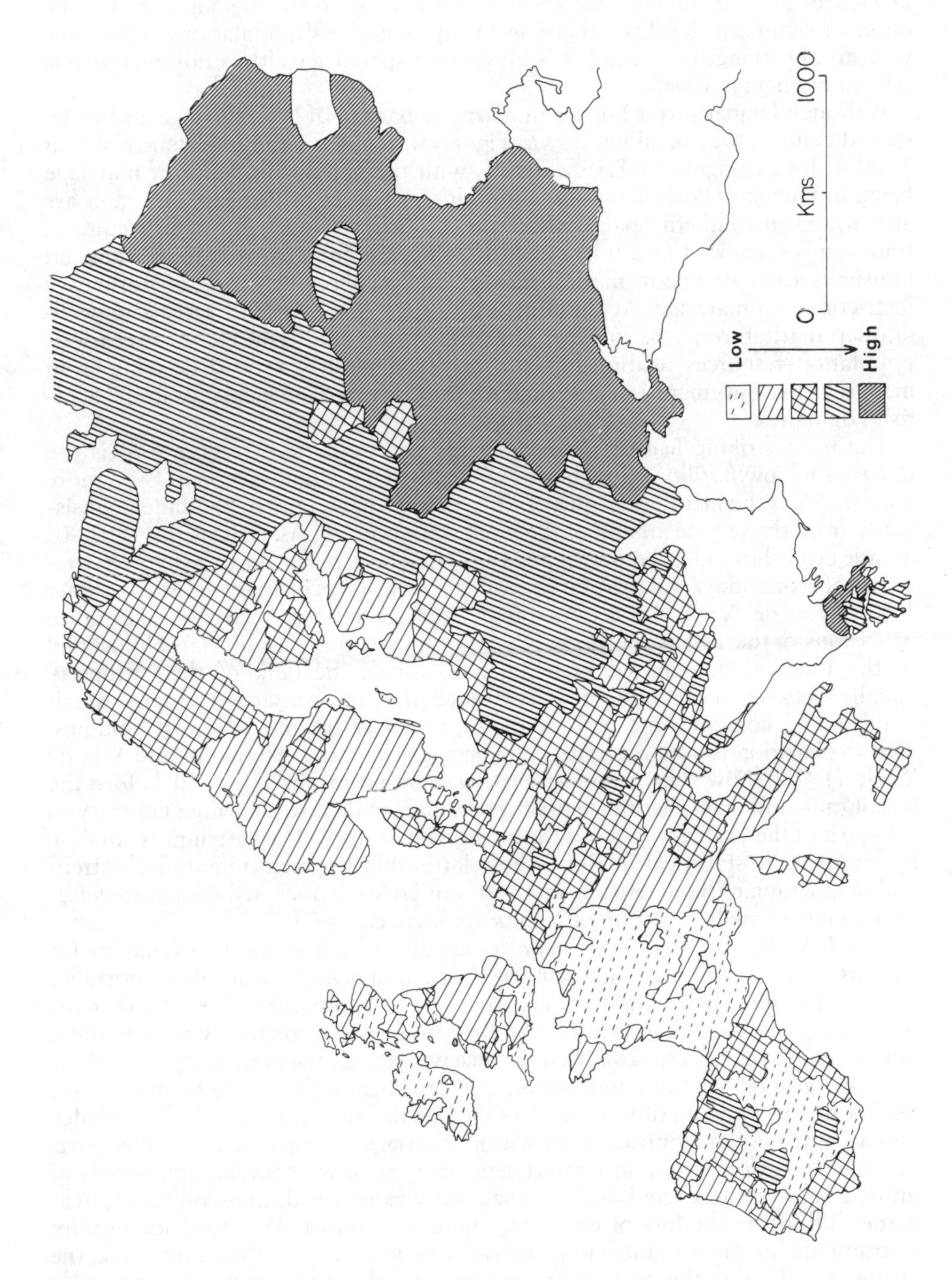

Figure 5.7
Index of overall fertility, 1900
Source: simplified from Coale (1969), map 3. By permission of The University of Michigan Press, © 1969 by The University of Michigan.

in Figures 5.5–5.7 for the year 1900 which is close to the median date for the onset of significant fertility decline in Europe's national populations. They provide an absorbing, but complex, snapshot of spatial fertility conditions at one critical moment in time.

Although Hajnal's west European marriage pattern of late marriage and widespread celibacy was breaking down, Figure 5.5 confirms the importance still at 1900 of his Leningrad to Trieste divide, with the highest incidence of marriage being in Hungary, Poland and the non-Baltic provinces of Russia. High rates are also found in southern Spain, where the Moorish element in Spanish cultural tradition is somewhat fancifully cited by Coale, and in parts of France, where an unusually early decline in marital fertility had enabled the lifting of normative restrictions on marriage. At the other end of the scale, Ireland reveals its well-known marital response to the famines of the mid-century. A precarious population–resources relationship and an imbalanced sex ratio deriving from male-selective out-migration also account for low marriage incidence in the Scottish Highlands.

The most striking feature of the marital fertility pattern (Figure 5.6) is the remarkably low fertility throughout France, Brittany alone excepted. Even more remarkable is the fact that fertility in this largely rural country had fallen consistently from the very beginning of the nineteenth century. As early as the 1801–10 decade crude birth rates below 25 per 1,000 had been reached by the five *départements* of Normandy, and a dozen other *départements* recorded rates of 25–29 per 1,000 (Van de Walle, 1978). An often cited reason has been the inheritance provisions of the 1804 Civil Code which, in line with the egalitarian philosophy of the Revolution, required Frenchmen to divide the bulk of their property equally between their children. It is argued that the peasantry adopted small families as a conscious means of avoiding excessive fragmentation of holdings. This explanation is now recognized, at best, as simplistic. Hermalin and Van de Walle (1977) show that primogeniture was by no means universal before the Revolution and that inheritance practices remained diverse and often contrary to the spirit of the Code throughout the nineteenth century. They attempt to analyse by multivariate statistical methods the relationship between inheritance pattern and demographic characteristics in a series of cross-sections, using *départements* as the units of analysis, but their results are inconclusive.

Van de Walle (1978) also uses *département*al data in a series of correlations for various years in the nineteenth century between indicators of fertility, mortality and rural wealth. He achieves some progress, by finding, for example, that the poorest areas with high mortality, like Brittany, preserve relatively high fertility, but he again concludes (p. 288) that 'France will remain the tantalizing puzzle'. At the most general level of explanation, one can suggest that there seems to have been a unique combination in rural France of the need for, and the knowledge and acceptability of, contraception within marriage. Compared with other parts of western Europe, agricultural improvements were slow to develop, the supply of unused land was restricted, industrialization was retarded, and emigration was made difficult by the loss of colonies. There was thus a clear need for fertility restraint among the peasantry to avoid demographic crises. At the same time, the means of achieving this were more readily available and acceptable in France, partly because its nobility had long been European leaders in the practice of contraception and partly because the Revolution had promoted rational attitudes towards the control of one's destiny.

Even if one excludes France as a very special case, it is still difficult to interpret the pattern of Figure 5.6 in relation to levels of regional economic development. One can find some support for the expected inverse relationship – low marital fertility, for example, in Catalonia, Piedmont-Liguria and the Stockholm area, and high marital fertility in Ireland, Greece, Poland and Russia. Yet major anomalies occur, like low fertility in the rural southern provinces of Hungary, which Demeny (1972) attempts with difficulty to explain, and high fertility throughout Belgium and the Netherlands.

Since ex-nuptial births were negligible, the map of overall fertility (Figure 5.7) shows essentially the combined effect of nuptiality (Figure 5.5) and marital fertility (Figure 5.6). High values for both components ensure very high overall fertility in Greece and in many parts of eastern Europe, particularly in Russia. The same pattern is shown to a lesser extent in Spain. Elsewhere in western and northern Europe there is a fairly uniform pattern of low or modest overall fertility, although it is often derived very differently – in the case of Ireland, for example, from low nuptiality and high marital fertility, and in the case of Aquitaine from high nuptiality and low marital fertility.

While the 600 or so province-type spatial units adopted in Figures 5.5–5.7 permit reasonably detailed analysis, they may still mask considerable demographic heterogeneity. Consequently, a few attempts have been made to provide

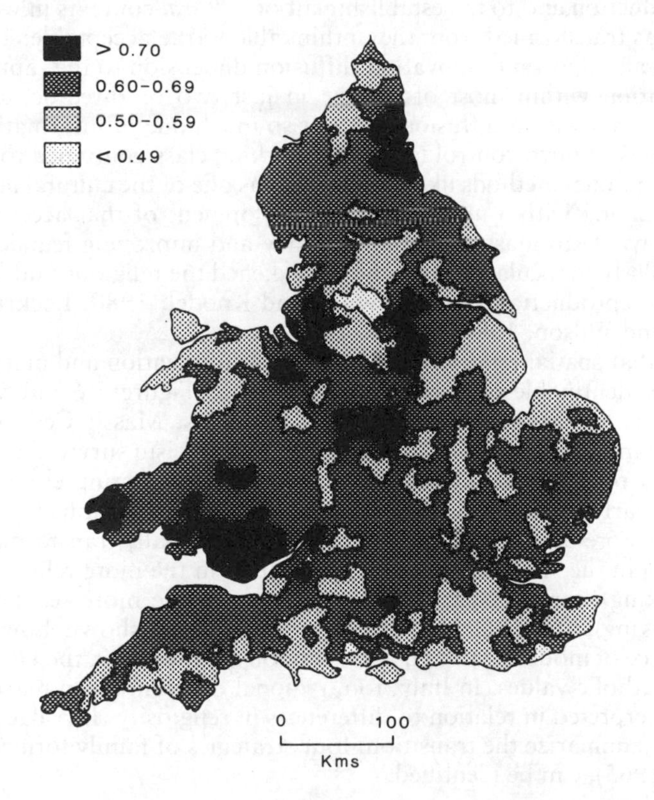

Figure 5.8
Index of marital fertility (see p. 134) for registration districts of England and Wales, 1891
Source: Woods (1986), Figure 3.7. Reproduced by kind permission of Unwin Hyman.

fertility indices for smaller and therefore more internally homogeneous areas. Woods (1982b, 1986) provides such a series of maps for late nineteenth-century England and Wales, based on some 600 registration districts. One of these maps (Figure 5.8) displays the pattern of marital fertility in 1891. It reveals the expected high fertility in the coalfields of the North East, South Yorkshire, Lancashire, Staffordshire, and North and South Wales, and low fertility in the textile belt of Lancashire and the West Riding of Yorkshire, where extensive female employment has long been thought to have promoted greater female autonomy and family limitation. On the other hand, there is no obvious pattern in the other low-fertility districts, since they embrace urban and rural, as well as northern and southern, districts. Multivariate analysis of the English registration county and district data (Teitelbaum, 1984; Woods, 1987) confirms that there is nothing approaching a clear-cut spatial co-variation of fertility with the social and economic indicators usually cited in demographic transition theory – a finding that also emerges from Lockridge's (1983) correlation analysis of small-area fertility data in nineteenth-century Sweden.

The basic argument, then, of classical demographic transition theory that fertility *adjusts* fairly routinely to socio-economic change is now being questioned. Except at very high levels of generality, the empirical evidence is not supportive. More explanatory prominence is now given to the replacement of the familial mode of production and to the establishment of *cultural* contexts in which family limitation was transformed from the unthinkable to the acceptable. There seems to have been an important innovation-diffusion dimension to the rapid spread of family limitation within most of Europe in just two or three decades around 1900, but the innovation-diffusion was not so much one of information on particular methods of birth control (since the working class continued to rely dominantly on traditional methods like withdrawal) as one of the cultural acceptability of family planning within marriage. The development of that acceptability has been related partly to mass education, literacy and improving female status but more particularly to secularization which weakened the religious underpinning of uncontrolled reproduction (Van de Walle and Knodel, 1980; Lockridge, 1983; Lesthaeghe and Wilson, 1986; Woods, 1987).

It follows that spatial relationships between secularization and marital fertility may well be identifiable at a regional level. Thus Figure 5.6 indicates higher marital fertility in three parts of France – the Alps, Massif Central and particularly Brittany – where strong adherence to Catholicism survived the secularizing impact of the French Revolution that was so devastating elsewhere in the country. Similarly, Figure 5.6 shows the higher fertility identified by Livi-Bacci (1971) in the more devoutly Catholic but less economically transformed north of Portugal than in the south, and the higher fertility in the more religious, Flemish (Dutch-speaking) northern part of Belgium than in the more secular, Walloon (French-speaking) south. Here Lesthaeghe (1977) has shown how education acted as a force of modernization in the Walloon area, while in the Flemish area it reinforced Catholic values. In Italy, too, regional differentials in marital fertility have been interpreted in relation to differences in religiosity (Livi-Bacci, 1977).

Finally, to summarize the transition, four strategies of family formation, based on Matras (1965), can be identified:

Strategy A Early marriage/natural fertility.
Strategy B Early marriage/controlled fertility.

Strategy C Late marriage/natural fertility.
Strategy D Late marriage/controlled fertility.

Strategy A is present in parts of the Third World today and characterized the whole of Europe until the sixteenth century, when strategy C was adopted in north-western Europe to form its pre-modern transition. The transition C to B in this area constitutes the modern transition, but it could have been accomplished via A or D. In some industrial areas a path via A may well have been followed, but the more common passage was through the intermediate stage D in which marital fertility was now controlled but marriage remained late. Thus, earlier marriage was made possible by substituting fertility control within marriage for delayed marriage and celibacy as the prime means of restraining population growth. Individual, deliberate control over fertility had replaced the unconscious control exerted by society norms. Thus the distinctive and repressive west European marriage pattern receded when it was no longer necessary. To the east of Hajnal's (1965) Leningrad–Trieste divide the move was directly from A to B, although generally delayed until the early decades of this century.

The transition outside Europe

In the developed countries of North America, Australasia and Japan the progress to low, controlled fertility provides interesting comparisons and contrasts to European experience.

In North America there is no equivalent to western Europe's pre-modern transition achieved through delayed marriage. In the United States the very high fertility rates up to the beginning of the nineteenth century are commonly attributed to ready availability of land in a frontier nation. Coale and Zelnick (1963) estimate that in 1800 the crude birth rate was 55 per 1,000 and the total fertility rate 7.0 per woman. There then ensued a long period of fertility decline until the 1940s. The crude birth rate fell to about 43 per 1,000 in 1850, 30 per 1,000 in 1900 and 18 per 1,000 in 1940, with the equivalent figures for the total fertility rate being 5.2, 3.2 and 2.0. The decline from the late nineteenth century is explicable in terms of increasing industrialization, urbanization and general modernization – standard demographic transition theory – but the earlier nineteenth-century decline in an essentially agricultural country is more intriguing. The popular modern explanation (e.g. Easterlin, 1976a) is that, as the nation's agricultural land filled up, farmers became increasingly concerned about their children's ability to acquire good farmland; hence they began to limit family size. Empirical evidence supports this view. Yasuba (1961) found that in the first half of the nineteenth century state and territory birth rates were strongly correlated negatively to population densities, suggesting that in the older settled areas relative scarcity of land contributed to fertility decline, while in the frontier regions of the west the availability of free or cheap land maintained high fertility. These conclusions are confirmed by more sophisticated regression analyses undertaken by Forster and Tucker (1972).

A similar relationship between human fertility and the density and nature of agricultural settlement is found by McInnis (1977) to have been operating in Upper Canada (the southern part of contemporary Ontario) in the mid-nineteenth century. On the basis of data from 55 counties he demonstrates a strong inverse correlation between fertility, measured by child:woman ratios, and population density, measured by population size in relation to maximum acreage ever cultivated. He also uses data from 1,200 farm households, extracted from manuscript returns of census enumerators, to demonstrate that fertility levels were at their highest in the most recently settled townships. In such frontier environments not only is farmland likely to be available to the subsequent generation but, in addition,

> Surrounding oneself with a large family is a rational response to the generally hazardous and risky life of the more isolated settlers. As a community grows up, people are able to depend upon the community at large and even institutions organized specifically to shoulder some of the risk of life.
>
> (McInnis, 1977, p. 208)

In Australasia, significant decline in marital fertility commenced in the 1880s, with a remarkable uniformity of timing and rate of decline exhibited by the individual Australian states and New Zealand (E. F. Jones, 1971). Despite considerable geographical isolation of states from one another, their common cultural heritage and common experiences led to very similar patterns of change in reproductive behaviour; there were also very few differences between socio-economic groups in the timing and pace of their adoption of new fertility strategies. It is unlikely, therefore, that there was any significant element of spatial or class diffusion in the spread of the small family norm. To Caldwell and Ruzicka (1978), the decisive factor in Australia was the way in which the development of mass education, and to a lesser extent secularism, undermined the morality of an austere, frugal and therefore cheap upbringing for children.

In pre-industrial Japan, as in Europe, there is now clear evidence from an accumulating series of local studies (Hanley and Wolf, 1985) that fertility levels were both socially and individually controlled to combat population pressure in subsistence agrarian societies with few opportunities for territorial expansion. Hanley (1977) shows that age at marriage was late and that significant proportions of women never married. She also demonstrates strong evidence for fertility control within marriage. Women typically terminated child-bearing in their early or mid-thirties after achieving an average family size of about three children. Such modest family sizes were regulated not only by fertility control, notably through abortion, but also by widespread resort to infanticide and child adoption. Adoptions were used negatively to divest families of excess children and positively to compensate for child deaths and, in particular, to ensure the existence of male heirs.

Japanese demographic evolution is also reminiscent of parts of Europe in that fertility almost certainly rose under the initial impact of economic modernization from the mid-nineteenth century. Thus Taeuber (1958, p. 55) concludes: 'The Japanese experience supports the hypothesis that it is the transitional society that manifests the highest fertility.' She also considers:

> the possibility that the fertility controls of an ancient society may decline simultaneously with the increase in the controls of an industrial society. If so, there may be a considerable period in which the demographic transition involves a change in the class incidence of family limitation but not in its total amount.
>
> (Taeuber, 1958, p. 55)

Certainly for a long time the overall decline in Japanese fertility was modest (crude birth rates of about 38 per 1,000 in the 1880s, 30 per 1,000 in the 1930s and 33 per 1,000 in the late 1940s), bearing in mind the rapid pace of industrialization and urbanization. This has been explained by both Taeuber (1958) and Wilkinson (1967) in terms of the retarded pace of social as opposed to economic transformation. They cite the maintenance of a paternalistic labour situation and a traditional family-centred social organization which left cultural milieu and values virtually unchanged.

Japanese resistance to the small-family ideal finally crumbled in the 1950s, when in a single decade the crude birth rate was halved, providing the most concentrated fertility reduction ever achieved in the world. This may be interpreted as the inevitable accompaniment of further modernization, but a special instigating factor was the desperate demographic plight of Japan immediately following the Second World War. At a time of economic chaos and national demoralization the country had to absorb 7 million Japanese from its lost empire and cope with a natural increase surge brought about by the rapid response of the death rate to medical and public health programmes. There was an immediate and widespread desire to limit family size in all social classes, and the prime means adopted was abortion, made available by the state on liberal grounds in 1948.

Post-transition trends

Demographic transition theory provides an adequate general account of the demographic evolution of developed countries until well into this century, and in particular of the sustained fertility fall which had lowered total fertility rates by the 1930s to 1.5–2.0 births per woman, some way below generation-replacement levels. But in the last half-century, in the period aptly characterized by Campbell (1974) as 'beyond the demographic transition', there have been appreciable, but

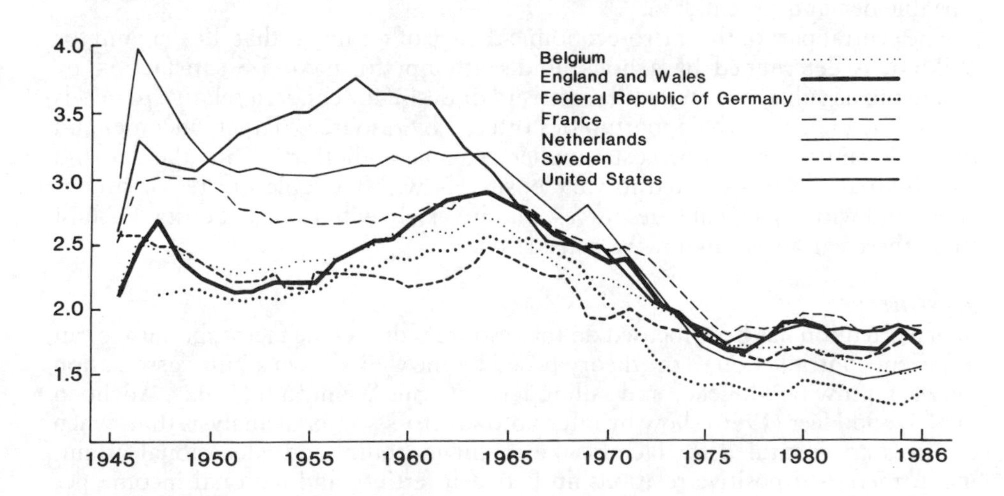

Figure 5.9
Total fertility rates (per woman) in selected developed countries, 1945–86

remarkably consistent fluctuations in fertility in developed countries, as Figures 5.2 and especially 5.9 demonstrate.

It is possible to argue that these fluctuations are entirely compatible with demographic transition theory, in that the postwar fertility increases, as measured by period indices, were a response to temporary and very special changes in the timing of marriages and births, with no appreciable impact on the long-term relationshp between modernization and fertility reduction and with only modest change in measures of *cohort* fertility. According to this view, the fertility falls throughout the developed world from the mid-1960s reflect a return to normality after the entirely exceptional postwar baby boom caused partly by the making up of marriages and births postponed during the Depression and the war and partly by a general trend towards younger age at marriage, at least until about 1960. In England and Wales, for example, the married proportion of women aged 20–24 rose from 34 per cent in 1939 to 59 per cent in 1961, before falling slowly to 55 per cent in 1977 and then very rapidly to merely 34 per cent in 1985. This recent decline in early marriage and the rise in cohabitation (particularly prominent in Sweden and Denmark, but widely observed in the Western world) are obviously linked, but which causes the other is arguable. It is also important to realize that the modern fertility downturn was well established *before* the modern trend towards deferred marriage had become significant. Thus only 10 per cent of the decline in Canada's total fertility rate between 1961 and 1981 can be attributed to declining marital proportions (Romaniuk, 1984).

More elaborate explanatory approaches centre on variants of the basic economic theory of fertility. Becker (1960, 1981) has led the way in suggesting that economic analysis can be applied beneficially to many aspects of human behaviour, not simply those that involve production or monetary transactions. In particular, he argues that fairly standard neo-classical theory of consumer behaviour can be profitably employed to analyse and explain fertility decisions. In developed countries at least, the influence of supply factors (fecundity, contraception) on fluctuations of fertility is minimal, so that the critical factor is the variable demand for children.

The central part of the micro-economic theory of fertility is that the demand for children is determined by a household's attempt to maximize satisfaction by balancing its subjective tastes (the intensity of desire for children relative to goods and services) against the opportunities offered by resources (largely income) and the constraints imposed by costs (monetary, time and effort). Thus the demand for children can be expected to vary positively with a couple's tastes or preferences and with its available resources, and inversely with expected costs. Each of these three variables can now be considered.

Resources

Much attention has been focused on the resources or income factor, because of an apparent contradiction to the theory posed by the well-known secular association between growing affluence and falling fertility. But Weintraub (1962), Adelman (1963) and Heer (1966) show by international cross-sectional analysis that, when controls are instituted for factors like infant mortality and educational attainment, there is a positive relationship between fertility and national income per head. Moreover, time-series data have been used to demonstrate that the relationship in the *short term* between income and both marriage and fertility is positive, as Malthus suggested. Evidence to this effect from pre-industrial and nineteenth-century Europe presented earlier in this chapter is confirmed by the work of Kirk (1942), Galbraith and Thomas (1941) and Silver (1965) on the relationship between business cycles and demographic trends in Germany and the United States in the twentieth century. They find that marriages and births increase following an improvement of business conditions and decline when business falls away. Economic fluctuations can thus be expected to advance or postpone births, but their effect on ultimate family size is more problematical.

The income effect on fertility helps to explain the enhanced fertility levels between the Second World War and the mid-1960s, since there were appreciable and continuous rises in per capita income in most developed countries at that time, in marked contrast to the depressed levels of the 1930s. One can also argue that the fertility falls of the late 1960s and the 1970s were to some extent a consequence of slower income growth and growing economic insecurity, expressed in inflation and unemployment. The relative prosperity of the child-bearing cohort is subject to several influences, but one important cyclical demographic influence identified by Easterlin (1980) is the echo effect of fertility on the age structure of the male working population. He demonstrates that the low fertility of the 1930s reduced the proportion of younger to older workers in the late 1940s and the 1950s, so that the consequent labour demand for young workers ensured a high level of prosperity (resources) for potential parents. The resultant high fertility led in turn to a flooding of the job market with young workers from the mid-1960s, so that the relative prosperity of the child-bearing cohort was undermined.

Tastes

The basic economic theory of fertility regards children in developed countries as consumption goods (certainly not the investment goods that they often are in the Third World). Consequently, in the allocation of scarce resources amongst competing uses there is a trade-off between goods and children. Several factors can influence the choice of trade-off. Easterlin (1976b) suggests that a major influence on material aspirations is the standard of consumption that young adults have inculcated from the households in which they were raised. Thus, young couples brought up by parents experiencing a low income and low material standard of living tend, in turn, to have low aspirations concerning their own consumption of goods and services. Easterlin argues that this was an important cause of the postwar baby boom, in that young couples at that time had only modest consumption aspirations because of their upbringing in the austerity era of the Depression, the war and postwar rationing. Consequently they felt they could afford several children, particularly in an era of buoyant labour demand and general prosperity. But the reverse pattern became evident by the late 1960s, when young couples had greatly increased aspirations for material consumption fashioned by their upbringing in the prosperous 1950s; hence the preference for goods and services over children in the late 1960s and 1970s.

Similar reasoning was used by Banks (1954) to explain the initiation of modern fertility reduction in England in the 1880s. He observes that the English middle classes were relatively prosperous between about 1850 and 1870, but a major depression in 1873 was the harbinger of several lean years. Middle-class incomes did not fall in real terms, but opportunities narrowed and the rapid growth of incomes ceased. Since, however, the aspirations of middle-class families continued to grow, parents perceived a worsening of their economic situation and thus restricted their family size.

But experience-based or socialized material aspirations can be only part of the explanation for changing tastes or preferences for children. Indeed, Blake (1968), in a memorably titled paper, 'Are babies consumer durables?', argues strongly that there is something qualitatively different in the decision to have a child from that to purchase a washing machine. To such sociologists the all-important factor governing fertility levels is that of group norms and peer-group pressure. Accordingly, preferences for children from the mid-1960s may well have been lowered by the then fashionable middle-class concern about global population pressure and, even more, by the burgeoning feminist movement's propagandizing of female fulfilment in spheres other than the domestic maternal one; the bell was beginning to toll for the *Child King*. In recognition of such influences, Easterlin (1978) and Lesthaeghe and Surkyn (1988) have attempted a reconciliation and synthesis of economic and sociological theories of fertility.

Costs

The decisive influence on recent fertility fluctuations in the West has probably been the opportunity costs (income forgone) involved in women leaving the labour force to raise children (Ermisch, 1982, 1983). There was little deterrence to high fertility from such costs as long as employment among married women was the exception, as in the 1950s. But married female participation rates rose

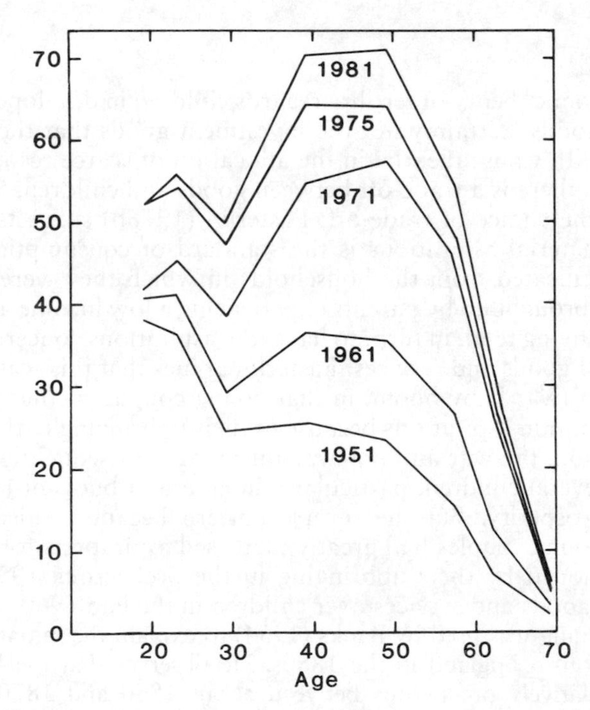

Figure 5.10
Labour force participation rates (%) among married women, Great Britain, 1951–81
Source: Eversley and Köllmann (1982), Figure 6.3.

steeply in the 1960s and 1970s (Figure 5.10), due primarily to structural employment changes which diminished the dominantly male-employing manufacturing sector and expanded the dominantly female-employing service industries and part-time employment. Under these circumstances the opportunity costs of leaving the labour force to rear children became an important disincentive to fertility, particularly as market forces and anti-discriminatory legislation enhanced the relative remuneration of women.

The interaction of the three key elements of resources, tastes and opportunity costs provides, therefore, an explanatory overview of post-transition fertility trends in the West. The standard of living aspired to by young couples in the 1950s and early 1960s was only modest, because of their upbringing in the prior era of scarcity. At the same time, there was appreciable economic growth, full employment and an alleviation of the acute postwar housing shortages, illustrated by British Prime Minister Macmillan's famous electioneering slogan in 1959: 'You've never had it so good.' Since there were very few married women in the labour force, this was obviously a prime period for child-raising. But from the mid-1960s very different conditions prevailed. Now there were high material aspirations among young people, based on their upbringing in the good times of the 1950s; there was a faltering and then a downturn in economic growth; and married female participation in the labour force became the norm – hence the rapid transition from the 'baby boom' to the 'baby bust'.

In this explanatory account there is no place for the more liberal provision of legal abortion facilities and improved contraceptive technology, notably the widespread adoption of the oral pill from the mid-1960s, essentially because they do not influence *demand* for children. In any case, the significance of their role in fertility reduction is problematical. Some have argued that the decline in unwanted births resulting from more widespread use of more effective birth control practices has been a significant factor in fertility decline in some developed countries since the mid-1960s. On the other hand, a historical perspective reveals that knowledge of, and access to, what might be regarded as at any time the more modern forms of birth control have never been important independent influences on fertility. Attitudes and values towards child-bearing have always been the critical factor. The limited influence of birth control technology can be inferred from the wide range of traditional and modern methods used in several developed countries in the early 1970s (Table 5.3). All of these countries were experiencing

Table 5.3 Percentage distribution of contraceptors by main method used, revealed by 1970–2 social surveys

	Denmark	Finland	USA	France	Czechoslovakia
IUD	4	4	9	2	14
Pill	37	26	41	17	4
Condoms	30	40	17	12	19
Diaphragm	9	–	7	1	–
Withdrawal	7	21	3	52	52
Rhythm	2	1	8	14	3
Other	11	8	16	2	8
Total	100	100	100	100	100

Source: Economic Commission for Europe (1976).

pronounced fertility decline, but were achieving it by the use of very different methods. In particular, one should note the heavy reliance in some countries on primitive methods like withdrawal.

The remarkably consistent pattern of post-transition fertility fluctuations in Western developed countries has not been followed by the USSR and eastern Europe. In the, initially, rather less developed countries of Yugoslavia and Poland, appreciable and sustained declines in total fertility rates (from about 3.5 births per woman in the early 1950s to 2.2 in the early 1970s) can be explained reasonably adequately in terms of general modernization, i.e. demographic transition theory. In the USSR fertility did not begin to decline significantly until after 1960, with its Central Asian republics being particularly resistant, but in Hungary, Czechoslovakia, East Germany, Bulgaria and Romania pronounced fertility reduction in the 1950s had brought about the lowest fertility levels in the world during the 1960s. In 1962, for example, the total fertility rate in Hungary was 1.8 births per woman, when the United States rate was still 3.5. Factors widely cited have been the housing shortages associated with intensive urbanward migration, and also the ready availability of government abortion services from the mid-1950s to promote maternal welfare and female employment at a time of manpower shortages in labour-intensive economies. In these countries there has been little further secular reduction in fertility in the 1970s and 1980s, partly because of the introduction of pro-natalist policies (see Chapter 11), enabling most developed Western countries to overtake them in fertility reduction.

Differential fertility

Sociologists study the ways in which fertility varies within developed societies by social patterning based on variables like class, income and religion. It might be argued that population geographers, on the other hand, should be more interested in the fashioning of spatial fertility variations by ecological patterning. Yet such patterning is itself largely the product of particular spatial mixes of social factors. Thus the attributes of an area which fashion its fertility level are essentially those discussed in sociological literature as differential fertility, so they must be of importance to population geographers.

The essence of differential fertility is that within national societies there exist social groupings with particular norms and values concerning family size that are maintained by peer-group pressure to conform. But since these groupings overlap, based as they often are on more than one variable, it is analytically difficult to assess the impact of any specific variable.

The significance of fertility differentials varies over time, particularly in relation to stage of demographic development. This is why differential fertility figures prominently in demographic transition theory, since fertility decline is often regarded as starting in particular social groups and then diffusing throughout society, with time-lags due to group differences in values and attitudes and in birth control practices. Therefore, as the transition gets under way, fertility differentials are at their widest. As the transition progresses, differentials can be expected to diminish and possibly disappear in post-transition societies of the modern world (Andorka, 1982; Hawthorn, 1982). This, of course, is very different from mortality differentials, since individual life expectancy is very far from being determined by individual choice. Fertility levels, on the other hand, are not only controlled almost entirely by demand or choice factors in developed countries, but such factors are themselves increasingly subject to the homogenizing role of powerful and universal influences like television. This social convergence is particularly evident in newer developed societies like Australia, which

> are at once less class-ridden; lack the inherited geography of the pre- and early industrial eras; were notable for the rapidity and universality of the uptake of the contraceptive pill during the 1960s; and remain distinctive for the extent to which the population of each state is both 'metropolitanized' and, for those living beyond the capital cities, 'metropolitan dominated' with respect to the dissemination of information, attitudes, entertainment, opinion, fad, and fancy.
>
> (M. G. Wilson, 1988, p. 9)

It has also been observed recently in Sweden – significantly, a thoroughly modernized society committed to equality of opportunity – that fertility differentials based on standard, traditional variables are rapidly being supplanted by more complex ones grounded in *individual* value systems (Hoem and Hoem, 1989). But in most developed countries, traditional fertility differentials, although declining, are persistent enough even in the 1990s to retain some analytical significance.

The major variables in differential fertility will now be reviewed, followed by a consideration of how such variables have been incorporated by population geographers in spatial analyses of fertility.

Social class

In the major period of fertility decline in developed countries between the 1880s and the 1930s there emerged a marked negative relationship between social class and fertility. This was because small-family norms and the practice of effective birth control were first adopted in higher status groups, before being diffused down the class gradient. But as diffusion is completed in contemporary societies, there has been an appreciable narrowing of differentials. Figure 5.11 shows this clearly enough for England and Wales during the 1970s, although it also indicates some modest widening of differentials in the early 1980s, suggesting that in societies like Britain, as opposed to Australia, there may well be limits to social and demographic convergence. In particular, it is hard to imagine that the lowest status groups, especially social class V, will not retain their higher fertility. They incorporate what some have termed an 'underclass', exhibiting in life-styles a 'culture of poverty' (Lewis, 1959), embracing attitudes of fatalism, apathy, lack of long-range planning and the like, which ensure the maintenance of high fertility and high mortality. Useful case studies in this area are by Askham (1975) and Blaxter and Patterson (1982) on fertility and health attitudes and behaviour among working-class families in Aberdeen.

Two components of social class which are thought to contribute particularly strongly to fertility differentials are educational attainment and family income. Using data from the 1-in-a-100 Public Use Sample census tapes in the United States, Rindfuss and Sweet (1977) have demonstrated that fertility differentials by female educational attainment were both appreciable and persistent in the 1950s and 1960s. One factor is likely to be the tendency for later marriage among more highly educated women, as well as their higher opportunity costs in leaving a career to raise children. Another concerns the tastes or preferences of such women, in that they are more likely to have major interests and commitments

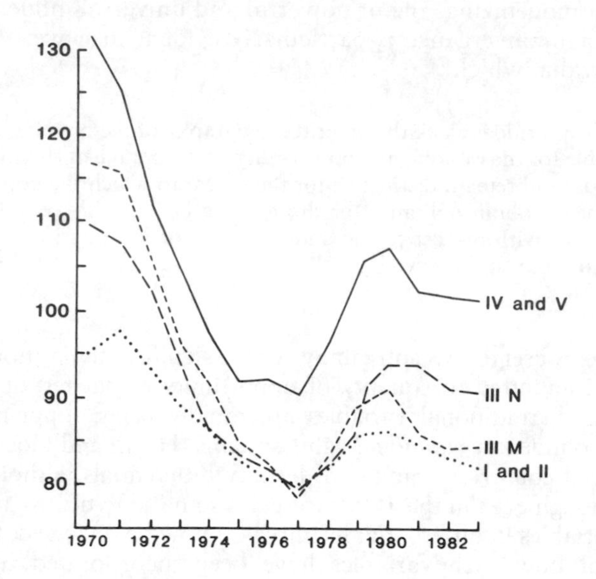

Figure 5.11
Nuptial live births per 1,000 married women aged 15–44 by social class of father, England and Wales, 1970–83
Source: Werner (1985), Figure 2(c), Crown Copyright.

outside family life. Then again, the better-educated couples are likely to be more efficient family planners.

The income variable is more complex to interpret (Freedman and Thornton, 1982). According to the basic economic theory of fertility, one should expect a positive relationship between fertility and family income, whether income is measured currently or in terms of long-term income prospects (Easterlin's 'potential income'); in other words, better-off couples can afford, and are likely to achieve, larger families. In reality, fertility usually varies inversely with income – 'the rich get richer and the poor get children'. Becker (1960) felt able to resolve the anomaly by pointing to the different abilities of groups to match desired fertility with achieved fertility. He argued that, when contraceptive efficiency in a population is extremely low or extremely high, fertility does indeed respond to income in the predicted fashion. Another explanation is that income is correlated with other fertility-influencing variables, so that the generally low fertility of high-status groups is due to high educational attainment, urban life-style and the like, *despite* a high-income level.

Perhaps the most acceptable expression of the income effect on fertility is in the so-called relative income hypothesis (D. Freedman, 1963; Stevens, 1981; Johnson and Lean, 1985). This suggests that fertility is positively related to relative income (the relationship between a couple's income and that of its peer group in terms of age, education, class, residence, etc.). Thus the more affluent a couple within its reference group, the higher will be its fertility.

Female employment

It is well known that participation by women in the labour force makes demands on time and energy at the expense of child-raising, thereby accounting for the inverse relationship consistently found within developed countries between fertility and married female employment; this is one reason for the relatively low fertility often found in textile areas. But it is far from proven that participation in the labour force influences fertility more than the reverse. Thus some women drop out of employment because they have children, while others seek employment because of sub-fecundity or sterility in themselves or their partners.

Much depends also on the conditions of employment. The conflict between roles of mother and worker is most acute in the case of full-time, career-structured employment outside the home. There is less incompatibility when employment is part time or based on the family home. This is how Federici (1968) accounted for the lack of any relationship between female employment and fertility in agricultural southern Italy, in contrast to the normal inverse relationship found in urban and industrial northern Italy. The role-clash faced by working mothers can also be reduced by supportive government policies, notably in Sweden (Hoem and Hoem, 1989), where 80 per cent of women with pre-school children are now in the labour force (including those on maternity leave).

Urbanization

Among the most persistent of fertility differentials has been the inverse relationship between urbanization and fertility evident from the early stages of demographic transition when urban social conditions proved more conducive than those in the countryside for innovations like small-family preferences and modern birth control practices. The differential seems to be maintained above all by the higher direct, indirect and opportunity costs of raising and educating children in cities.

Religion

Within Western developed countries the fertility of Roman Catholics is invariably higher than that of the rest of the population. The proscription by the Roman Catholic Church on 'unnatural' methods of birth control is well known, although survey data reveal an ever-increasing majority of Catholic women deviating from the birth control teaching of their Church. More important is thought to be the higher value attached to a larger family in Catholic doctrine than in Protestant teaching, which tends to emphasize more the quality rather than quantity of children.

Protestantism itself covers a wide spectrum of denominations, and there is plenty of evidence to show that the more fundamentalist groups like Mormons, Nazarenes, Pentecostals and Jehovah's Witnesses in the United States have fertility levels as high as Catholics. It is significant that the Mormon state of Utah exhibits by far the highest fertility rate of American states. In 1980 its total fertility rate of 3.2 births per woman was almost double the national rate of 1.8, despite its population being heavily urban (Salt Lake City).

It has sometimes been argued that in Western societies religion does not have a significant independent effect on fertility, in so far as fertility differences between religious groups are essentially due to the different compositions of these groups by occupation, education and the like. While Protestants have a socio-economic structure close to the national average, Catholics are over-represented in low-status groups, and Jews, who have the lowest fertility of all, are over-represented in high-status urban groups. Yet whenever atempts are made to standardize for socio-economic structure (Feeedman, Whelpton and Smit, 1961; Compton, 1978; Day, 1984), the Protestant–Catholic fertility differential does survive, not least because the normal inverse relationship between fertility and class structure does not always occur among Catholics; higher-status Catholics, it seems, are more likely to heed Church views on family size. Some demographers (Jones and Westoff, 1979) have been predicting the imminent disappearance of the Catholic–Protestant differential in the United States, but some of the more refined surveys of fertility intentions suggest otherwise (Coombs, 1979; Blake, 1984; Mosher and Hendershot, 1984; Goldscheider and Mosher, 1988).

A provocative hypothesis is that Catholic fertility will be particularly elevated in those societies in which Catholics constitute an appreciable minority (van Heek, 1956–7; Day, 1968). In such circumstances there may well be a stricter adherence to basic Catholic principles and, particularly if there is a feeling of discrimination, there might be an incentive to increase the demographic strength of one's group within the national community. This is how Day explains that fertility is higher in the important Catholic minorities in the United Kingdom, Switzerland, the Netherlands, the United States, Canada, Australia and New Zealand than in the Catholic majorities of France, Belgium, Luxembourg, Italy and Austria.

An expression of this theory at work in the contemporary world is surely to be found in Northern Ireland, given the long-standing hostility between the Protestant and Catholic communities and the 'appreciable minority' status of Catholics (now almost 40 per cent of the population). There are, indeed, appreciable differences in the 1980s between the two communities in achieved fertility, for the country as a whole and for each of its major ecological zones (Table 5.4), as well as in attitudes towards legal abortion provision (Table 5.5). It is elevated Catholic fertility that gave to Northern Ireland the highest total fertility rate of all developed countries in 1985 (2.45 births per woman), surpassing even the traditional leader, the Irish Republic (Werner, 1988). More important, Compton (1982) has shown that a projection of growth in the two communities, based on their respective fertility, mortality and emigration schedules in the late 1970s, would lead to a Roman Catholic majority in Northern Ireland by the middle of the next century. The social and political implications are immense. But in a new assessment, Compton and Coward (1989) conclude that modern falls in Catholic fertility are making the attainment of a majority by demographic means much less likely.

Table 5.4 Number of children born to ever-married women aged 45–49, 1983, in Northern Ireland by area and religion

	Roman Catholic	Non-Roman Catholic
Belfast	4.0	2.4
Belfast suburbs	3.7	2.6
Belfast fringes	3.3	2.2
Intermediate zone	5.2	2.7
Outer zone: west of the Bann	5.4	3.2

Source: Northern Ireland Fertility Survey, 1983, reported in Coward (1986a), table 3.8. By permission of Croom Helm.

Table 5.5 Percentage of women in Northern Ireland, by religious denomination and religiosity, 1983, opposed to liberalizing currently restrictive abortion legislation

	Importance of religion	
	Very/Quite	Not very/Unimportant/Don't know
Roman Catholic	80	59
Fundamental Protestant	74	*
Protestant	43	26

* Too small sample.
Source: Northern Ireland Fertility Survey, 1983, reported in Compton Coward and Power (1986), table 5. By permission of *Irish Geography*.

Canada provides another intriguing case, where the French-speaking, almost entirely Catholic population of Quebec comprises just over one-quarter of Canada's population. Perhaps nowhere in the world was the Catholic large-family ideal more effectively realized, since it was reinforced by nationalist propaganda of the *revanche des berceaux* type. It was the high fertility of French Canadians that ensured their political survival in face of a flood of English-speaking immigrants. But in recent times there has been a startling turnabout. Provincial fertility rates indicate that Quebec's premier position was already being eroded in the 1930s and that from about 1960 Quebec has consistently shown the *lowest* provincial fertility in Canada. By 1988 its total fertility rate had fallen to a remarkably low 1.4 births per woman, resulting in the provincial government's announcement that year of appreciable financial incentives for higher-order births. It is easier to chart than to explain Quebec's fertility fall, but it can be argued that it does not necessarily invalidate the van Heek–Day hypothesis. This is because French-speaking Catholics, although comprising an appreciable minority within Canada, do comprise a large majority within Quebec, which has considerable autonomy within a federal state.

Ethnicity

The only way in which the biological attribute of ethnic or racial group could possibly affect fertility is through fecundity, and there is simply no evidence for this. But since different ethnic groups within a country are likely to occupy different social situations, whether or not caused by discrimination, their fertility levels are likely to vary.

Figure 5.12 shows the remarkable persistence over time of the fertility differential between American whites and blacks. The similar temporal trend of black and white fertility does, however, mask some very important differences between the two groups in the pattern of child-bearing. By the early 1980s the proportion of ex-nuptial births had climbed to just over 50 per cent for blacks, but to only 10 per cent among whites, and the proportion of all births to teenage mothers was 30 per cent among blacks and 15 per cent among whites (Reid, 1982; Evans, 1986).

There is considerable concern in the United States about an inner-city black underclass incorporating large numbers of one-parent households where young black women are stranded alone, on welfare, with small children, little education, few skills and inadequate family planning knowledge and motivation. In contrast, among the most highly educated American women, fertility has for some time been lower among blacks (Rindfuss and Sweet, 1977); these are the middle-class

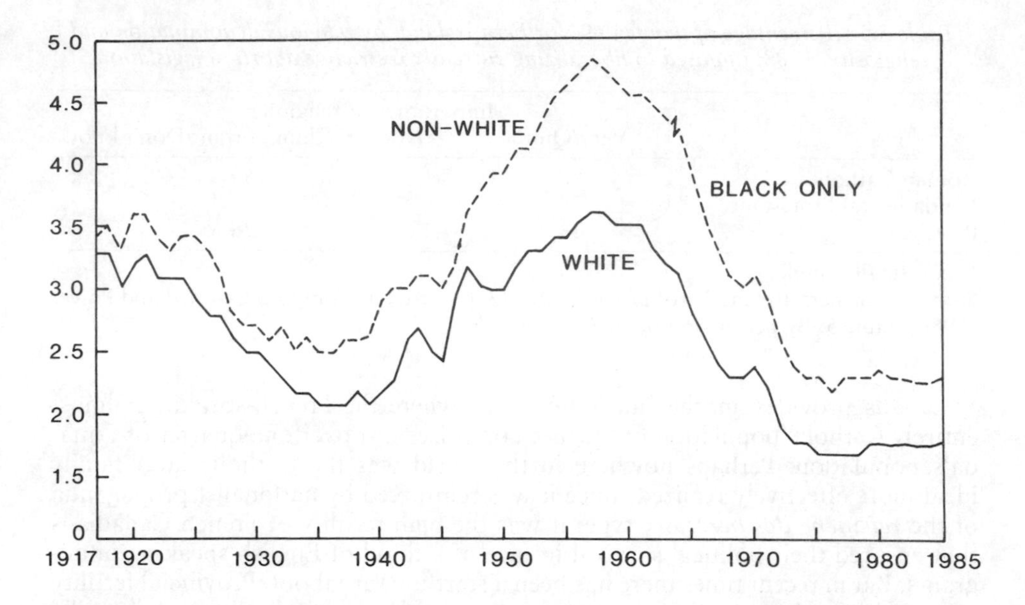

Figure 5.12
Total fertility rates (per woman) for white and black women, United States, 1917–85
Source: Reid (1982), figure 4, by permission of Population Reference Bureau Inc.; and data from *Statistical Abstract of the US 1988*, US Bureau of the Census.

blacks who have joined the formerly white flight to the surburbs and who are increasingly alienated from inner-city blacks.

Spatial analyses

A spatial perspective may well be the central concern of geographical analysis in population and other fields, but this is no justification for inferring that spatial analyses of fertility are the distinctive preserve of population geographers. There has, in fact, been considerable use by demographers of territorially disaggregated data in their attempts to understand the factors which influence fertility. The goal of such work is often to throw light on the dynamic relations between fertility decline and socio-economic change, particularly when time-series data are lacking, although it has been emphasized (e.g. Janowitz, 1973) that cross-sectionally observed relationships between variables are not necessarily applicable to longitudinal interrelations.

Reference has already been made to the way in which cross-sectional fertility data have been profitably subjected to cartographic methods (Coale, 1969; Braun, 1978; Coale and Watkins, 1986) and to statistical analysis (Yasuba, 1961; Forster and Tucker, 1972; McInnis, 1977; Van de Walle, 1978; Teitelbaum, 1984). Additional work by non-geographers in this field is discussed extensively by Andorka (1978, chapter 4.3). It is clear, then, that some demographers have made detailed spatial studies of fertility, invariably with a view to testing hypotheses on the relationship between fertility and socio-economic factors. Such studies complement much more numerous studies using time-series data for national units and social survey data for individuals. There is, however, an interpretation problem associated with territorial data, known as the 'ecological fallacy' (Robinson, 1950; Duncan, Cuzzort and Duncan, 1961, chapter 3). The essence is that one cannot crudely infer behaviour relating to individuals from aggregate population data relating to places. Andorka (1978) provides a good example from 1930 regional data in Hungary. In both bivariate and multivariate analysis there was a significant positive correlation between fertility and the percentage of Protestant population. But it would be premature to conclude on the basis of these results that Protestants had a higher fertility than the rest of the population. In fact, Protestants had the *lower* fertility in each region, the point being that in predominantly Protestant regions *each* denominational group had higher fertility. Much has been made of the ecological fallacy in sociological literature, but examples like the Hungarian one are rare, and by way of balance readers can be referred to a study by Duncan (1964) which, in assessing the relationship between fertility and house rents in an American City, found a very close accordance in results between an analysis using census tract data and one using individual survey data. He concludes (p. 88): 'This exploration, therefore, seems to bolster our confidence in the general validity of the findings of studies in differential fertility where areal variation was the sole source of information on socio-economic differentials.'

From this short review it is obvious that population geography cannot claim spatial analysis of fertility as its own preserve, yet its practitioners have made significant contributions. As in so many fields of geography, traditional cartographic methods have been supplemented and in some cases supplanted, by statistical analyses of varying degrees of sophistiction. An excellent example of the traditional approach is provided by Heenan (1967) in his consideration of the evolving spatial pattern of fertility in South Island, New Zealand. The portrayal and description of fertility patterns are meticulous, but interpretations are tentative. A good deal of the explanatory evidence, notably on the religious composition of the population, is presented in map form, so that there are subjectivity problems in comparing patterns.

Methodological evolution is illustrated by Compton's study (1978) of the spatial pattern of fertility in Northern Ireland. For the province as a whole, and separately for Belfast, he maps two indices of fertility by electoral wards. One is a period measure, the general marital fertility rate (nuptial births in the year 1970–1 per 1,000 ever-married women aged 15–49), while the other is a cohort measure, completed family size (children ever born to 100 ever-married women aged 50–59 in 1971). Correlation analysis aids the interpretation of the resultant patterns. The correlations by ward between fertility and Catholicism are impressively high, despite the lower marriage rate of Catholics in the prime child-bearing age group of 20–29 years. Compton notes that the positive correlations are stronger for completed family size ($r = 0.73$ for Northern Ireland and 0.80 for Belfast) than for current marital fertility (0.68 and 0.50 respectively), suggesting that there might be a weakening over time of the relationship, particularly in Belfast. He also notes statistically significant positive relationships between fertility and social classes IV and V, after controlling for religious denomination. A similar spatial analysis of fertility in the Republic of Ireland (Coward, 1978) indicates that the pattern of overall fertility is the outcome of two very different component patterns: those of nuptiality (highest in the west) and marital fertility (highest in the east).

A more complete, technically sophisticated reliance on multivariate statistical methods in geographical studies of fertility is provided by M. G. Wilson in a series

Table 5.6 Total fertility rate by country of birth of mother, England and Wales, 1981 and 1986

	1981	1986
United Kingdom	1.7	1.7
Caribbean	2.0	1.8
India	3.1	2.9
Pakistan and Bangladesh	6.5	5.6

Source: OPCS, *1986 Birth Statistics*, HMSO, table S15, Crown Copyright.

of papers analysing recent spatial variations in age-standardized fertility in Scotland (1978a) and Australia (1971, 1978b, 1984). The basic interpretative approach is to regard spatial differentials in fertility as being reducible to the spatially clustered effects of the variables recognized in differential fertility; and, indeed, the multiple regression analyses generally identify a small number of independent variables like Catholicism, female employment and non-agricultural occupations that account for a sizeable proportion of the fertility variation. More recently, however, Wilson (1988) has led the way in questioning the continued utility of such work in contemporary, post-transition societies characterized by appreciable demographic convergence. His argument has two strands: first, that there has been a widespread and substantial narrowing of traditional fertility differentials in response to the homogenizing of norms around the two-child family; and second, and more fundamentally, that spatial differentials, where they survive, have little more than surrogate status, being simply the horizontal reflection of the structural differentiation of society. The argument is compelling, and confirms an important theme of this text that the elucidation of spatial patterns should not be regarded by population geographers as an end in itself.

Wilson recognizes that his views, fashioned as they are in the metropolitan-dominated space economy of Australia, may have lesser relevance in older and more complex industrial societies. Indeed, Coward (1986a) in a review of recent spatial fertility differentials in the various countries of the British Isles, was convinced that the differentials, although declining, were still significant enough to warrant continued study. The decline in spatial variation from the first half of the century in England and Wales was regarded as being caused by the narrowing of class differentials in fertility and by the spatial convergence in key compositional factors. The most important of these factors is married female participation in the labour force. The traditional pattern, now being eroded, has been for high participation rates in Greater London (clerical positions), the Potteries, and West Yorkshire and south-east Lancashire (textiles), contrasting with the lower rates in agricultural, mining and heavy industrial regions.

Nevertheless, despite modern convergence, Coward could still demonstrate that a quarter of all local authority districts in England and Wales in 1981 differed from the national level of age-standardized fertility by at least 10 per cent. There were two major areas of relatively high fertility. First, there are areas which have attracted young couples aspiring to parenthood by their supplies of new, relatively affordable housing (e.g. Peterborough, Milton Keynes, Crawley, Luton, Maldon). Second, there are those areas with high proportions of ethnic minorities, particularly Asians (parts of inner London, Leicester, Derby, Birmingham, Bradford, Blackburn, Bolton, Burnley, Rossendale, Oldham, Rochdale). Interestingly, the high fertility of relatively recently arrived immigrants (Table 5.6) in several of these areas in Lancashire and Yorkshire has turned around their traditionally low fertility. The areas of lowest fertility in 1981 are concentrated in parts of inner London, the suburban fringes of Greater London and several medium-sized towns in southern England. Most are characterized by high housing costs, while some are also university towns which obviously attract large numbers of young single people.

There are also some appreciable differences between English regions in the overall pattern of age-specific fertility. Thus Figure 5.13 indicates that regions of the North and the Midlands have fertility rates well above the national average in the younger age groups and well below the average in the post-30 groups. The pattern is reversed for Greater London and the South East. Three factors seem to underlie these regional disparities. First, the greater availability of well-paid employment opportunities in the South East encourages deferred motherhood. Second, such deferment has become particularly fashionable in social classes I and II, which are again concentrated in the South East. Third, there is the role of net migration of young single persons from north to south.

It is clear, then, that spatial variations in the overall level and age-specific pattern of fertility are still very much alive in some post-transition societies. Yet Wilson's fundamental concern remains valid: that such variations rely essentially on explanations that are 'space or place neutral', with little explanatory role for any regional cultural tradition and with space simply providing a muted, inefficient representation of the different fertility levels of different social categories. There is a clear contrast to spatial variations in mortality, where the intrinsic nature of place *continues* to matter.

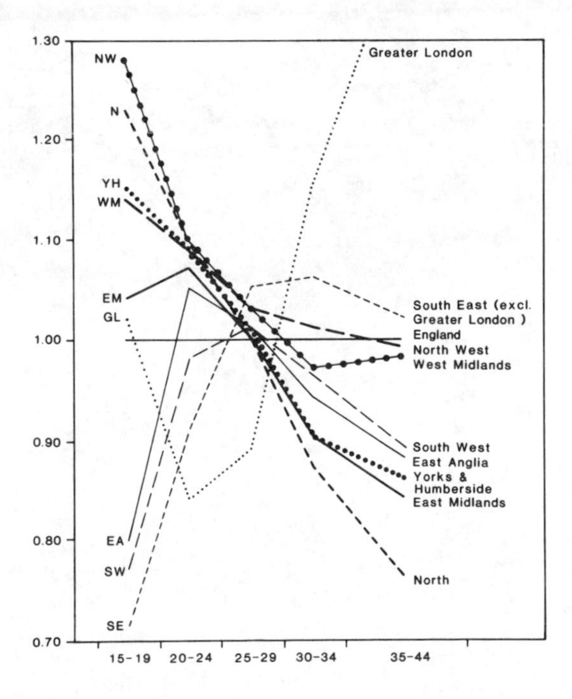

Figure 5.13
Ratios of regional to national age-specific fertility rates, England and Wales, 1982–5
Source: Armitage (1987), Figure 1(b), Crown Copyright.

Note

1. For example, the index of marital fertility for any area is constructed from:

(a) vital statistics records of the number of births to married women in that area during a particular period;
(b) census data of the number of married women in that area in each age group;
(c) a known age-specific fertility schedule of a population with extraordinarily high fertility – married Hutterite women in North Dakota in the 1920s.

The index is calculated by comparing the actual number of nuptial births in the particular area and period with the number that would be expected if married women there were experiencing the Hutterite age-specific fertility schedule. It is a ratio of observed to expected births. The indices are described fully in Coale and Watkins (1986), pp. 153–62, and Newell (1988), pp. 44–9.

6

FERTILITY PATTERNS: LESS DEVELOPED COUNTRIES

Fertility increases

Uncritical acceptance of the demographic transition model has fostered a widespread impression that there was a fairly uniform pattern of stable high fertility in less developed countries until about the 1960s. It was also thought that the disturbingly high population growth rates in the Third World in the third quarter of this century were entirely due to rapidly falling rates of mortality. Some analysts began to question these assumptions by pointing to significant rises of fertility in the 1940s and 1950s associated with early modernization in South America (Collver, 1965) and the Caribbean–Central American region (Byrne, 1972) and, in the 1970s, in parts of sub-Saharan Africa (Valentine and Revson, 1979; Romaniuk, 1980). Others have gone further, claiming convincingly that such fertility increases are a common, and perhaps universal, feature of incipient modernization (Nag, 1980; and especially Dyson and Murphy, 1985, 1986).

A major problem, as always in less developed countries, is the quantity and quality of demographic data. The major agencies which collect, evaluate and adjust data to produce estimated fertility rates are the UN Statistical Office, the US Bureau of Census, the Population Council and the UN Population Division. But there are often appreciable differences between their estimates, so that 'using various rates from different sources . . . it is possible to "prove" that a given country's rate has greatly declined, declined only moderately, has stayed the same or has increased depending on what one wants to "prove" ' (Cavanaugh, 1979, p. 292). Rises in the crude birth rate may sometimes reflect simply an improvement in the level of birth registration over time. However, when such rises can be confirmed by measures or estimates of fertility independent of the births registration system, their credibility is enhanced. Thus an important part of Dyson and Murphy's work is the confirming of birth rate increases, like those shown in Figure 6.1, by measures of fertility calculated retrospectively from maternity histories of women questioned in national fertility surveys.

The level of fertility increase in the 1945–60 period in the countries represented in Figure 6.1 is even more impressive when one considers that the proportion of population 'at risk' to child-bearing was declining during this period due to

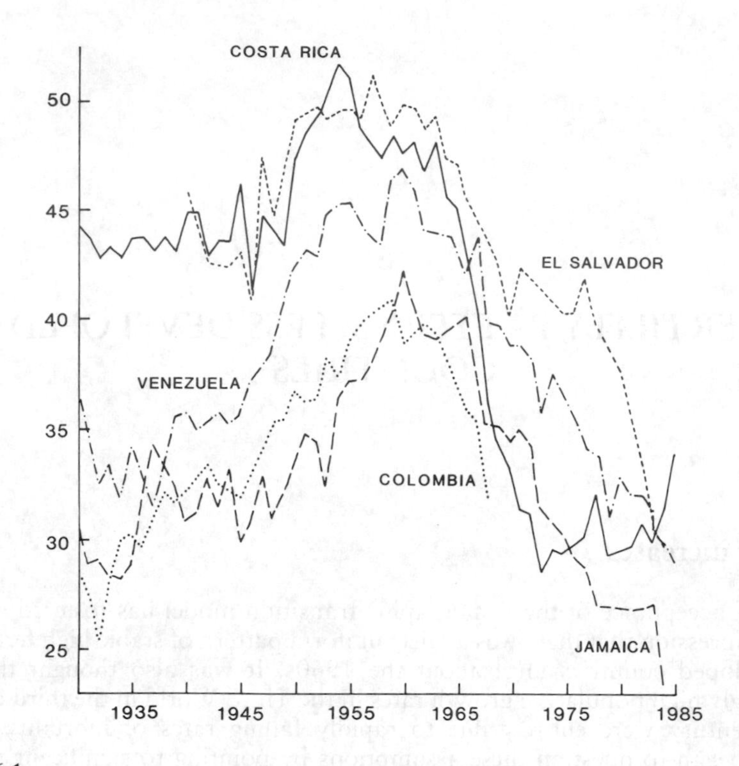

Figure 6.1
Crude birth rates (per 1,000) for selected countries in the Caribbean–Central American region, 1930–85
Source: Dyson and Murphy (1986), Figure 5.4. Reproduced by kind permission of Unwin Hyman Ltd.

appreciable mortality declines which particularly improved the survival of infants, children and the elderly. For most countries in the Caribbean–Central American region, Dyson and Murphy show that rises in the birth rate were responsible for between 30 and 60 per cent of the 1945–60 population growth, indicating the very real importance of the phenomenon.

Explanation for the fertility increases can now be given in relation to changes in several of the intermediate or proximate determinants of fertility specified in Figure 5.1. There are interesting regional differences in the contributions that different determinants make to overall fertility increase.

Age at marriage

Dyson and Murphy (1986) present age-specific nuptiality data to demonstrate that a trend towards younger and more universal marriage was the major determinant of fertility increases in the Caribbean–Central American region in the 1940s and 1950s. It is not easy to explain this phenomenon, although a comparable pattern in Mauritius is almost certainly related to the prosperity of the island's all-dominant sugar industry. Table 6.1 shows a substantial rise in the proportion of young women currently married between 1932, when the industry was suffering from the worldwide depression, and 1952, when the Mauritian

Table 6.1 Percentage of females currently married by age group, Mauritius*

Age group	1931	1944	1952	1962	1972	1983
15–19	30.7	35.9	39.9	27.8	12.4	10.5
20–24	62.8	65.4	72.3	68.2	49.7	47.9
25–29	74.9	74.4	82.9	83.1	76.0	70.8
30–34	76.4	76.1	83.8	85.3	83.6	76.8
35–39	75.0	74.0	81.8	83.9	84.2	79.1
40–44	68.1	67.9	76.5	78.5	80.0	78.9
45–49	63.3	61.6	69.0	71.1	74.4	75.0

* The data refer to legally, religiously and consensually married women.
Source: Calculated from data in censuses of Mauritius.

population was benefiting from a combination of high sugar prices and good harvests (always a variable factor in an island highly vulnerable to cyclone damage). Figure 3.11 shows clearly the associated fall in the birth rate in the early 1930s and the substantial rise in the early 1950s, so that there are obvious parallels with the marriage and fertility fluctuations in relation to living standards observed in Chapter 5 for pre-industrial European populations.

More intriguing and debatable is the extent of any demographic parallel between Mauritius and the not dissimilar plantation economies of the time in several parts of the Caribbean and Central America. One special factor in that region was the fertility increase in Cuba in the early 1960s associated with the conditions, real or assumed, that led young adults to foresee a much more promising future after the 1959 revolution (Diaz-Briquets and Perez, 1982).

Widowhood

Reductions in mortality mean that married couples survive longer as child-bearing units. In addition, in some societies remarriage of widows has become more socially acceptable. This is cited by Srinivasan, Reddy and Raju (1978) as contributing to increased fertility in the Indian state of Karnataka between 1950 and 1975, despite appreciable family planning programme effort. Both factors would seem to underpin the increases in female nuptiality in the older child-bearing age groups that are illustrated well by the Mauritius data (Table 6.1).

Coital frequency

The Caribbean region and parts of sub-Saharan Africa have long been characterized by an informal polygamous mating system embracing several types of socially accepted non-marital as well as marital unions. Opinions differ as to the effect of non-marital unions on fertility. It can be argued that women in such unions may well feel impelled to have children to make the relationship more permanent and that men are unlikely to show sexual restraint when they are not strictly responsible for their offspring. Nevertheless, the favoured demographic view (G. Roberts, 1955; Blake, 1961; Stycos, 1968; Nag, 1971) is that fertility is lower in less stable unions, essentially because sexual activity is reduced in the often extensive periods between unions. Consequently, any increase in the stability of sexual unions (and this has been widely observed in the Caribbean region) is likely to have had a positive effect on fertility.

Lactational amenorrhoea and post-birth abstinence

Following a pregnancy, a woman is unable to conceive until the normal pattern of ovulation and menstruation is restored. The length of delay is determined primarily by the duration and intensity of breastfeeding. A reinforcing inhibition to fertility is the custom, particularly prevalent in sub-Saharan Africa, of abstaining from sexual relations for a post-birth period of up to three years, a practice designed specifically to protect infant and maternal health (Page and Lesthaeghe, 1981).

Any shortening of the periods of breastfeeding and abstinence for whatever reason (e.g. Westernization) must, in the absence of contraception, lead to higher fertility. This is thought to have occurred widely in the 1970s and 1980s in sub-Saharan African. In Kenya, where the estimated total fertility rate rose from between six and seven births per woman in the 1960s to well over eight in the 1980s, despite the mean age at first marriage for women rising from about 18.5 to 20 years, the most important contributory factor has been the rapid reduction of the post-birth abstinence period to a mere four months (Frank and McNicoll, 1987).

Pathological sterility

The prevalence of venereal disease, particularly gonorrhoea, resulted in remarkably high levels of childlessness (20–40 per cent of women aged 45–49) and in total fertility rates of below 5.5 births per woman in the 1950s in a large area of central Africa, including Zaire, Congo, Gabon, Central African Republic and Equatorial Guinea (Frank, 1983; Lesthaeghe, 1986; Doenges and Newman, 1989).

Two factors promoting venereal disease in that region have been the multiplication of sexual partners in polygamous mating systems and the use of prostitutes encouraged by customs of post-birth abstinence. Antibiotics have reduced the extent of sterility and have therefore increased fertility, although the central African region can still be observed in the mid-1980s as having lower fertility than neighbouring countries (Figure 6.2).

Whether sterility will continue to decline in the region is uncertain because of the recent spread of gonorrhoea strains that require more expensive antibiotics than penicillin. Another sexually transmitted disease, AIDS, thought to be widely present among heterosexuals in parts of sub-Saharan Africa, has little direct influence on fecundity, although it is possible that it could influence fertility through modification of sexual behaviour in attempts to limit exposure to the virus (Frank and McNicoll, 1987).

Spontaneous abortion

Foetal loss is likely to be reduced with improvements in nutrition and also with the reduction of malaria. The latter relationship was probably significant in the fertility increases in the early 1950s in parts of Sri Lanka, where analysis of district data shows that fertility increases were concentrated in the formerly malarial, high mortality districts of the Dry Zone (Langford, 1981). Malaria, as a debilitating disease, is also thought to depress fertility through its effect on coital frequency.

These, then, are changes in some of the proximate determinants that are likely to have increased fertility at early stages of economic and social transformation,

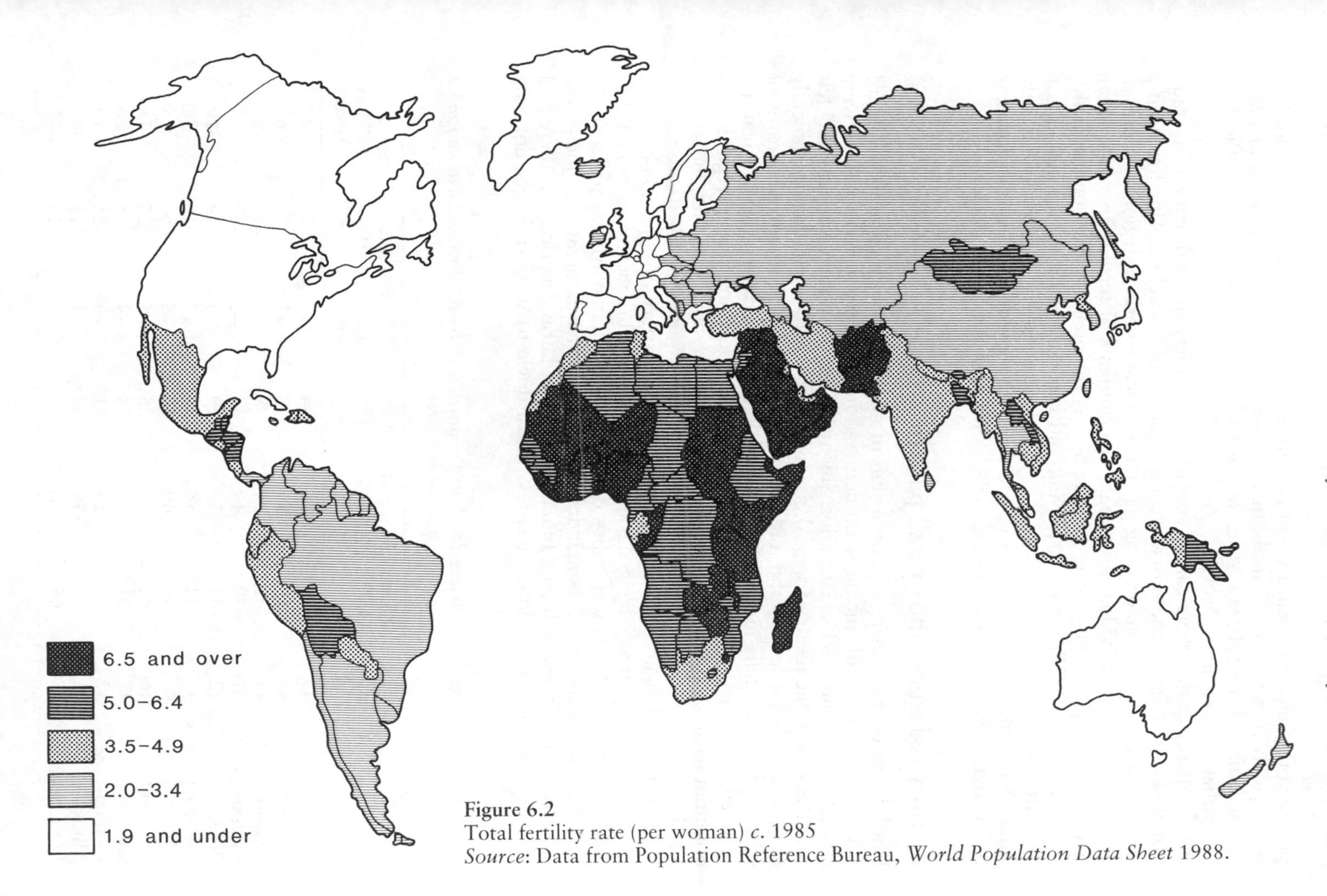

Figure 6.2
Total fertility rate (per woman) *c.* 1985
Source: Data from Population Reference Bureau, *World Population Data Sheet* 1988.

provided that no compensatory contraceptive action is adopted. Such conditions were widely present in the Caribbean and Latin America in the 1940s and 1950s and in much of sub-Saharan Africa in the 1970s and 1980s, but it is difficult to generalize on the pattern in Asia.

It is also clear that there can be considerable variety of fertility levels in traditional societies *before* the onset of significant fertility decline. This has been a fairly recent appreciation, deriving particularly from work on regional variations in the strength of individual proximate determinants of fertility in sub-Saharan Africa. Of particular interest to geographers is the preliminary mapping in this field (Adegbola, 1977; Bongaarts, Frank and Lesthaeghe, 1984), the description of culturally and ecologically based reproductive systems in the various sub-Saharan regions (Lesthaeghe, 1986), and some ethnographic case studies focusing on demographic strategies (Kraeger, 1982; Newman and Lura, 1983).

Fertility reduction: the overall pattern

Until the mid-1960s there was little sign of any fertility reduction in the Third World associated with incipient modernization. Even when fertility falls were recognized in countries like Barbados, Singapore and Taiwan, it was generally felt that these were the result of special circumstances operating in demographically insignificant countries. By the mid-1970s, however, the widespread nature of sustained fertility falls in the Third World had become an established feature, even encompassing several of the most populous countries long regarded as potential seedbeds for population-based catastrophes, notably China, India, Java and Egypt.

Table 6.2 and Figure 6.2 suggest that the global pattern of reduction has been characterized by appreciable geographical variability. In fact, by the mid-1980s there had been little or no fertility decline in sub-Saharan Africa, parts of the Middle East, Pakistan and Bangladesh; there had been moderate decline in the majority of countries, including India and most of Latin America; but there had been appreciable decline, in some cases almost down to Western levels, not only

Table 6.2 Crude birth rate in less developed countries, 1950–85, by continent and most populous countries

	Crude birth rate (per 1,000)				Percentage change		
	1950–5	1965	1975	1985	1950–5 to 1965	1965 to 1975	1975 to 1985
Africa	49	48	46	44	–1	–4	–4
Nigeria	49	50	49	46	+1	–1	–6
Egypt	45	42	35	38	–6	–17	+8
Ethiopia	51	50	49	46	–2	–2	–6
The Americas	42	41	36	29	–3	–14	–19
Brazil	41	42	38	28	+2	–10	–26
Mexico	47	44	40	30	–6	–9	–25
Colombia	46	44	33	28	–5	–25	–15
Asia	42	41	35	28	–4	–15	–20
China	37	34	26	21	–9	–24	–19
India	42	43	36	33	+1	–16	–8
Indonesia	45	46	40	27	+2	–13	–33

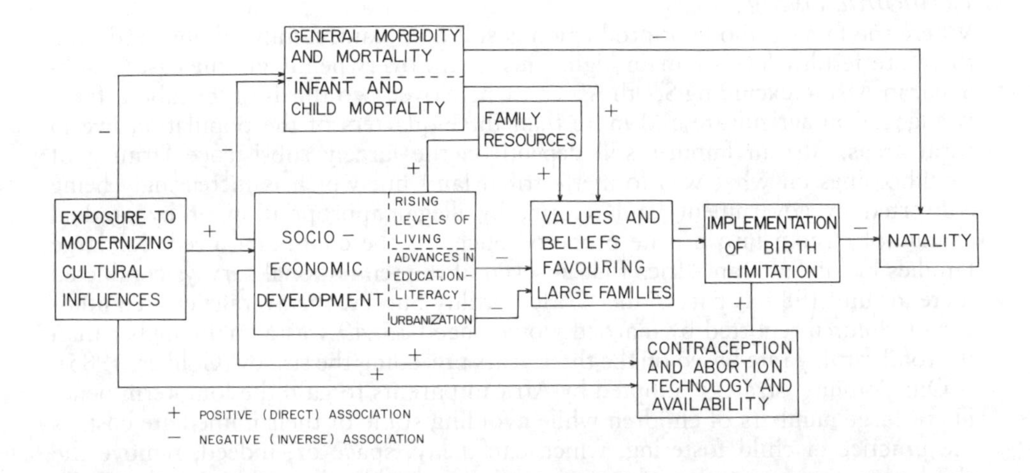

Figure 6.3
Causal diagram of influences on natality with particular reference to less developed countries. Reverse relationships and weak relationships are not shown. An example of a reverse relationship is how high infant mortality may reflect, through neglect and abuse of children, parental reaction to high natality generally and unwanted births specifically; such a relationship has been postulated for both pre-industrial Europe (Van de Walle and Knodel, 1980) and modern Latin America (Scrimshaw, 1978)
Source: adapted from S. E. Beaver (1975), Figure 4.1.

in some small crowded islands like Barbados, Mauritius and Singapore, but also in some major Asian countries, notably China, Taiwan, South Korea and, most recently, Thailand and Indonesia. There are also appreciable variations within countries, illustrated by a total fertility rate of under three births per woman around 1980 in the Brazilian state of São Paulo, compared with a rate of over six in north-eastern Brazil (Daly, 1985).

The rest of this chapter attempts to explain this variability in relation to three fundamental influences – development, culture and family planning programmes. Clearly, these three broad influences on fertility behaviour are deeply interwoven, so that it is very difficult, and perhaps technically impossible, to untangle interrelationships in such a complex bio-social system. A useful introduction to the basic influences and interrelationships is provided by Figure 6.3. It should also be borne in mind that the key proximate determinants (Figure 5.1) in fertility reduction have been those relating to age at marriage and to contraceptive use.

Development influences

The introduction to fertility decline theory in Chapter 5 explains how high fertility in traditional societies is based on well-founded needs that parents have for the labour, security and status provided by children under a system in which inter-generational benefits flow mainly from the young to the old. Fertility decline is associated with the erosion of that system, particularly through two broad elements of transformation widely but variably present in the Third World: economic change and education.

Economic change

Where the familial mode of production is still dominant, family labour needs and therefore fertility levels remain high. This is still overwhelmingly the case in sub-Saharan Africa (excluding South Africa), where over two-thirds of the labour force is engaged in agriculture and more than three-quarters of the population live in rural areas. African families still depend on the largely subsistence farming of smallholdings on what was formerly tribal land but which is increasingly being converted by government land reform, by illegal appropriation or by gradual change of custom into private property. Such are the continuing needs for large families in sub-Saharan Africa that in seven of the ten national surveys conducted there around 1980 as part of the World Fertility Survey, the average desired number of children reported by married women aged 15–49 was actually higher than the total fertility rate based on the three years preceding the survey (Goliber, 1985).

One common strategy adopted by African parents to gain the long-term benefits of large numbers of children while avoiding some of their immediate costs is the practice of child fostering, which can delay, space or, indeed, remove the child-rearing consequences of reproduction (Isiugo-Abanihe, 1985; Frank, 1987). Most fostering takes place within a kinship network, since children are generally regarded as belonging not simply to biological parents but also the lineage or kinship group. The willingness of relatives to foster children is indicative of the value of children to them. Fostering is thus a potent means of reallocating resources within the extended family and a far cry from the crisis form of agency-controlled fostering prevalent in the West. So important is the practice that the World Fertility Survey shows the proportion of children under 15 not living with their mothers as between 10 and 20 per cent in the various African countries surveyed, although in much of West Africa the proportion exceeds 30 per cent (Caldwell and Caldwell, 1987).

In much of Latin America and Asia the pro-natalist familial mode of production of the African type has been undermined increasingly by the commercialization of agriculture, economic diversification and urbanization. Not only is the need for a family production team removed, but the market penetration of the countryside, the material aspirations raised by media propagandizing of conspicuous consumption, and the monetization of daily life, all lead to a growing appreciation of the economic costs rather than the benefits of large families.

Where such economic transformation and associated social change have been greatest, there can be found the most appreciable fertility falls – in South Korea and Taiwan, for example, rather than in India and Pakistan, and in urban areas much more than in rural areas.

But it would be wrong to make a simple association between increasing living standards and lowered fertility. Recently, there have been several reportings of declining fertility caused by *deteriorating* real income in *monetized* Third World societies – in Brazil (Merrick and Berquo, 1983), Cuba (Diaz-Briquets and Perez, 1982), Ghana (Caldwell and Caldwell, 1988), Mauritius (Jones, 1989), Java (Freedman, Kooh and Supraptilah, 1981) and even Bangladesh (Phillips *et al.*, 1988) – which, although posing a challenge to conventional wisdom, is entirely consistent with the basic economic theory of fertility. Demand for reduced fertility in these circumstances is induced by the poverty–material aspirations gap.

Education

The growth of formal education systems in less developed countries reduces fertility in two major ways. First, and most immediately, there is a greater appreciation by parents of the costs, rather than the benefits, of children, as expenditure is required on uniforms, books, stationery and other fees, while school attendance reduces the household services contributed by children. Second, educated children have their material aspirations for adult life raised, they become familiar with the Western small-family norm, and they tend to marry later than their parents. Later marriage comes partly from individual choice, but also many parents demand that, after 'investing' in the secondary school education of their children, these children should postpone marriage to help with family finances.

The impact, therefore, of education on fertility is obvious, but what is not so clear are the reasons for the rapid expansion of primary and, more selectively, secondary school education in large parts of the Third World in the last three decades. Supply-side factors include the growing wealth and therefore capability of some nations to provide basic services, while a more universal factor has been the influence of development theorists in emphasizing the infrastructural benefits of mass education. There are also the demand factors, which are vital in ensuring the use of education facilities when it is impossible, as in most of the Third World, to enforce school attendance.

The demand for schooling seems to have been particularly high in crowded agricultural societies where land division on inheritance is accentuated by declining infant and child mortality, so that holdings are no longer capable of supporting large families. Caldwell, Reddy and Caldwell (1985) examine these circumstances in a part of the southern Indian state of Karnataka, which is also a dry farming area and therefore subject to periodic drought-induced disaster. They find that an important security strategy has been for some of the family to seek off-farm employment, particularly in the expanding regional city of Bangalore. Schooling is regarded as an essential qualification for acquiring a decent job outside agriculture, especially in civil service occupations, which are widely seen as providing security, steady income, social welfare benefits and prestige. The education of daughters is also increasingly favoured in rural Karnataka, so that they become more eligible as wives to educated men working outside agriculture – a more indirect strategy of providing parents with risk insurance. In Bali

> the concern over agriculture as a viable occupation for the next generation has coincided with a dramatic increase in emphasis on education. This is partly government development strategy and partly an almost desperate attempt by parents to ensure secure employment for their children. At this time secure employment very largely means a Civil Service job . . . what this requires is a large investment in just a few children.
>
> (Streatfield, 1986, p. 139)

Almost identical processes have been observed at work in several other crowded agricultural societies – in eastern Nigeria (Ekpenyong, 1984), in parts of Thailand, where 'education has come to be seen as a substitute for land' (Knodel *et al.*, 1987, p. 306), and in Mauritius (Jones, 1989). In these circumstances the net inter-generational wealth flow may *not* have changed decisively, since educated children will ultimately benefit their parents, but there is such a financial crunch during the schooling period that numbers of children must be restricted.

It is clear, then, that economic transformation and education provision are two potent, fertility-reducing elements of modernization that generally accompany each other, as data from Taiwan illustrate (Table 6.3). However, in an important

Table 6.3 Indicators of economic, social and demographic development, Taiwan, 1952–84

	1952	1964	1970	1984
Per capita real national income index	100	172	263	668
Labour force in agriculture, forestry, fishing (%)	56	49	37	18
Enrolment in junior high school (%)	34	55	80	99
Automobiles per 1,000 population	1	3	7	65
Married women aged 22–39 currently practising contraception (%)	*	22	44	77

* No data.
Source: Chang, Freedman and Sun (1987), tables 9 and 19.

minority of settings, the provision of mass schooling has set off fertility decline in advance of much economic change. Outstanding examples are Sri Lanka, the Indian state of Kerala and some Caribbean territories like Barbados.

Cultural influences

These comprise the normative beliefs that guide behaviour in each society. Such beliefs have often been regarded – certainly in the classical demographic transition model – as by-products of the pre-eminent socio-economic determinants. However, in the previous chapter it has been observed that more recent analyses of the West's fertility transition have given more prominence to the semi-autonomous influences of ideational factors like secularism. Certainly, particular cultural environments do seem to have accelerated or delayed demographic transformation in the West. So, too, has been the case in the contemporary Third World, where the geographically differentiating influences of religion and kinship systems are particularly evident.

Religion

Among the world's major religions and philosophies, Roman Catholicism, Islam and traditional African religions are generally regarded as buttressing high fertility, whereas Buddhism and Confucianism have more liberal attitudes to legitimizing the notion of family limitation.

In Africa Islam has been spreading for centuries through the savanna belt of West Africa and along the East African coast, while the association of Christianity with colonialism has been a potent conversionary force. Consequently, most Africans are now Christians or Muslims. Yet, many features of traditional African religions have survived to dominate belief systems. Caldwell and Caldwell (1987) show how African religion focuses essentially on the reproduction of the lineage, which is seen as a descent group stretching infinitely far back and reaching indefinitely into the future. Thus, both for ancestral spirit survival and lineage survival, there is a premium on living descendants. Fertility is affected in two ways. First, fertility is equated with virtue and spiritual approval, while reproductive failure and self-imposed fertility restriction are seen as punishment and evil respectively. Second, the upward inter-generational wealth flows that are common in traditional societies are particularly prominent in sub-Saharan Africa because they are religiously expected and, indeed, guaranteed by the fear of rebuke or curse from ancestors living or dead.

The pro-natalist role of Islam operates more indirectly. There are no theological objections to contraception (Ghallab, 1984). More important is the code of ethics, morality and behaviour which ensures a highly subordinate position for women in Islamic societies. Women are expected to marry early after very little education, not to engage in employment, and to be subservient to their often much older, decision-making husbands. All of these conditions promote high fertility, as does the ease with which men can repudiate, abandon and divorce their wives.

Such lack of security for women encourages their desire for several sons as risk insurance. Gender inequality widened in the 1970s and 1980s with the growth of Islamic fundamentalism, orthodoxy and sometimes militancy, which is particularly evident in Iran but by no means confined to it. In Pakistan, for example, the common law and other reasonably egalitarian socio-cultural legacies of British rule have been subject to appreciable modification along the lines of the Shariah, the body of Islamic theological doctrine and legal practice, which many think discriminates against women (Sathar *et al.*, 1988).

Figure 6.4 shows the global distribution of Islamic populations, reflecting the diffusion from a Middle Eastern heartland into Turkey, Central Asia, northern Africa and coastlands of the Indian Ocean. There are some considerable demographic differences within this Islamic realm (Clarke, 1985; Weeks, 1988), particularly between the more broadly developed countries like Tunisia and Egypt and the poorest countries like North and South Yemen, Mali and Niger. Nevertheless, the important point is that, for any given level of development, Islamic countries invariably show higher fertility than other countries.

The most obvious contrast to Islam in demographic influence is Buddhism, with Sri Lanka and Thailand being outstanding cases of Buddhist societies which have achieved appreciable fertility reduction despite remaining fairly poor agrarian societies. Buddhism seems to pose no major religious barriers to the notion of family limitation or to the use of contraception, although abortion is certainly proscribed. Knodel, Chamratrithirong and Debavalya (1987, p. 163) comment that religion was almost never mentioned in their field discussions on family size in Thailand. Important concerns in Buddhism that are relevant to fertility behaviour are the seeking of enlightenment, which has promoted the spread of formal education, and the primacy of individual action and responsibility. Females also enjoy more autonomy than in most less developed societies, this being particularly evident in Thailand. Finally, the role of an unmarried person is socially acceptable; indeed, many young men are expected to spend some time in monastic sexual seclusion.

More debatable is the demographic influence of Confucianism in the Chinese culture realm that embraces China, Taiwan, Singapore, Hong Kong and significant parts of the population in other South East Asian countries. The most profound assessment is that of Greenhalgh (1988). She shows that although Confucianism as a system of thought is fundamentally pro-natalist (evidenced in sayings such as 'more sons, more happiness'), it has also promoted pragmatism, 'costs and benefits' rationality, achievement orientation, competitiveness for upward social mobility and respect for education. In the free-market economies of

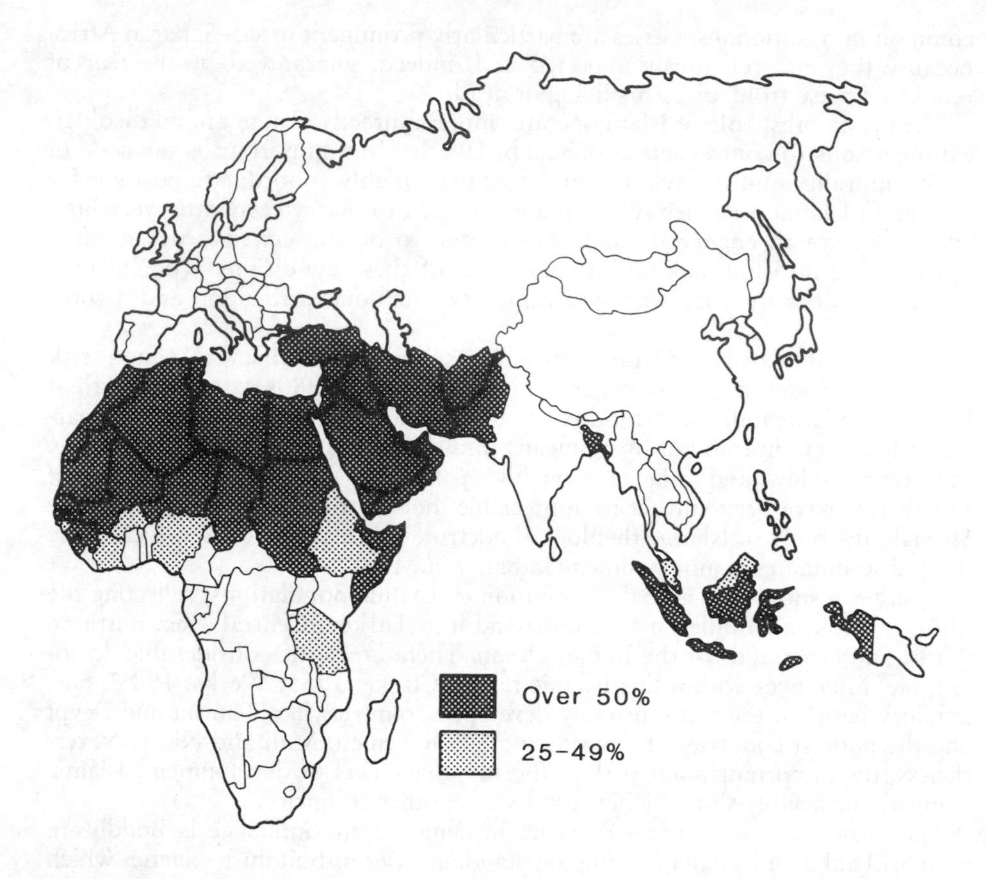

Figure 6.4
The Islamic World; percentage of total population Muslim
Source: Weeks (1988), Figure 1. By permission of Population Reference Bureau, Inc.

eastern Asia these qualities have underpinned both economic growth and fertility reduction, while in China itself they permitted the likely economic benefits from large rural families to be overruled in fertility decision-making by the potential costs of deviating from the state's insistence on low fertility. Therefore, an appreciable fertility reduction in China in the 1970s and early 1980s was not simply due to state coercion of individuals, but rather to the carefully calculated strategic behaviour of these individuals. Their strategies are now being modified. The decollectivization of agriculture has increased the economic benefits of rural children while undermining the power of local cadres and their capacity to reward and punish households. Not surprisingly, rural fertility in China is thought to have shown modest increases by 1990.

Kinship systems

An important macro-level spatial relationship of long standing between demographic regimes and kinship systems has been identified in India by Dyson and Moore (1983) in their north–south division of the country, although they are well aware of additional vertical differentiation by caste. It also needs to be said that any short summary, such as this, inevitably highlights striking contrasts and fails to capture the subtlety of regional differences, which, in truth are essentially those of degree rather than of kind.

In the northern Indian kinship system a dominant element is the co-operation and often co-residence of males closely related by blood. Their wives are chosen from outside the kinship network and often, therefore, from external districts. A wife takes up residence with her husband's family and has very little status and influence within it. Emotional ties and communication between husband and wife are limited by the need to maintain the solidarity of the patrilineal group. Mindful of family honour, females are secluded from other males by not working outside the household and often by practising *purdah*. Indicative of their low status, women generally cannot inherit property for their own use.

Clearly, women are subjected to strong pro-natalist pressures in such a socio-cultural system, which has many similarities to an Islamic society. Women marry early, have little choice in the male-arranged marriage, and enter a strange household with the prime task of producing male heirs for their husband's descent group. Their social isolation further encourages them to produce their own security group in the form of sons. With little access to education and employment and with severe restriction on their independence of movement and initiative, they are difficult to access by family planning personnel; in any case, the recruitment of that personnel is itself compromised by the system.

In southern India there is a rather different pattern of gender relations. Women there are much more likely to marry men they know. Thus, in the population studied by Caldwell, Reddy and Caldwell (1982, 1985) in Karnataka, 30 per cent of all marriages were between persons closely enough related to specify the exact relationship (half of the marriages were between first cousins). A wife is able to maintain contact with her own family, and little importance is placed on her re-socialization into the husband's kin group; indeed, strong nuclear families are often established at marriage, preventing the northern pattern of strict segregation between male and female spheres of activity. The better social position of southern women is reflected in their more favourable inheritance rights, in greater participation in schooling and in employment outside the household, and in greater freedom of movement and action. All of these conditions promote later marriage than in the north, more participation by wives in household decision-making, even in reproductive matters, and greater access to family planning services.

This important north–south socio-cultural division, deeply rooted in Indian society, is reflected clearly in the spatial pattern of fertility in recent decades. Figure 6.5 shows the geographic consistency of the north–south division as the Indian total fertility rate fell from an estimated 6.1 births per woman in 1961–6 to 4.7 in 1976–81. The maps show, however, that the state of Punjab has begun to deviate from the high fertility pattern of northern India, and this may well be related to the fact that about 60 per cent of Punjab's population is Sikh.

> While Sikh patterns of marriage and family organizations follow most of the principles of north Indian kinship, the status of women within the institutional framework has been conditioned by the prescripts of a relatively egalitarian religion whose leaders have sometimes protested against the position of women and have modified some of the disabilities they suffered under the old codes.
>
> (Dyson and Moore, 1983, p. 53)

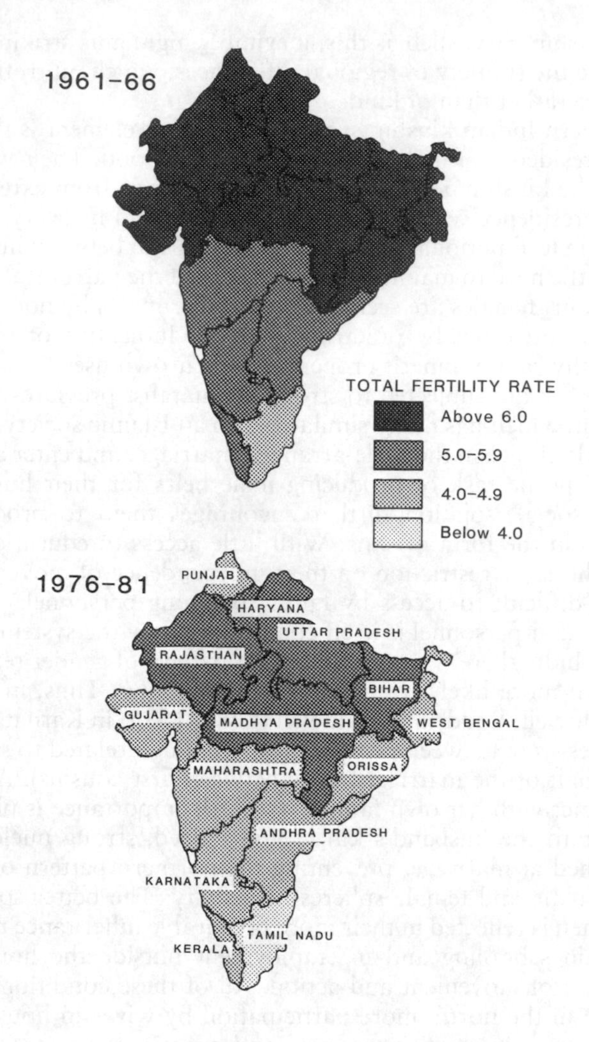

Figure 6.5
Estimates of the total fertility rate for the most populous states of India, 1961–6 and 1976–81
Source: Rele (1987), Figure 2.

The lowest total fertility rate in India, 3.1 in 1976–81 in the southern state of Kerala, has been widely discussed (Nayar, 1984; Zachariah, 1984), not least because of its possible prescriptive policy relevance. It is intriguing that low fertility has been achieved in a state which has well below the per capita income of India as a whole – itself a very low one. Kerala's progressive demographic status, demonstrated in Chapter 3 in relation to mortality (Table 3.12), is attributable to appreciable social, rather than economic, development, and, in particular, to the remarkably high education and literacy status of women. The causes of this seem to lie partly in a long fortuitous line of benevolent and paternalistic rulers and, more recently, of leftist governments committed to major expenditure on public welfare, and partly in a particularly strong and long-established manifestation of the higher status that southern Indian women enjoy, expressed traditionally in Kerala in matrilineal descent and polyandry. All this is reflected in the female mean age at first marriage – 23 years in Kerala in 1981 compared with 18 for India (Nayar, 1984).

Programme influences

The origins, growth and changing nature of family planning programmes will be discussed fully in the concluding chapter on population policies, but it is appropriate to consider at this juncture two questions. Have these programmes had any significant independent effect on fertility reduction? If so, what factors have caused different levels of programme effort and effectiveness between and within less developed countries?

Impressive claims were made in the 1960s for the actual and potential role of intervention programmes. There seemed to be a close temporal and therefore, it seemed, functional relationship between the growth of programme effort and the decline of fertility in some of the early decline countries, notably Taiwan, Singapore and Hong Kong. Moreover, many KAP surveys (knowledge, attitude and practice of family planning) seemed to suggest that women desired smaller family sizes than they were actually achieving, pointing to an apparent unfulfilled need for the supply factors of contraceptive information, goods and services.

By the 1970s these claims had been appreciably qualified, and sometimes rejected. More careful analysis revealed that many of the early claims for programme influence in South East Asia had failed to control adequately for continuing modernization, as the data in Table 6.3 for Taiwan suggest. A more sensitive approach to survey interviewing also revealed misleading responses in some of the early KAP surveys when excessive deference was being given by poor peasants to educated interviewers and programme personnel. In a classic study of the failure of a village family planning programme in India despite favourable survey returns, Mamdani (1972, p. 23) quotes the explanation of a villager for accepting foam tablets without having any intention of his wife using them: 'But they were so nice, you know. And they come from distant land to be with us. Couldn't we even do this much for them? Just take a few tablets? . . . If they are happy writing my name, let them do it.' Some eminent demographers were particularly critical:

> The idea that a family planning program can bring fertility decline to a country without social and economic modernization . . . is not even plausible, much less proven.
>
> (Kingsley Davis, 1967, p. 730)

> The power of family planning programs to achieve more than limited demographic objectives . . . is questionable. Since people can be expected to find the means to control fertility if they want to, and since without such a desire the availability of . . . contraceptive technology is inconsequential, it is unlikely that family planning can do much more than accelerate a process that should occur in any case.
>
> (Demeny, 1974, p. 158)

> Policy success, where it can be claimed, usually turns out to refer to a marginal acceleration of a change already underway.
>
> (McNicoll, 1978, p. 96)

Such criticisms, in turn, may have been overstated, since there is now firm evidence of appreciable fertility falls in the 1980s in some countries which are relatively poor and dominantly agrarian but which have strong family planning programmes, notably China, Thailand and Indonesia. The modern consensus view is that programme effort can, and often does, play a significant role over and above that of modernization in promoting fertility reduction (Freedman, 1987). A perceptively early and still entirely plausible compromise view is that:

> a program tends to have the greatest impact at medium stages of development; at lower stages the motivation for smaller families, which is a consequence of development, is not great, and at later stages the natality decline will take place regardless of the program. In the middle ranges an active family-planning program diffuses birth control more rapidly than otherwise would be the case.
>
> (Oechsli and Kirk, 1974–5, pp. 416–17)

Given, then, the likelihood that family planning programmes have *some* demographic significance, what are the factors that determine their geographically variable presence and relative strength? Preliminary examination, in advance of the chapter on population policies, suggests that there are four major contributory factors:

1. For ideological reasons, some national governments are much less committed than others to intervention programmes. The reduced commitment, and sometimes complete opposition, is most evident in many, although not all, Islamic, Catholic and socialist countries.
2. Programme funding, which, outside China and India, is substantially from external Western sources, can vary significantly. Not surprisingly, per capita funds available to programmes in small island societies like Barbados, Jamaica and Mauritius have been much higher than in the likes of Pakistan and Bangladesh.
3. Since successful programmes are dependent on an infrastructural capability (roads and media networks, health centres, administrative skills, etc.), it is clear that there is a functional link between development and programme effectiveness.
4. There are also explicitly geographical factors that provide logistical advantages for programmes in some settings. A nucleated settlement pattern is advantageous for client recruitment and provision, while in small crowded islands, environmental limitations and oceanic isolation often encourage community accountability and collective demographic restraint since inhabitants can at least partially identify their own interests with those of their fellows. Thus, in Singapore:

   > lack of natural resources, inadequate water supply, and the virtual absence of agricultural land convey to residents a sense of precariousness and an awareness of environmental limitations. The pressure of population on resources is reinforced by frequent references to this problem in the press and in government policy announcements.
   >
   > (Fawcett and Khoo, 1980, p. 551)

We have now discussed the variability in the overall global pattern of fertility reduction in less developed countries. Three fundamental influences have been identified: development, culture and intervention programmes. The interaction between them will now be assessed more fully in relation to results from different types of analytical study.

Table 6.4 Components of crude birth rate decline in selected less developed countries

	Crude birth rate			Percentage of decline due to:		
	Beginning	End	% change	Age structure	Marriage pattern	Marital fertility
Fiji						
1966–75	36	28	−22	−21	29	93
South Korea						
1966–74	35	25	−28	0	9	91
Mauritius						
1962–72	38	24	−36	−25	53	72
1972–83	24	20	−17	−114	40	174
Singapore						
1957–70	43	22	−48	−13	53	60
Sri Lanka						
1963–71	34	30	−13	−42	89	53
Taiwan						
1965–75	32	23	−28	−42	49	93
1980–5	23	20	−13	−17	50	68
Thailand						
1965–75	42	37	−12	−22	19	103
Turkey						
1960–75	45	32	−29	8	23	69

Assumptions: Women recorded in consensual unions are regarded as married; all births are attributed to married women.
Source: Mauldin and Berelson (1978), Chang, Freedman and Sun (1987), Jones (1989).

Decomposition studies

Following the method of Retherford and Cho (1973) it is possible to decompose crude birth rate change into three component parts, but one does need data on age- and marital-specific birth rates which are available only in a minority of Third World countries. The pattern revealed by such analysis in Table 6.4 is illuminating. For nearly every country shown, it is clear that changes in marital structure (primarily a trend to later age at first marriage) more than compensate for increases in the proportion of women of reproductive age that result from the effect of incipient fertility decline on age structure. In addition, there have been appreciable reductions in marital fertility, although, contrary to general opinion, these *marital* fertility rates were never as high as those in pre-industrial Europe.

Macro-analytical studies

Appreciable literature and numerous technical analyses have been devoted in the last three decades to the problem of disentangling the interactive effects on Third World fertility of development, cultural background and family planning programmes. The most common studies are cross-sectional regression analyses, using countries as units of analysis, that employ a fertility measure as the dependent variable and several modernization indicators as independent variables. There are obvious problems of analysis and interpretation, not always confronted openly by practitioners, stemming from the unreliability of crude data, particularly on births. Moreover, cross-sectional studies reveal associations, rather than causes as such, so that one must beware of attaching significance to theoretically unsound relationships. Consider, for example, the way in which De Castro (1952) used the statistically impressive inverse relationship between national fertility and protein consumption to suggest that hunger and protein deficiency stimulate sexual activity and production of female hormones – a view refuted by physiologists.

Cross-sectional studies, which will now be reviewed selectively, may be divided conveniently into those using pre- and post-1970 data. Generally it is only among the latter that family planning programme variables have been incorporated, as a response to their growing strength and likely influence.

Studies of pre-1970 data

As early as 1950 the Latin American nations with more modern socio-economic characteristics were more likely to exhibit lower fertility. Stycos (1968) was able to demonstrate correlations of –0.61 and –0.80 between mid-century crude birth rates and levels of literacy and urbanization respectively, but the correlation between crude birth rates and GNP per capita was only –0.26, suggesting that within a fairly narrow development range (Latin America in 1950) there are more important influences on fertility.

Using 1960–4 data, Kirk (1971) examined natality–modernization relationships within three broadly homogeneous cultural regions of the less developed world: Latin America, Islamic countries and southern Asia. The relationships were much stronger within each of the groups than for all countries combined, suggesting that gross cultural differences, however crudely defined, are significant influences on fertility. For the 25 Latin American countries, modernization variables of urbanization, literacy, non-agricultural employment, life expectancy, and telephone, newspaper and hospital provision gave individual correlations with the crude birth rate of between –0.68 and –0.94, so that collectively these measures accounted for as much as 90 per cent of the variance in birth rates. Associations for the other two regions were lower, and the best fits of natality to modernization were achieved with rather different independent variables: education in the Islamic nations and GNP per capita in southern Asia.

Support for the cultural region approach is provided by Janowitz (1971) in a study of gross reproduction rates and modernization variables in 57 countries about 1960. She found that use of five regional dummy variables considerably enhanced the predictive power of her regression model.

A problem in all multivariate regression analysis is multicollinearity. When there is appreciable intercorrelation among the independent variables, as there invariably is within the modernization package, it becomes difficult to assess the relative importance of individual variables. This problem was clearly recognized by Oechsli and Kirk (1974–5) in their analysis of crude birth rate variation among 25 Latin American and Caribbean countries in 1962 and 1970. In place of several individual indicators of development, they adopt an overall index, which is an equal-weighted average of ten socio-economic indicators commonly used in fertility analysis. The method of index construction is to standardize indicators by giving a mean of zero and a standard deviation of one to each indicator's data set; then the average of the ten standardized indicator values for any particular country is its development index. They also recognize that continuous, unlagged variables may not be the best way of relating natality to development, since transition theory suggests that the values and attitudes governing family size take some time to respond to a changing social environment. Accordingly Oechsli and Kirk adopt a lag of about eight years, relating 1970 birth rates to 1962 development index values.

In cross-sectional analyses one must always beware of glibly deducing temporal changes from cross-sectional data. Each country is, in fact, 'on its own trajection through time and the development process. The cross-section picture can be generalized to the probable time series only if there are great similarities in the trajectories of the countries through time and modernization' (Oechsli and Kirk, 1974–5, p. 409). Recognizing this objection, S. E. Beaver (1975) employed a pooled cross-sectional and longitudinal multivariate analysis of fertility and modernization for a set of 24 countries in Latin America and the Caribbean over the 1950–70 period. Each country at a given time was treated as a separate observation, and since four time periods were used this resulted in a total of 96 (4 × 24) observations. Seven independent variables, all of them indicators of modernization and most of them lagged 7–10 years before natality, together accounted statistically for 65 per cent of the variance in standardized birth rate and 77 per cent of that in crude birth rate.

Beaver regards his analysis as providing strong empirical support for the theory of demographic transition. But, more distinctively, he also draws attention to the significant role of cultural factors, unrelated to development, in explaining the pattern of residuals – that part of natality variance 'unexplained' by the regression model. By adopting a cultural-racial grouping of countries, he finds that the European countries (Argentina, Uruguay, Cuba and Puerto Rico) and the African–East Indian countries of the Caribbean tend to have lower natality than predicted by the regression model. On the other hand, the mestizo countries of mainly Amerindian racial background but with some mixture of European ancestry (Mexico, Peru, Venezuela, El Salvador and Honduras) have higher than predicted natality. The Amerindian countries (Bolivia, Ecuador, Guatemala and Paraguay) show only slight departures, in no systematic direction, from predicted levels.

Beaver suggests that differences between the groups in family structure and sex roles hold the key to interpretation. Southern Euorpean countries, which supplied the bulk of European immigrants to Latin America, subscribed to chauvinistic sex roles, with a strong ideal of male dominance and very different standards of sexual behaviour for men and women. On the other hand, nuclear families were stable and the kinship system exerted fairly strict controls over sexual relations and marriage. In the mestizo countries the chauvinsim of southern Euorope becomes what is popularly termed 'machismo'. But having adopted the sexual ideals of southern Europe, these societies lack the social controls to restrain the resulting behaviour, since the social structure of the indigenous Amerindian society has been disrupted. Many births are ex-nuptial, and even married men may have few obligations to their wives. In the African–East Indian countries ex-nuptial births are also common, but the southern European pattern of sex roles is absent. Fertility reduction is promoted here by the instability of sexual unions and by the greater freedom women enjoy to engage in many activities, including fertility control. Beaver's study, therefore, confirms the view that cultural background can facilitate or delay demographic change independently of socio-economic development.

Modern cross-sectional studies

The most comprehensive and most cited cross-sectional study has probably been that of Mauldin and Berelson (1978). The analytical kernel of this study is a series

Table 6.5 Coefficients of correlation between selected variables, 94 developing countries, around 1970

	1	2	3	4	5	6	7	8	9
1. Adult literacy	1.00								
2. School enrolment	0.80	1.00							
3. Life expectancy	0.87	0.76	1.00						
4. Infant mortality rate	−0.78	−0.71	−0.86	1.00					
5. Non-agricultural male employment	0.65	0.73	0.80	−0.73	1.00				
6. GNP per capita	0.23	0.38	0.40	−0.37	0.62	1.00			
7. Urbanization	0.45	0.58	0.58	−0.54	0.78	0.57	1.00		
8. Family planning programme effort	0.64	0.52	0.70	−0.64	0.52	0.07	0.32	1.00	
9. Crude birth rate change (1965–75)	0.70	0.60	0.76	−0.71	0.61	0.13	0.42	0.89	1.00

Source: Mauldin and Berelson (1978), pp. 99, 104, 105.

of correlations for 94 developing countries between crude birth rate change, 1965–75, and variables measuring modernization level and family planning programme effort about 1970. Seven variables were selected to represent the education, health, economic and urbanization dimensions of modernization, while programme effort was measured by an index based on fifteen programme criteria.

The correlations are presented in Table 6.5. Predictably there is appreciable intercorrelation among the modernization variables, and programme effort is also correlated, although to a lesser extent, with the modernization variables. Crude birth rate change is correlated significantly with all modernization variables apart from GNP per capita. The modernization variables in combination achieve a multiple correlation coefficient of 0.81 with crude birth rate change, thus accounting statistically for 66 per cent of the latter's variance (coefficient of determination, $r^2 = 0.66$). When the modernization and programme variables are considered together, the r^2 value rises to 0.83. The addition, therefore, of the programme effort variable to the set of modernization variables raises the 'explanation' of variance in crude birth rate decline from 66 per cent to 83 per cent.

These findings are confirmed, even to the extent of remarkably similar r^2 values, when the 1965–75 crude birth rate decline is regressed against:

1. changes in modernization variables, 1960–70, and programme effort, 1970–2;
2. modernization variables, 1960 (for lag effect), and programme effort, 1970–2.

With respect to the 'development versus family planning' debate, Mauldin and Berelson conclude that the joint impact is more effective than either alone, and that programme effort makes a significant difference, not merely a trivial one. This overall conclusion is generally supported by other cross-sectional analyses of international data (Tsui and Bogue, 1978; Cutright and Kelly, 1981; Lapham and Mauldin, 1984). One study concludes:

> the important implication is not the amount of total explanation rendered by the model but the verification of family planning's independent effect on birth rates. Having allowed other factors to reign freely in explaining 1975 fertility levels, we find that, by

including family planning effort last, it is able to make an independent and significant impact on 1975 fertility levels.

(Tsui and Bogue, 1978, p. 32)

Having considered overall relationships in less developed countries between fertility and associative factors, it is now of interest to observe how closely *individual* countries conform to the overall pattern. Table 6.6 presents this information in an economical manner, although the dependent variable illustrated is not fertility decline itself but rather the dominant proximate determinant of fertility decline – the contraceptive prevalence rate. The quality of data on contraceptive use should always be carefully considered. Particularly suspect are prevalence rates provided by some national family planning programmes which are based on cumulative numbers of first-time acceptors, without adequate correction for drop-out or discontinuation. It is now widely recognized that cross-sectional surveys provide more accurate information on contraceptive prevalence than programme statistics, partly because they avoid the problem discussed above and partly because they can more readily detect and include traditional, non-programme methods, as well as contraceptive supplies from the private sector. Accordingly, the rates in Table 6.6 are all derived from national sample surveys between 1977 and 1982, mainly from two standardized international programmes: the World Fertility Survey and the Contraceptive Prevalence Survey.

The vertical axis of Table 6.6 represents an index of development calculated from the seven modernization variables specified in Table 6.5. The horizontal axis represents strength of family planning programmes measured by an index based on thirty criteria (e.g. public statements by leaders, use of mass media, availability of particular methods, presence of home-visiting fieldworkers); each criterion is scored on a 0–4 range, so that index scores for countries can vary from 0 to 120.

One can ignore initially the contraceptive prevalence figures and concentrate on the relationship between modernization and programme strength. That a close relationship exists is evident in the clustering of countries in the four cells along the diagonal from bottom right to top left. The countries to the left of these cells all have stronger family planning programmes than their development level would suggest: China, Sri Lanka, Indonesia, India, etc. Conversely, countries to the right have weaker than predicted programmes. They fall almost entirely into three groups – African, Islamic and Latin American – where culturally based pronatalism has resisted fertility regulation policies.

Considering now the mean contraceptive prevalence by cell, there are clear gradients both by row and column, showing the graded influence of both modernization and programme effort. One might note, however, that at the high modernization level, there is very little differential impact on contraceptive prevalence of weak, moderate and strong programmes, but at the lower-middle level of modernization, there are appreciable differences by programme strength. This would seem to confirm the view, quoted on p. 150, of Oechsli and Kirk (1974–5).

Finally, within each cell there is a substantial degree of uniformity in contraceptive prevalence, although there are individual anomalies. The overall pattern of Table 6.6 is, therefore, a clear demonstration of the essential regularity of pattern of contraceptive usage and, through it, of fertility decline in the Third World.

Table 6.6 Contraceptive prevalence rates (%), by 1970 development level and 1982 programme effort index, in 73 less developed countries

1970 Development Level	1982 programme effort: Strong		Moderate		Weak		Very weak or none	
High	Hong Kong	80	Cuba	79	Costa Rica	66	Paraguay	36
	Singapore	71	Panama	63	Brazil	50		
	Taiwan	70	Jamaica	55	Venezuela	49		
	Korea	58	Trinidad/ Tobago	54	Peru	43		
	Colombia	51	Fiji	38	Chile	43		
	Mauritius	51						
	Mexico	40						
		(60)		(58)		(50)		(36)
Upper middle	China	69	Thailand	58	Ecuador	40	Iran	23
	Sri Lanka	57	Philippines	45	Turkey	40	Syria	20
					Honduras	27	Ghana	10
			Dominican Republic	43	Egypt	24	Nicaragua	9
			Malaysia	42	Morocco	19	Zaire	3
			El Salvador	34	Guatemala	18	Zambia	1
			Tunisia	31	Algeria	7		
		(63)		(42)		(25)		(11)
Lower middle	Indonesia	48	India	32	Haiti	19	Bolivia	24
			North Vietnam	21	Zimbabwe	14	Nigeria	6
					Kenya	7	Lesotho	6
					Pakistan	6	Burma	7
							Cameroon	2
					Papua New Guinea	5	Uganda	1
					Senegal	4	Kampuchea	0
					Liberia	1		
		(48)		(27)		(8)		(6)
Low			Bangladesh	19	Nepal	7	Benin	18
					Tanzania	1	Sudan	5
							Sierra Leone	4
							Ethiopia	2
							Somalia	2
							Yemen	1
							Burundi	1
							Chad	1
							Guinea	1
							Malawi	1
							Mali	1
							Niger	1
							Burkina Faso	1
							Mauritania	1
				(19)		(4)		(3)

Note: Mean prevalence, at each level of programme effort and development level, is shown in parentheses

Source: Lapham and Mauldin (1985), Table 4.

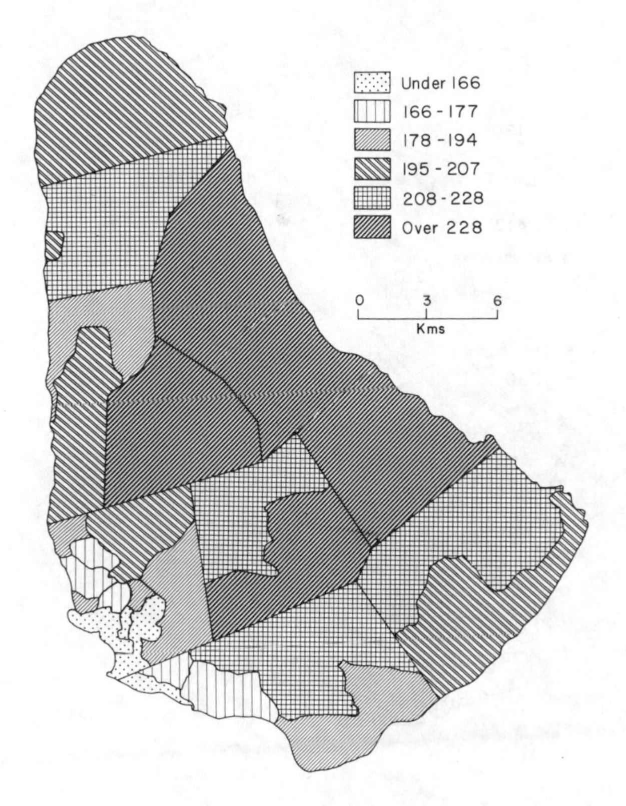

Figure 6.6
Number of children under two years of age per 1,000 females aged 15–44 years, Barbados, 1970
Source: H. R. Jones (1977), Figure 3.

District studies

Multivariate analysis of areal data has also been adopted to examine district variations in fertility within individual countries. The approach and methodology have been reviewed by Hermalin (1975), while recent case studies include India (Jain, 1985), Sri Lanka (Langford, 1981), Thailand (Foggin, 1983) and Trinidad (Coward, 1986b).

Barbados also provides interesting evidence for the interactive influence of development and programme activity on spatial differences in fertility. A comparison of Figures 6.6 and 6.7 suggests that the 1970 fertility pattern may be interpreted through the ecological theory of metropolitan dominance, by which cities exert an organizing influence on the social structure of their hinterlands by the diffusion of urban norms and values, so that gradient patterns in socio-economic characteristics are established. Fertility rates, represented here by a modified child:woman ratio, are clearly lowest in and around Bridgetown, with intermediate rates in the suburban coastal extensions to Speightstown and Oistins. Almost two thirds of the areal variation in fertility can be 'explained' by the use in multiple regression of several modernization and demographic structure variables ($r^2 = 0.62$). However, the pattern of residuals does indicate the

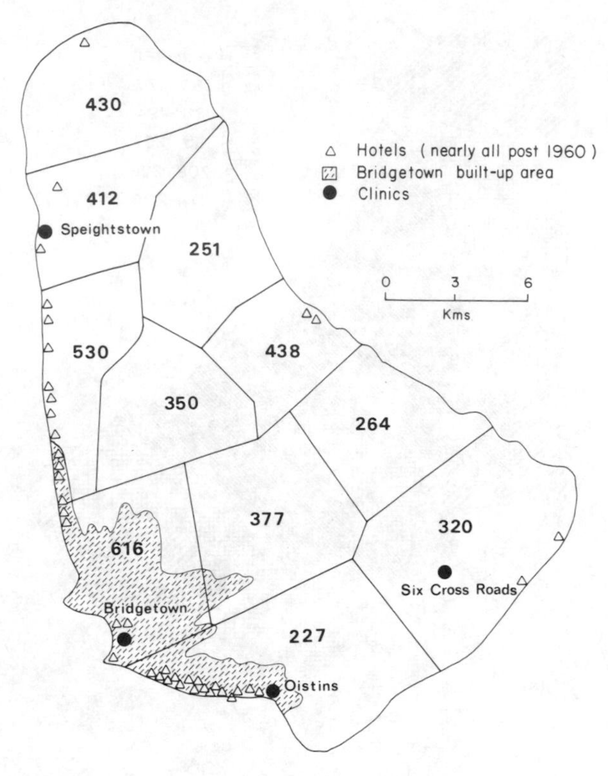

Figure 6.7
Selected features of settlement and of family planning activities in Barbados *c.* 1970. The figure in each parish refers to the number of women (per 1,000 women aged 15–44 in 1970) contacted by fieldworkers of Barbados Family Planning Association in an eighteen-month period, 1970–1
Source: Data from Barbados Family Planning Association.

spatial role of the family planning programme, which has been the most successful in the Western hemisphere. The areas of significant positive residuals, where the regression equation overestimates 1970 fertility levels, are all close to clinics or in parishes with the highest fieldwork density.

Clearly, important regional fertility differentials do exist in an island even as small as Barbados, and metropolitan dominance in family planning activity contributes actively to their retention. This suggests an obvious need for some spatial reallocation of programme resources within the island, along the lines of the successful, but rather exceptional, programmes in South Korea (Foreit, Koh and Suh, 1980) and Mauritius (Jones, 1989) where programme input has always been concentrated in rural areas.

Survey analyses

An important criticism of the macro-analytical cross-sectional studies of fertility discussed above is that causes are inferred remotely and crudely from aggregate-level association, without adequate examination of the detailed mechanisms that may be at work. Indeed, modernization variables should more properly be regarded, not as the fundamental determinants of fertility behaviour, but rather as surface, identifiable expressions, or causal surrogates, of decision-making.

Some progress to analytical refinement is achieved in social surveys where information is collected on individuals rather than on populations or places, thus enabling aggregation to levels and groups specified by analysts. This permits the identification of fertility differences within countries between groups differentiated on the basis of a range of demographic, social and economic characteristics.

This approach is represented well in internationally organized and funded programmes which attempt to ask a standard set of survey questions from a representative national sample of women of child-bearing age in a large number of less developed countries. The World Fertility Survey is the best known of these programmes, encompassing surveys in 42 less developed countries between 1974 and 1982 (Cleland and Hobcraft, 1985).

The most recent trans-national programme has been the Demographic and Health Surveys of the Institute for Resource Development of Westinghouse. Selected late-1980s data from the first of these national surveys are presented in Table 6.7. They reveal the expected gradation in fertility by community of residence and by maternal education, but there are some interesting differences between continents and countries. Gradations in sub-Saharan Africa are modest, whereas in Latin America they are substantial. In Africa the promotion of high fertility by traditional religious belief-systems remains a powerful, socially universal feature, while the fertility-inhibiting effects of both breastfeeding and pathological sterility are reduced among more educated urban dwellers (Cleland and Rodriguez, 1988). In Latin America rural areas remain dominated by an agro-social system of large estates, small tenants, landless labourers, and 'master and man' relationships. There has been little of the rural development and land reform that has benefited many Asian countries and given their peasants an incentive not to fragment their owner-occupied holdings. In Latin America, the only security for the mass of the rural population remains that provided by family members.

Table 6.7 also highlights the remarkable recent falls of fertility in Thailand, a country which remains overwhelmingly agrarian (83 per cent of the population rural and 71 per cent of the labour force in agriculture in 1980) but where a

Table 6.7 Total fertility rates for selected countries, derived from births data reported in 1986–7 Demographic and Health Surveys

	Rural	Urban	Education of mother None	Primary	Secondary
Latin America					
Brazil	5.0	3.0	6.5	5.1	3.0
Colombia	4.9	2.8	5.4	4.2	2.3
Peru	6.6	3.2	7.0	5.1	2.9
Asia					
Thailand	2.6	1.7	3.5	2.5	1.5
Africa					
Liberia	6.8	6.1	6.7	7.0	4.8
Senegal	7.1	5.5	6.8	5.7	3.8
Burundi	6.9	5.3	6.8	7.2	5.5

Source: *Studies in Family Planning* (1988 and 1989), Vols. 19 and 20.

Buddhist culture has permitted appreciable progress in family planning activity. The 1987 Demographic and Health Survey recorded an astonishing 65 per cent of married women aged 15–44 as being current users of contraception, with the leading methods, in order, being female sterilization, the pill, injection and IUD – significantly, 'female' methods in a society where gender relations ensure a substantial degree of female autonomy.

Community studies

Large-scale surveys can undoubtedly take analysts some way down the road to identifying and explaining fertility patterns, but the detailed behavioural mechanisms are often missed because of rigidities imposed by a common-format questionnaire and by the lack of trust often accorded to unfamiliar interviewers.

An important response to these problems has been the micro-demographic approach of community-level studies. Anthropological-type methods have thus been used in selected localities where analysts reside for at least some months. Participant observation and long, leisurely, wide-ranging discussions can allow a fuller appreciation of the all-important economic and cultural contexts which fashion fertility behaviour, and also help to gain a suffficient degree of rapport to permit reliable responses to the more intimate questions.

Rich as these studies are (e.g. Caldwell, Reddy and Caldwell, 1982, in Karnataka; Tilakaratne, 1978, and Caldwell *et al.*, 1987, in Sri Lanka; Collins, 1983, in Andean Peru), there remains the problem of the extent to which one can generalize from what are essentially local studies. Possibly their greatest value is in the way they can tease out and properly explain from field experience the associations and apparent causal relationships that have been identified in more aggregated analyses. For example, two village studies in the Punjab region of India have provided particularly good insights into how fertility behaviour responds to modernization.

Das Gupta (1977) studied a village 15 miles from Delhi which had been the scene of an earlier sociological study in 1953. Many changes had taken place in the village between these dates. Subsistence agriculture had given way to cash cropping, and the village had become integrated into a wider economy. There was a much greater dependence on outside income, particularly from employment in Delhi. New schools and a health centre had been built, electricity introduced, and a proper road built to give direct access and a new bus service to Delhi. Different caste-based groups within the village reacted quite differently in their fertility behaviour response to these modernization changes, and in particular to external employment opportunities:

> The upper strata find it easier to secure the better paid permanent jobs, simply because they are in a position to meet the heavy but necessary investment in education and bribes. The volume of investment thus required puts a premium on lower fertility and helps to explain the increasing interest among this group to limit their family size. Of the lower strata, only some have struggled to make these investments and thereby managed to get a foothold on the upward spiral to economic success. However, the bulk of the poorer strata cannot afford such heavy expenditure and have to rely on their traditional village occupations and casual jobs for their subsistence. These people hope to maximize their potential income by having large numbers of children, which in turn makes it increasingly difficult for them to get on to the upward spiral.
>
> (Das Gupta, 1977, p. 119)

In a similar study Nag and Kak (1984) revisited the Punjab village of Manupur, where Mamdani (1972) had explained the failure of an early family planning programme in terms of what he saw as the universal *need* within the community for large families. By the early 1980s Nag and Kak were able to show how various elements of modernization had considerably modified attitudes towards family size, so much so that about 50 per cent of Manupur couples with women of reproductive age were using a contraceptive method, primarily female sterilization. The critical new factor in Manupur seems to have been the transformation of agricultural activities and technology associated with the new high-yield wheat varieties of the Green Revolution, Punjab being the leading innovative state in India in this respect. An increase in credit facilities through co-operative societies and the nationalization of major banks in the 1970s have allowed an unusually wide range of farmers here to exploit the potential of the new technology, which includes, for example, the extensive owning or renting of tractors.

The new agricultural system impacts on the traditional need for child labour in Manupur in several ways. Reduction of fallowing practices and increase in multiple cropping have led to a virtual disappearance of grazing land in the village and, with it, of the traditional role of children in the tending of cattle and the collecting of cowdung for fuel and manure. Chemical weedkillers have similarly reduced the field role of children in the growing season. The new technology and the modern credit system have made most parents conscious of the need for their sons to have secondary school education, while many also felt that the amount of dowry demanded by prospective in-laws would be less if their daughters, too, were educated. The appreciable changes in attitudes to family size in Manupur in just ten years serve to remind us that demographic change in the contemporary Third World *can* proceed very rapidly, but only under favourable enabling circumstances.

7

PROBLEMS OF POPULATION GROWTH AND AGE COMPOSITION

Global growth problems

In the last three decades in affluent Western countries there has been a huge surge of concern by scholars and the general public, although somewhat less by governments, about global population growth and its associated problems. This can be related to three factors.

1. After the increases in fertility in the 1950s and 1960s in developed countries and the reductions in mortality in less developed countries, population growth has reached unprecedented levels, with the world's population doubling every 40–45 years, raising ever more vividly the spectre of 'standing room only'. At no point in their demographic transitions did European nations experience the growth in population-related demand for basic necessities that now confronts the Third World. Medium-variant projections of the UN Population Division in 1988 suggest that the world's population will have grown from about 4.1 billion in 1975 to 6.3 billion in 2000 and 8.5 billion in 2025. Figure 7.1 indicates that nearly all of this increase will occur in the less developed countries, which comprised 68 per cent of the world's population in 1950, 76 per cent in 1985 and an expected 84 per cent in 2025. The most dramatic change will be in Africa, where continued resistance to the current global trend of reduced fertility will increase its proportion of the world's population from 11 per cent in 1985 to 19 per cent in 2025; in absolute terms its population will triple.
2. The snowballing role of population momentum has become more widely appreciated. An analogy is the momentum of a speeding train that cannot be brought to an immediate halt even with full application of brakes. An enormous momentum for growth has been created by the very young age structure of today's less developed countries, the legacy of high fertility in the immediate past. The UN medium-variant projection assumes that world fertility will fall by 2035 to the replacement level of 2.1 births per woman (compared with an actual level of 3.5 births in 1990) but this would still enable the population to expand to just over 10 billion before stabilization occurs around the year 2100.

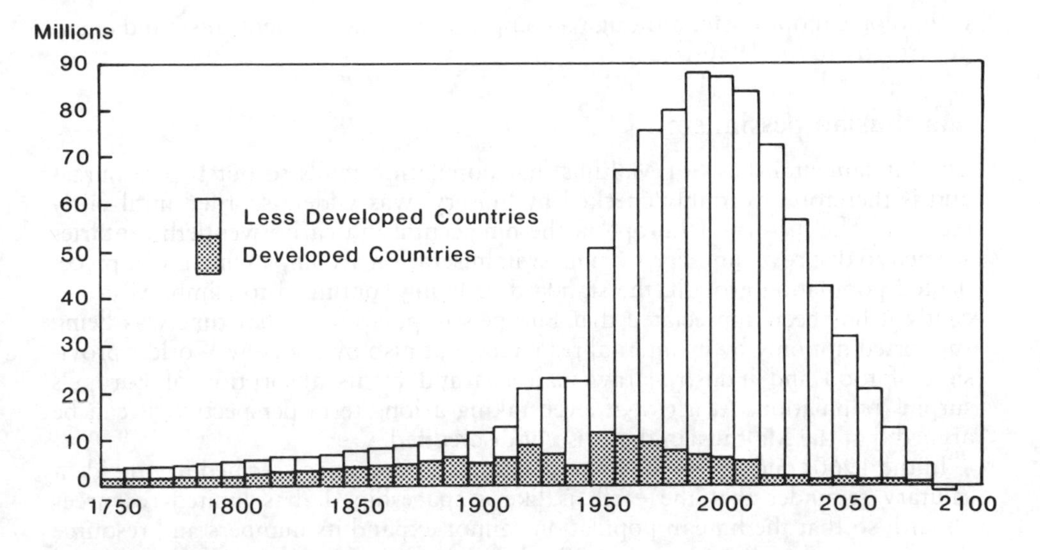

Figure 7.1
Average annual increase in population per decade
Source: *Population Bulletin* (1986), Vol. 41(2), Figure 1. By permission of Population Reference Bureau, Inc.

3. The publication in 1962 of Rachel Carson's *Silent Spring*, highlighting the decimation of wildlife by pesticides like DDT, set off major concern in the developed world about environmental deterioration. Well-publicized disasters like the *Torrey Canyon, Exxon Valdez* and other oil-spills, Minimata Bay mercury poisoning, defoliation in Vietnam, chemical poisoning at Seveso and Bhopal, radioactivity release at Chernobyl and desertification in the Sahel have fuelled concern about humanity's ability to contain pollution and effectively husband life-support systems. This concern expressed itself in the formation and rapid growth of pressure groups like the Conservation Society, Friends of the Earth, Green Peace and Zero Population Growth, and in major world conferences on environmental issues.

The potentially harmful role of population growth in environmental management is expressed in a simple equation:

$$I = PF$$

where I is the total impact of a society on the ecosystem, P is population size, and F is impact per capita (Ehrlich and Ehrlich, 1972, p. 259). Some scientists like Barry Commoner have argued convincingly that F is the more critical factor, embracing the high, and some would say profligate, material consumption of the West, but there is little doubt that when P and F increase simultaneously, as has generally been the case in the modern world, the effect on I can be massive.

However, despite the continuing concern about the consequences of population growth, there has been remarkably little research that explicitly addresses the issue. Indeed, it is a subject in which preconception and ideology, rather than empirical analysis and assessment, tend to dominate. Attitudes have tended to polarize into standpoints which can be termed Malthusian pessimism and technological optimism, although a compromise stance of neutralism did emerge strongly in the 1980s.

Malthusian pessimism

The fundamental thesis of Malthus that population tends to outstrip resources and is therefore invariably checked by 'misery' was widely scorned until fairly recently. The history of Europe in the nineteenth and early twentieth centuries seemed to disprove him. Far from mass immiseration accompanying the unprecedented population growth, the standard of living continued to climb. More recently it has been appreciated that Europe's population at that time was being supported not only by its internal resources, but also by the New World's provision of food and industrial raw materials and by its absorption of Europe's surplus population. At a *global* level, taking a long-term perspective, it can be argued that the Malthusian thesis is *not* discredited.

In the 1960s and 1970s the first images from outer space seemed to provide a salutary reminder that the earth is like a spaceship. It has limited resources aboard, so that the human population cannot expand its numbers and resource demands indefinitely, or spaceship Earth would simply run out of fuel, food and other vital support systems. Fears and forecasts of global eco-catastrophe became common, even fashionable, in literature (*The Population Bomb, Born to Starve, The Hungry Future, End to Affluence, People: An Endangered Species*). The arch-prophet of doom was the American biologist Paul Ehrlich (1971; Ehrlich and Ehrlich, 1972), haunted by an apocalyptic fear that spaceship Earth would be blown up by its internal population bomb: that humanity would breed itself into oblivion. In much quoted phrases he asserted that the battle to feed all of humanity was lost, that India had twenty minutes to famine, and that hundreds of millions of people would starve to death in the 1970s and 1980s.

The high-water mark of modern Malthusianism was reached with the publication of *The Limits to Growth* (Meadows *et al.*, 1972) – a landmark book which commanded worldwide attention and debate. An informal international association with a restricted, prestigious membership, the self-styled Club of Rome, had embarked on a breathtakingly ambitious task, nothing less than a Project on the Predicament of Mankind. It had been impressed by the system dynamics work of Jay Forrester and colleagues at the Massachusetts Institute of Technology, so that model-based work was commissioned on the global interaction of five basic factors: population, agricultural production, natural resources, industrial production and pollution.

The basis of the simulation conducted in *The Limits to Growth* was a series of calculated or assumed relationships between the five basic factors and their component elements. Given these relationships, computer runs were made to project recent trends into the future. These runs invariably spell out disaster, since one cannot indefinitely accommodate exponential growth in a finite system. Sooner or later, negative feedback loops of famine, pollution and resource depletion result in a system overshoot and collapse. Stripped of jargon, these are the positive checks of Malthus.

Figure 7.2 is the 'standard' run, based on the projection of historical growth values and the assumption that there is a 250–year supply of non-renewable resources at 1970 usage rates. The system collapses disastrously well before 2100, as is the case with subsequent runs using what the study team regards as even the

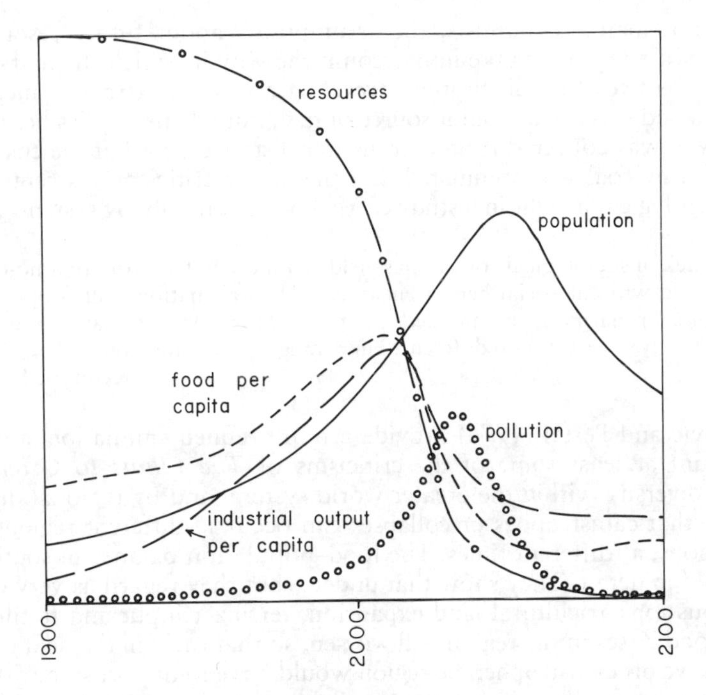

Figure 7.2
The Limits to Growth: world model standard run. The standard run assumes no major change in the relationships that have historically governed the development of the world interaction system, and all plotted variables follow historical values from 1900 to 1970. Food, industrial output and population grow exponentially until rapidly diminishing resources brake industrial growth. Population growth is finally halted by a rise in the death rate due to decreased food and medical services
Source: Figure 35 from *The Limits to Growth: A Report for the Club of Rome's Project on the Predicament of Mankind*, by D. H. Meadows, D. C. Meadows, J. Randers and W. Behrens. A Potomac Associates Book published by Universe Books, New York, 1972.

most optimistic assumptions about resource expansion, pollution control and fertility reduction. Such assumptions prolong the period of population and industrial growth, but cannot remove their eventual limits.

The social advocacy of the study and its supporters is clear. Since unlimited growth in a finite world is impossible, two other options are left: an unthinkable nature-imposed limitation of growth, involving massive famine, disease, pollution and the like, or a self-imposed limitation – an extension of the preventive checks of Malthus to embrace birth control programmes (Neo-Malthusianism), capital shifts from manufacturing to agriculture, environmental management, and the adoption of ecologically harmonious life-styles (Schumacher's 1973 *Small is Beautiful* and the 1972 Blueprint for Survival of *The Ecologist* magazine).

The Limits to Growth was favourably received by many biologists and ecologists well versed in the exponential growth of animal and plant populations up to a critical environmental 'carrying capacity'. But reviews by social scientists have generally been harsh (Cole *et al.*, 1973), with the major criticism being summed up by the old computer maxim: Garbage In, Garbage Out. In other words, computer studies are only as good as the quality of information and assumptions adopted. For many economists, the assumptions about future resource availability do not adequately take into account the way in which the market system provides incentives for substitution, recycling and new extractive methods; for example, wood was once a major source of fuel, but when supplies became scarce near cities, it was collected from ever greater distances, so that the cost rose and substitution by coal was stimulated. But the major criticism has centred on the lack of sociological insight in a study carried out essentially by systems engineers:

> Its utter lack of sociological content has yielded a model of a world that no-one knows. It is a world without social heterogeneity . . . Thus all nations, all people move to a catastrophe or harmony at an identical pace. There are no haves and have nots simultaneously. There are no differences in demographic transition.
>
> (McGinnis, 1974, p. 299)

Mesarovic and Pestel (1974) provide a more refined simulation study taking into account at least some of the criticisms of *The Limits to Growth*. They recognize diversity within the greater world system, and by regional disaggregation show that catastrophes or collapses can occur in different regions, for different reasons, at different times. The food–population balance in southern Asia is examined in detail. They show that under what they regard as very optimistic assumptions for agricultural land expansion, fertilizer input and fertility reduction, the food crises in the region will worsen, so that early in the next century, in order to stave off catastrophe, the region would have to import some 500 million tons of grain a year. Even if this were available, the cost and logistics of shipping and internal distribution are staggering. It is at the regional level, then, that the Malthusian scenario of catastrophe becomes most credible, evidenced by the famines of the 1970s in Bangladesh, Bihar, Ethiopia and the Sahel.

The legitimation of modern Malthusianism was demonstrated clearly in *The Global 2000 Report to the President* (Barney, 1980), when several US government agencies responded to a presidential directive to assess the likely changes in global population, resources and environment as a foundation for long-term planning. It concluded:

> If present trends continue, the world in 2000 will be more crowded, more polluted, less stable ecologically, and more vulnerable to disruption than the world we live in now . . . For hundreds of millions of the desperately poor, the outlook for food and other necessities of life will be no better. For many it will be worse.
>
> (Barney, 1980, p. 1)

Technological optimism

In assessments of humanity's survival prospects, the arguments of Malthusian doom-mongers are opposed by the 'technical fix cheer-mongers'. Their views are neatly represented in a poem, 'The Technologist's Reply to a Conservationist's Lament', written by the American economist Kenneth Boulding.

> Man's potential is quite terrific
> You can't go back to the Neolithic.
> The cream is there for us to skim it
> Knowledge is power, the sky's the limit.
> Every mouth has hands to feed it.

Food is found when people need it.
All we need is found in granite
Once we have the men to plan it.
Yeast and algae give us meat,
Soil is almost obsolete.
Men can grow to pastures greener
Till all the earth is Pasadena.

Moral:
Man's a nuisance, Man's a crackpot.
But only man can hit the jackpot.

Faith is thus put in man's innate ability to adapt, invent, substitute and recycle, so that a breathing space is provided until the Third World populations complete their demographic transitions.

There are two major strands to the technological optimists' case. First, the great majority of economists are sceptical about the possibility of natural resources being a constraint to the growth of population or living standards. Neo-classical economists have an abiding faith in the pricing system, where the 'invisible hand' of the market generates adjustments which bring population, resources and environment back into balance after any dislocation. Thus, as a resource is depleted, rising prices reduce consumption, stimulate conservation and recycling, and also speed the search for alternatives in an 'Age of Substitutability' (Goeller and Weinberg, 1976); some even argue that capital is a near perfect substitute for resources. Simon (1981) believes he had 'proved' that resources are not being depleted by demonstrating that the prices of most resources in recent history have fallen, rather than risen, in real terms.

A second contention is much more controversial. Drawing upon the pre-nineteenth-century Mercantilists, who believed that national population growth promoted vigorous and beneficial commerical expansion, and on writers like Boserup (1965, 1987), who believe that population growth precedes and provokes agricultural progress, Simon (1981, 1986) and Simon and Khan (1984) stand Malthusianism on its head by arguing that population growth, far from being a problem, is a golden opportunity. In this 'population push' view of history, technological progress overcomes Malthusian diminishing returns because population growth *causes* progress. It does this partly by demand ('challenge and response', 'necessity the mother of invention') and partly by an alleged (but, to critics, incredulous) functional relationship between population size and numbers of inventions.

> The ultimate resource is people – skilled, spirited, and hopeful people – who will exert their wills and imaginations for their own benefit and so, inevitably, for the benefit of us all
> (Simon, 1981, p. 348)

> There will be a higher standard of living, at first despite the greater numbers and then later because of them.
> (Simon, 1986, p. 171)

Neutralism

The contrasting extremist scenarios of doom-mongerers like Ehrlich and population boosters like Simon have encouraged a broad consensus to emerge, embracing several positions on the scale between alarmism and complacency and several very different ideologies, but all sharing the belief that population growth is only one of several key influences on humanity's future survival and prosperity. The dominant Neo-Malthusian paradigm in Third World development thinking in the 1960s and 1970s had undoubtedly exaggerated the negative functional link between population growth and development. Even under the unrealistic assumption that other things are equal, it is now widely believed that prospects for future advancement in less developed countries are not precluded by rapid population growth, although they may well be enhanced by more modest growth.

A growing appreciation of two factors underpins the new consensus. First, empirical analysis has shown that there is no significant correlation, positive or negative, between national population growth rates and various measures of development, like per capita income growth, over recent decades among less developed countries (Kuznets, 1973; Bloom and Freeman, 1986). Variations in economic growth and development are determined much more by a range of structural and institutional factors.

Second, both Malthusian and technical-fix viewpoints ignore the critically important social organization of production, distribution and consumption, which determines that the basic problems of humanity derive not from scarcity of resources but from relationships between individuals and groups. Marxists observe that only the poor go hungry, that scarcity is engineered for profit maximization by capitalism and that 'poverty is due to maldistribution of resources both internationally and within nations and not to the physical limits of producing the resources themselves' (Sandbach, 1978, p. 27). The widespread malnutrition and occasional famines in the Third World are rarely due to global or even national food deficits, but essentially to the inability of poor families to purchase the food that is potentially available. A much-quoted case is Brazil, where the considerable growth in GNP per capita is not thought to have benefited the vast bulk of the population, because the benefits of growth have been siphoned off by powerful élites; there has been no trickling down of benefits to the great mass of landless rural poor and shanty-town dwellers.

Three influential assessments in the 1980s of the global population–resources issue will now be reviewed briefly. They illustrate the differences of emphasis and interpretation that are accommodated within the broadly neutralist position.

The *World Development Report 1984* of the World Bank discusses population change and development as its special topic. The report has been described aptly by economist Colin Clark as Malthusianism in retreat. For the foreseeable future at the global level, it argues that population pressure on natural resources will not be an important obstacle to development, and any food shortages should not be acute. Yet it is convinced that rapid population growth, young age structures and high dependency ratios slow the accumulation of vital human capital skills (especially education and health) and exacerbate income inequalities. Consequently, it concludes:

> No one would argue that slower population growth alone will ensure progress; poor economic growth, poverty, and inequality can persist independently of population change. But evidence described in this Report seems conclusive: because poverty and rapid population growth reinforce each other, donors and developing countries must co-operate in an effort to slow population growth as a major part of the effort to achieve development.
>
> (World Bank, 1984, p. 185)

An authoritative 1986 report, *Population Growth and Economic Development*, was commissioned by the National Research Council of the US National Academy of Sciences. Its overall conclusion – that prospects for development are not precluded by rapid population growth, although they are generally enhanced by more modest growth – does not differ much from the *World Development Report*, although the general tone had moved nearer to complacency.

A third assessment is the Policy Statement of the US government at the 1984 UN International Conference on Population at Mexico City:

> First, and most important, population growth is, of itself, a neutral phenomenon. It is not necessarily good or ill. It becomes an asset or a problem only in conjunction with other factors, such as economic policy . . . The population boom was a challenge; it need not have been a crisis . . . It provoked an overreaction by some, largely because it coincided with two negative factors . . . The first of these factors was governmental control of economies . . . As economic decision-making was concentrated in the hands of planners and public officials, the ability of average men and women to work towards a better future was impaired . . . Under such circumstances, population growth changed from an asset to a peril . . . Localized crises of population growth are, in part, evidence of too much government control and planning, rather than too little. The second factor . . . was confined to the western world. It . . . attacked science, technology, and the very concept of material progress . . . The combination of these two factors . . . led to demographic overreaction in the 1960s and 1970s . . . and too many governments pursued population control measures without sound economic policies.
> (from the Policy Statement reprinted in *Population and Development Review* (1984), pp. 574–9)

Reflecting the ideological stance of the New Right, this is essentially a case for 'free market demographics' where the best contraceptive is seen as unfettered entrepreneurial initiative (Demeny, 1985; Finkle and Crane, 1985). In a remarkable turnaround, the analysis and prescriptions of *The Global 2000 Report to the President* were explicitly rejected.

Sustaining life-support systems

As concern about the *independent* influence of population growth on humanity's future has receded in the 1980s, it has been replaced by growing worries about the environment's ability to sustain life in the long term. These worries are not the Malthusian ones, fashionable in the 1970s, of depletion and possible exhaustion of non-renewable natural resources like fossil fuels and minerals. Rather, they concern the vital renewable resources of air, water, forests, topsoil and genetic diversity, which are being threatened in such a way as to prejudice sustainable development (Repetto, 1987).

There are two threats to environmental systems. First, exploitation can proceed past the point of regeneration. Growing population pressure certainly contributes (e.g. in the degradation of pastures and woodland in the Sahel), but equally important is the common property or public domain nature of many resources (McCay and Acheson, 1987) so that ecological limits are rarely protected by rising prices. Second, there is the ecologically destructive element in many of our modern technical fixes, particularly involving the release of toxic fluids, gases and substances into the water supply, the atmosphere and the soil, as well as the dislocating and potentially catastrophic effect of carbon dioxide emissions on global climatic patterns. Free markets are incapable of dealing with these threats; indeed, they exacerbate them.

Recent trends in the population–food balance

The preoccupation of human beings has always been to obtain enough food, the basic life-support resource. This is now less evident in the developed world, where less than one-fifth of disposable income is spent on food, but in many Third World countries the proportion is between 60 and 90 per cent. Any assessment of the world food situation should recognize the importance of three dimensions: supply, distribution and consumption.

Supply

At a global scale the evidence indicates clearly that the population–food balance has improved significantly in recent decades. Despite unprecedented population growth, global food production has expanded at an even faster rate, so that the price of most food grains has fallen in real terms. Until about 1960 increases in production were largely the result of bringing new lands into cultivation, since yields per hectare were fairly stable. Subsequently, the situation has reversed (Tables 7.1 and 7.2). The Second Agricultural Revolution in the developed world, based on mechanization, chemical fertilizers and pesticides, has permitted major increases of production in the West from the 1950s, while, more recently, similar technological developments in association with the breeding of more productive cereal strains have brought the higher yields of the Green Revolution to appreciable parts of the Third World.

But the food:population ratios that matter for human welfare are not the global ones, but those at national, regional, class and household levels. Figure 7.3 and Table 7.2 illustrate some of the important macro-regional variations in staple food production. One should always remember the intrinsic unreliability of such agricultural data, because of estimates that have to be made for subsistence production and for unregulated trade, as well as downright falsification of data for political reasons. For example, the scale of crop failure and famine in China in 1958–62 (see Figure 7.3) associated with the forced and mismanaged collectivization of the Great Leap Forward has only relatively recently been documented (Ashton *et al.*, 1984; Jowett, 1989). Nevertheless, it is clear from Figure 7.3 and Table 7.2 that the basis of the apparently healthy global position is the appreciable production growth in North America, western Europe and Australasia.

The situation in the Third World is much more variable. In Latin America the continuing availability of new cropland in the sparsely peopled interior has been the major factor in the modest, but sustained, increases in grain production per capita. Southern and especially eastern Asia has been the biggest success story, disproving the dire predictions of chronic food shortages for this densely peopled

Table 7.1 Average grain yields (tonnes per hectare)

	1934–8	1952–6	1967–70	1981–3
Developing countries	1.15	1.15	1.41	2.01
Industrialized countries	1.15	1.37	2.14	2.67

Source: FAO data from Schuh (1987), table 3.2.

Table 7.2 Increases in crop area and yields by region, 1964–6 to 1982–4

	Increase in cropland (%)	Increase in yields (%) Cereals	Roots and tubers
Africa	14	13	22
Asia	4	77	58
North and Central America	8	44	23
South America	35	42	–1
Europe	11	76	19
USSR	1	35	13
Oceania	24	25	13
World	9	58	21

Source: Repetto (1987), table 6. By permission of Population Reference Bureau, Inc.

region that were being made at the 1974 World Food Conference. Although there are some lagging countries, notably Bangladesh, where under-nutrition remains rife and famine sometimes threatens, most Asian countries have expanded their food output faster than their population – not through any significant increase in cropland in these already crowded countries, but because of yield increases. There were early gains in Japan, Taiwan and South Korea, countries which benefited in the 1950s from the development of land reform, agricultural research, comprehensive rural infrastructure and mass education. Many of the more intensive agricultural methods pioneered there were extended to other parts of Asia, allowing a rapid take-up, particularly in India, Pakistan and the Philippines, of the new

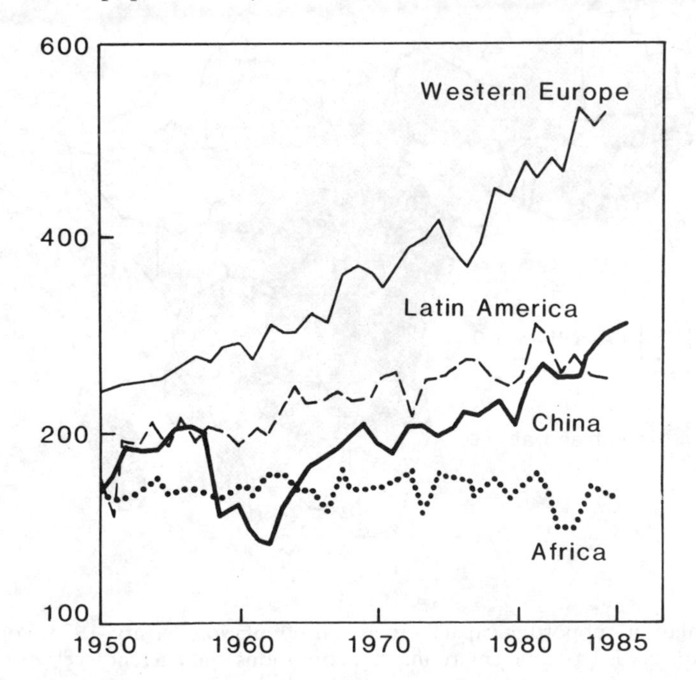

Figure 7.3
Grain production, kilograms per capita, 1950–85
Source: L. Brown (1985). Copyright Worldwatch Institute.

high-yielding varieties of wheat and rice which became available from the mid-1960s.

Some influential early assessments of the Green Revolution were hostile (Griffin, 1974; Fraenkel, 1971), cast as they were in theories of dependency and underdevelopment. Not only was the new technology seen as benefiting exclusively the wealthy farmers, who could afford the necessary package of seeds, fertilizers, pesticides and water input, but it was also perceived as positively disadvantaging the poor by promoting land consolidation and landlessness. However, most recent assessments (Farmer, 1986; Glaeser, 1987; Pinstrup-Anderson and Hazell, 1987; Hayami, 1988; Rigg, 1989) are more optimistic about the social consequences. They show that small farmers quickly followed the lead of innovators once they saw the success of the new technologies under farm conditions and that there have been beneficial multiplier effects in the rural economy. It now seems that government direction and controlled irrigation are much more important than farm size and wealth in determining adoption patterns.

Africa is now the continent of crisis for food production. As Figure 7.3 suggests, food production per head fell during the 1970s and 1980s in response to severe production problems and the highest rates of natural increase in the world. From near self-sufficiency in food in 1970 there has been a decline to a position in which imports provided about one-fifth of the continent's cereal needs in 1985. There is now an immense literature (reviewed by Watts, 1989) on the complexity of Africa's agrarian crisis. Five key factors are involved:

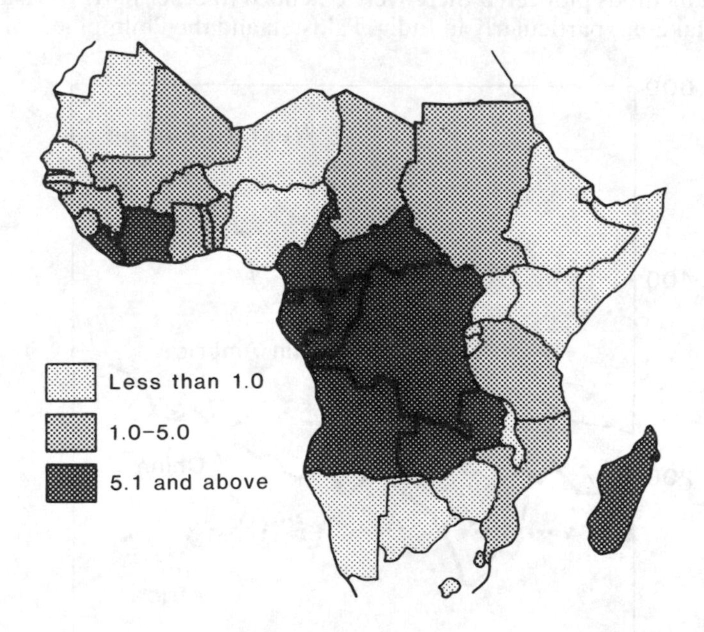

Figure 7.4
Ratio of population–supporting capacity to actual population density, 1975. Population–supporting capacity is based on environmental conditions and current levels of farm technology
Source: FAO estimates reported in World Bank (1984), p. 164. Copyright © 1984 The International Bank for Reconstruction and Development/The World Bank. By permission of Oxford University Press, Inc.

1. It is almost impossible to find a country in sub-Saharan Africa where agriculture has not been disrupted by political and military strife, but armed conflicts in Ethiopia, Somalia, Sudan, Mozambique and Angola have been particularly devastating.
2. The agricultural potential of different African environments varies considerably, and Figure 7.4 provides a useful introduction. It shows that fourteen countries did not have adequate environmental resources, assuming subsistence farming methods, to achieve self-sufficiency in food for their 1975 population levels. Most of these countries are in the ecologically fragile, semi-arid zones afflicted by occasional severe droughts, and by long-standing deforestation and desertification. On the other hand, there are eleven countries, largely in humid, forested central Africa, that obviously possess extensive areas of under-used land capable of supporting at least five times their 1975 population even with traditional farming methods.
3. Development strategies in much of Africa have assigned a very low priority to agriculture. Despite the great bulk of African populations being engaged in agriculture, central government expenditure (capital and recurrent) on agriculture in 1978–80 in a set of fifteen African countries ranged from a mere 2 per cent of the total budget in Nigeria to just 12 per cent in Ghana (Mellor, 1987, table 9.5). Even that small investment goes largely to export crops.
4. Obsessive state centrism in Africa has been widely cited as the root of appreciable waste, inefficiency, mismanagement and corruption. In a generally fruitless search for scale economies, peasant agriculture in substantial areas has been converted to large mechanized enterprises – whether capitalist, co-operative or state farms – often tied to agro-industry processing. In addition, the monopolistic nature of state marketing boards invariably restricts commodity prices and therefore incentives to farmers. Consumer prices for basic foods are also strictly controlled by most African governments, usually to benefit vocal and politically volatile urban consumers to the detriment of rural producers. Moreover, food aid and subsidized imports of wheat and rice shift consumer preferences away from traditional local food like sorghum, millet and other rough grains, further reducing incentives to local production.
5. With the one exception of hybrid maize in Kenya, Zimbabwe and Zambia, Africa has benefited little from the Green Revolution. Research on improved varieties of traditional African food crops is still at an early stage, while the adoption of high-yielding strains of wheat and rice has been deterred by the need for controlled irrigation systems and for institutional back-up. In addition, external attempts to stimulate improvements in agricultural training, credit and land reform often fail to recognize that it is women who are the real farmers, certainly of food crops, in sub-Saharan Africa.

Distribution

The sequence of activities from harvest to ingestion, involving transport, storage, processing and marketing, receives comparatively little policy attention, yet post-harvest food losses in most Third World countries invariably exceed 10 per cent of production and often reach 30 per cent in Africa (Pariser, 1987). In the midst of his many excessive claims, Paul Ehrlich did convincingly argue that the soundest means, ecologically, of addressing the world's food problem was to tackle the problem of preventable food loss.

Another critical issue in food distribution is the aid and trade in food for deficit countries. The precarious balance between food and population would not be so potentially catastrophic if the huge potential surpluses of North America and western Europe could be made readily available to counter Third World under-nutrition and famine. But the complicated mix of commercial, diplomatic and humanitarian considerations among potential donors has thwarted the development of a system of internationally co-ordinated reserves that was optimistically programmed at the 1974 World Food Conference.

A particular problem has been that donor interests have been dominated by the needs of internal surplus disposal and of external diplomatic leverage. A former US Agriculture Secretary, Earl Butz, has been quoted (*Nation's Business*, June 1973) as saying: 'Food is a weapon. It is now one of the principal tools in our negotiating kit.' Indeed, Tarrant (1980) has shown that US food aid in the mid-1970s was targeted on countries like South Korea, Chile, Egypt, Jordan and Cyprus, not on the basis of objectively assessed need, but because of their importance to American foreign policy. It is also widely believed that Western food aid has been unresponsive to leftist governments even at times of critical shortages (Chile under Allende in 1971, Bangladesh under Mujib in 1973–4, Marxist Ethiopia and Mozambique in the 1980s).

Consumption

Demand factors are now seen to be just as important as supply factors as causes of famine and under-nutrition. Access to resources is always mediated through the social structure, and the plain fact is that regardless of global or national food:population ratios there are always some individuals with inadequate access to food because of their poverty. Consequently 'hunger has increasingly been classified as an injustice rather than a misfortune' (Watkins, and Menken, 1985, p. 665). Indeed, Sen (1981) has shown that some of the world's worst famines have occurred with no significant aggregate decline in food availability per head – Bengal in 1943, Ethiopia in 1973 and Bangladesh in 1974. Similarly, the relatively healthy overall food situation in Latin America (Figure 7.3) masks the inability of pauperized rural and urban proletariats to purchase adequate food.

The policy response that emerged in the 1970s among radical development theorists was an advocacy of 'basic needs' approaches or 'growth with equity', with models being sought in the relatively egalitarian development experience of China, Cuba and Sri Lanka. Even under the free-market philosophies more fashionable among funding bodies like the IMF and World Bank in the 1980s a place generally remains for targeted interventions in food policy. Thus, while general food subsidies are frowned upon, subsidies targeted at low-income groups are approved – for example, the subsidizing of a staple that is consumed particularly by the poor.

What this review of population and food resources has shown is the complexity of interaction between food supply, distribution and consumption, as well as the very different situations that are evident at global, national and household levels. Food is perhaps the most basic human necessity, but it is far from being thought of as a basic human right.

Problems of changing age composition

When age-specific birth rates and death rates are constant over time in a closed population (one without external migration), a constant age structure is achieved and the population grows or declines at a constant rate. Such a population, termed by demographers a stable one, has obvious advantages for planning the allocation of age-specific resources. But when there are rapid changes in fertility and/or mortality, the altered age structure can badly dislocate the resource provision in view of the very different needs of different age groups.

In the poorer, least developed Third World countries, a pattern of rising fertility and falling mortality (Latin America in the 1940s and 1950s, sub-Saharan Africa in the 1970s and 1980s) raises in the short term the already large proportion of population in the dependent 0–14 age group. In 1990 this proportion in most African countries was in the 44–48 per cent range, although in Kenya it exceeded 50 per cent. Immense pressure is exerted by such proportions on government expenditure for education, and there is now clear evidence (World Bank, 1984; Najafizadeh and Mermerick, 1988) that after considerable progress in educational provision in the 1960s, public spending per schoolchild fell in real terms in many less developed countries in the 1970s and 1980s. The other major implication of such youthful demographic structures is the considerable potential

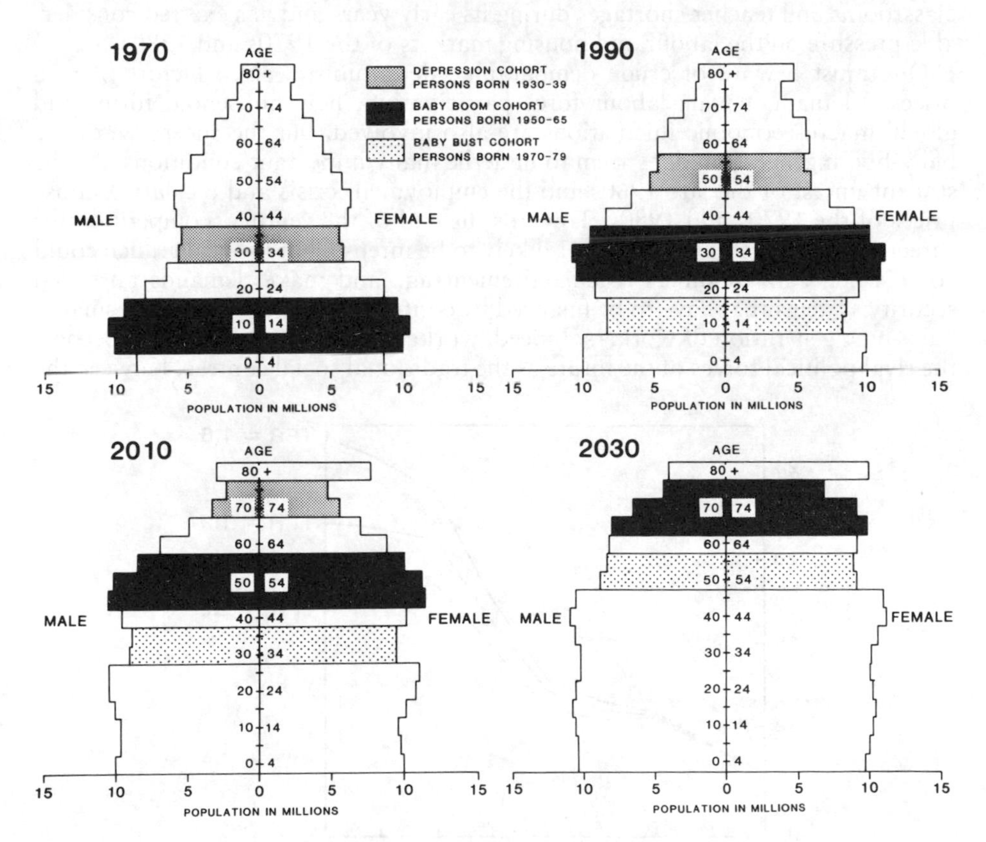

Figure 7.5
Age–sex pyramids, United States, 1970–2030. The projections assume: total fertility rate constant at 2.0; life expectancy at birth rising to 72.8 years for males and 82.9 years for females by 2050; net immigration constant at 750,000 per year
Source: Based on Bouvier (1980), Figure 4.

for overall population growth as the 0–14 cohort proceeds into the child-bearing age groups, regardless of any fall in fertility rates.

In developed countries the impact of changing age structure has been more immediately dramatic because of the volatility of fertility in the last half-century, embracing the low fertility of the 1930s Depression era, the prolonged postwar baby boom and the subsequent baby bust of the 1970s. Mortality trends have had a much lesser impact because changes have been gradual and in one direction. A major planning problem in the West has been, and will continue for some time to be, the provision of resources to meet the needs of the baby-boom cohort, sandwiched as it is between two much smaller, low-birth cohorts (Bouvier, 1980; Eversley and Köllmann, 1982; Ermisch, 1983).

Figure 7.5 shows the projected progression through the life-cycle of the United States baby-boom cohort born between 1950 and 1965 (see Figure 5.9). During this progression, age-specific institutions experience the strain of rapid, costly expansion and, a decade or so later, the often painful task of retrenchment and contraction. Thus the baby-boom cohort has had to cope with overcrowded classrooms and teacher shortages during its early years and has exerted considerable pressure on the labour and housing markets of the 1970s and 1980s.

One must beware of crude demographic determinism, since factors like oil prices, Vietnam, female labour force participation, new household forms and global macro-economic fluctuations are also involved, but the sheer size of the baby-boom generation does seem to underlie many important conditions like the student unrest of the late 1960s and the employment crisis and escalating house prices of the 1970s and 1980s. Towards the end of the century, competition for career advancement in middle age is likely to be intense, but the real crunch could come when baby boomers reach retirement age and make demands on social security systems that have to be financed by contributions from the much smaller baby-bust generation of workers. Indeed, workers and retired could well become the rival political forces of the future as the traditional social contract between the

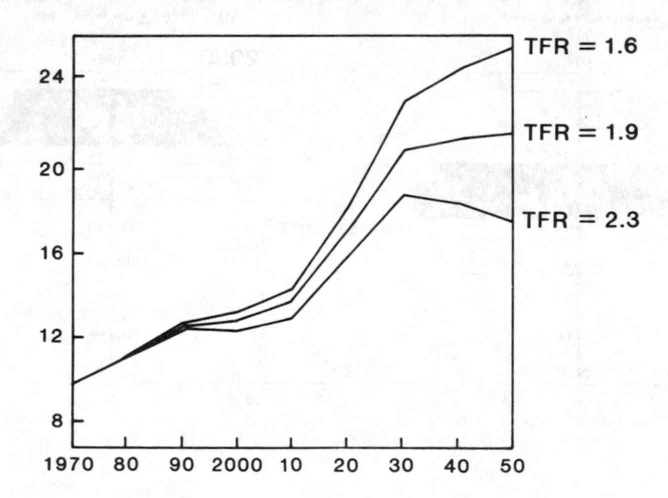

Figure 7.6
Proportion of the US population aged 65 and over, 1970–2050, projected by alternative total fertility rate assumptions and intermediate mortality and migration assumptions
Source: Weaver (1987), Figure 1.

generations is threatened (Johnson, Conrad and Thomson, 1989); this issue is considered further in the concluding chapter on population policies.

Figure 7.6 corrects a prevalent view that the proportion of the elderly will continue to increase steadily and appreciably. In fact, in most developed countries the major surge, particularly by comparison with those in the labour force (Figure 7.7), will not come until the baby-boom cohort reaches retirement age around 2015. However, any complacency is misplaced, because the numbers at the younger and older ends of the elderly range will change very differently in the short term. Those in their 60s will not increase significantly since they will be the survivors of the low-birth cohort of the Depression, but there will be appreciable increases among the super-elderly (80+), who are particularly demanding of state resources for care and support at a time when the family unit seems increasingly ill-adapted to shoulder the burden of the aged (Soldo and Agree, 1988). The trend shown by the US projections in Figure 7.6 is typical of developed countries, although the actual proportions of the elderly are somewhat higher in European populations not subject to American levels of net immigration of younger population. Thus the proportion of population over 65 years in Sweden had reached 17 per cent in 1985 and is projected, under medium-variant assumptions, to reach 23 per cent in 2025. Moreover, migration patterns on retirement produce particularly high proportions of the elderly in coastal areas and scenically attractive rural areas (Warnes and Law, 1984; Rees *et al.*, 1989).

A final effect of the baby boom is in the 'echo' of increased births (despite stable fertility rates) in the 1980s, evident in Figure 7.5 in the widening of the 0–9 cohort in 1990 and the 20–29 cohort in 2010. The subsequent echo in the next generation is also discernible – just – in the profiles for 2010 (0–4 cohort) and 2030 (15–24).

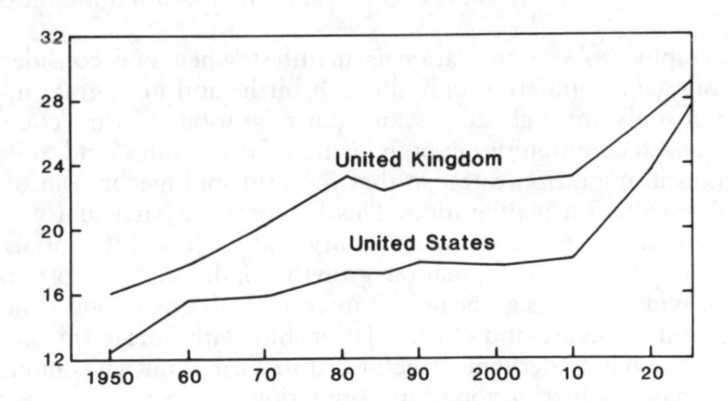

Figure 7.7
Number of persons aged 65 and over per 100 persons aged 15–64 projected by UN intermediate assumptions
Source: Weaver (1987), Figure 2.

8

THE ANALYSIS OF MIGRATION

Who moves? Where to and from? How many? Why? And with what consequences for the areas of origin and destination? These are the questions that have sustained a huge literature in the social sciences on migration; for recent surveys by geographers, see Clark (1982), Lewis (1982) and Cadwallader (1986). Convention distinguishes migration within countries (internal migration) from migration between countries (international migration). This distinction, which is observed in the arrangement of the following chapters, is not simply one of scale. More fundamental is the extent of government intervention in the migration process, which has been appreciable in the case of international migration in this century.

The demographic role of migration is manifest when one considers that a community can gain population only through births and in-migration, and can lose population only through deaths and out-migration. Within countries the differences between communities in both birth and death rates tend to be smaller than differences in migration rates, so that the principal mechanism of internal population redistribution is migration. This has become particularly evident in recent decades with the reduction of mortality and fertility differentials and the establishment of quasi-zero population growth conditions in most developed countries. The wider social significance of migration (discussed in the next chapter) derives from its cause-and-effect relationship with social transformation. Every society that has undergone structural transformation has simultaneously experienced a major redistribution of its population.

Disciplinary approaches

The major distinctions between migration studies are in relation to scale, theoretical perspectives and analytical methods. Such distinctions often have a disciplinary basis.

In formal demography there has been remarkably little interest in migration, reflecting its traditional lack of concern for spatial, as opposed to temporal, perspectives. Even in the field of spatial demographic accounting, where migration is considered as part of a system of stocks and flows, demographers have stood back in favour of planners, regional scientists and geographers (Rogers, 1968; Alonso, 1973; Rees and Wilson, 1977; Congdon and Batey, 1989). The approach of economists to migration is typically macro-analytical, using aggregate data to consider migration as an adjustment to labour market mechanisms. Sociologists and social anthropologists have relied more on field-survey, community-based data and have focused their interest on migration differentials, motivation of migrants, the relationship between physical and social mobility, and the adaptation of migrants to host societies.

Migration, as a spatial reallocation of human resources, is of central interest to the spatially orientated discipline of geography, particularly with the modern emphasis in human geography on spatial processes and spatial interaction. Among the demographic components of change, it is certainly migration rather than mortality or fertility that has attracted most analytical attention by population geographers. Earlier, cartographically based, descriptive work on the spatial structure of migration flows has given way to the interpretation of such patterns by statistical analyses of co-varying factors and to the behavioural investigation particularly of intra-urban movement, where the human geographer's familiarity with small-area data as well as survey work has been helpful. Increasingly, however, there has been disciplinary convergence in migration research, so that it now represents a truly interdisciplinary field of investigation, indicative of modern holistic tendencies in the social sciences.

What is migration? Temporal and spatial dimensions

Of the three major components of population change, migration is the most difficult to conceptualize and measure. The definition of birth and death, at least for statistical, if not ethical, purposes, is clear cut, but migration is 'a physical and social transaction, not just an unequivocal biological event' (Zelinsky, 1971, p. 233). The fact that a migrant is a person who travels 'is the only unambiguous element in the entire subject' (Barclay, 1958, p. 243).

The term 'mobility' is the most general concept in the field. *Spatial mobility*, embracing all sorts of territorial movements, should be distinguished from *social mobility*, a term extensively used by sociologists for changes in socio-economic status. But all forms of spatial mobility cannot be regarded as migration. Of the four categories shown in a simple typology of spatial mobility (Table 8.1), only category D would universally be accepted as migration. Category A largely comprises commuters, involving no change in residence. Category B includes the movement of seasonal or temporary workers, some seasonally nomadic pastoral groups, and also students moving termly between family home and college; such movements are often designated as *circulation*, which covers 'a great variety of movements, usually short-term, repetitive, or cyclical in character, but all having

Table 8.1 A classification of spatial movements of population

	Recurrent	Non-recurrent
Local	A	C
Extra-local	B	D

Source: Duncan (1959), p. 699. For a more elaborate typology, see Gould and Prothero (1975).

in common the lack of any declared intention of a permanent or long-lasting change in residence' (Zelinsky, 1971, p. 226). Category C embraces, above all, intra-urban residential relocation, the dominant category of residential movement within developed countries like Britain, where the 1981 census indicated that 69 per cent of those moving residence within Great Britain in 1980–1 had moved less than six miles.

Whether or not intra-urban residential movement should be regarded as migration is a moot point. There are those like Lee (1966, p. 49) who define migration broadly 'as a permanent or semi-permanent change of residence. No restriction is placed upon the distance of the move.' More typically a distinction is drawn between local movers and migrants by the erection of a migration-defining distance or boundary. Thus one analyst reserves the term migration

> for those changes of residence that involve a complete change and readjustment of the community affiliations of the individual. In the process of changing his community of residence, the migrant tends simultaneously to change his employers, friends, neighbours, parish membership, and many other social and economic ties. The local mover, by contrast, may simply move across the street to a house a few blocks away. Very likely he retains his same job, breaks no community ties, and maintains most of his informal social relationships.
>
> (Bogue, 1959, p. 489)

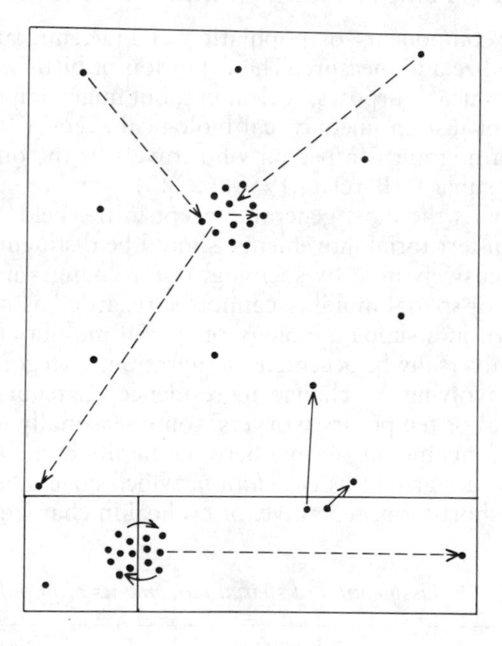

Figure 8.1
The effect of areal size, shape and population distribution on the classification of spatial mobility as migration (continuous line) or local movement (broken line)

Table 8.2 Moves of residence within country per 1,000 population by distance of move: United States, Great Britain and Sweden

Distance moved	United States 1975–6	Great Britain 1980–1	Sweden 1974
Under 50 km	125	75	*
Over 50 km	46	15	24

* Not available.
Source: Long, Tucker and Urton (1988), table 1.

Many analysts have found this distinction intuitively attractive, although it does raise thorny questions about the nature, areal extent and hierarchical ordering of communities.

The crossing of a civil boundary of some kind is a crude means, and often in terms of data availability the only means, of distinguishing a migrant from a local mover. In many countries there are three levels of territorial unit whose boundaries may be used: large provinces or regions, intermediate counties or districts, and minor townships, parishes, wards or communes. In most cases the migration-defining boundaries adopted are those of the intermediate units – a compromise between too gross and too detailed extremes. It is essential, therefore, that in any study of migration care should be taken to define the precise nature of spatial mobility that is to be classified as migration, with a clear specification of both temporal and spatial criteria.

Geographers have emphasized that migration numbers and rates are strongly influenced by the size, shape and internal population distribution of the areal units employed in migration analysis (Kulldorff, 1955; Willis, 1972; White and Mueser, 1988). The larger and more compact an area, the fewer will be its migrants and the greater its internal movers, other things being equal. A concentration of population around the centre of the unit has a similar effect (Figure 8.1). The point is significant because the areas commonly adopted for migration tabulations vary widely in size, shape and population distribution. In the United States, for example, the smallest county is a mere 5 square kilometres while the largest is the size of the Netherlands. This is why Long, Tucker and Urton (1988) used 'distance moved' as the most objective indicator of migration for an international comparative study. Some of their results (Table 8.2) suggest that the rate of migration (regarded as movement over 50 kilometres) in the United States is about three times the British rate and twice the Swedish rate, confirming the American reputation for high mobility.

Data sources

'Just as in physical and biological sciences theory cannot develop without experimentation, in social sciences there can be no development without a systematic observation of social events' (Elizaga, 1972, p. 127). 'In any society, knowledge and understanding of migration patterns is largely determined by the quality and detail of the data available and, only after that, by the precision of the analytical approach and the insight of the theoretical concepts employed' (Woods, 1979, p. 196). Therein lies the justification for a fairly extensive review of data sources on migration.

Population registers

The most detailed, yet comprehensive, migration data are usually provided in those countries which maintain continuously updated records on each member of the resident population; they include the Scandinavian countries, West Germany, the Netherlands, Japan and most eastern European countries (UN Dept. of Economic and Social Affairs, 1970; Verhoef and van de Kaa, 1987). The recording at a local level of demographic events including migration may be a primary purpose of some registers, as in Sweden (Hofsten, 1966), or may simply be a by-product of registration for security and a range of administrative reasons, as in Japan (Kono, 1971).

The spatial organization and tabulation of register data vary. Some files are arranged by place of birth, most by place of residence, but others by place of residence of family head. Only in some countries are the data from local registers aggregated regionally, so that

> the real difficulty in developing adequate models [of migration] from population register data may well end up being the overabundance rather than the lack of data. Disaggregation is not a totally unmixed blessing since it creates more and more statistics, and the need for some summary organization becomes imperative.
>
> (Clark and Moore, 1978, p. 274)

But the outstanding merit of population registers is that they generally record all migrations, even of chronic movers, as well as some socio-economic characteristics of migrants. The richness of register data may be illustrated by Ginsberg's (1978) study of time intervals between residential moves in Norway. His data file covered all men aged 16–67 in 1971 resident in Norway throughout the period 1965–71. The month, year, regional origin and regional destination of each move within this population could be recorded, as well as the age, marital status, income, wealth and employment of the mover in 1970.

Social researchers have often made pleas for the introduction of population registers in countries like Britain where migration data have been notoriously deficient. Strong opposition has focused on possible infringements of individual liberties, so that it was only under the exceptional circumstances of the Second World War and its immediate aftermath that a national register operated in Britain.

Population censuses

A national census is the only other source of migration data covering the whole population, although unlike a continuous registration system it is tied to particular dates and can provide only a 'snapshot' of an essentially dynamic phenomenon. Several forms of migration data may be provided by, or calculated from, national censuses (Rees, 1977).

Net migration

The balance between in- and out-migration can be estimated by indirect, residual methods for administrative areas which are areally stable between censuses.

By the so-called vital statistics method, net migration during an inter-censal period may be obtained as a balance between total population change and natural increase:

$$M_n = (P_{t+n} - P_t) - (B_n - D_n)$$

where M is the net migration during the inter-censal period n; P_t and P_{t+n} are the total populations at two successive censuses; and B_n, D_n are the total births and deaths (corrected for place of usual residence) during the inter-censal period. This method can provide only an estimate of net migration, because registration systems do not distinguish births and deaths in the original population from those among migrants. More critically, an emphasis on net movement can give a misleading impression, since the balance between a migration stream and its counterstream is invariably small. There are no such beings as 'net migrants'; rather, there are people who arrive at places and people who leave them, often almost

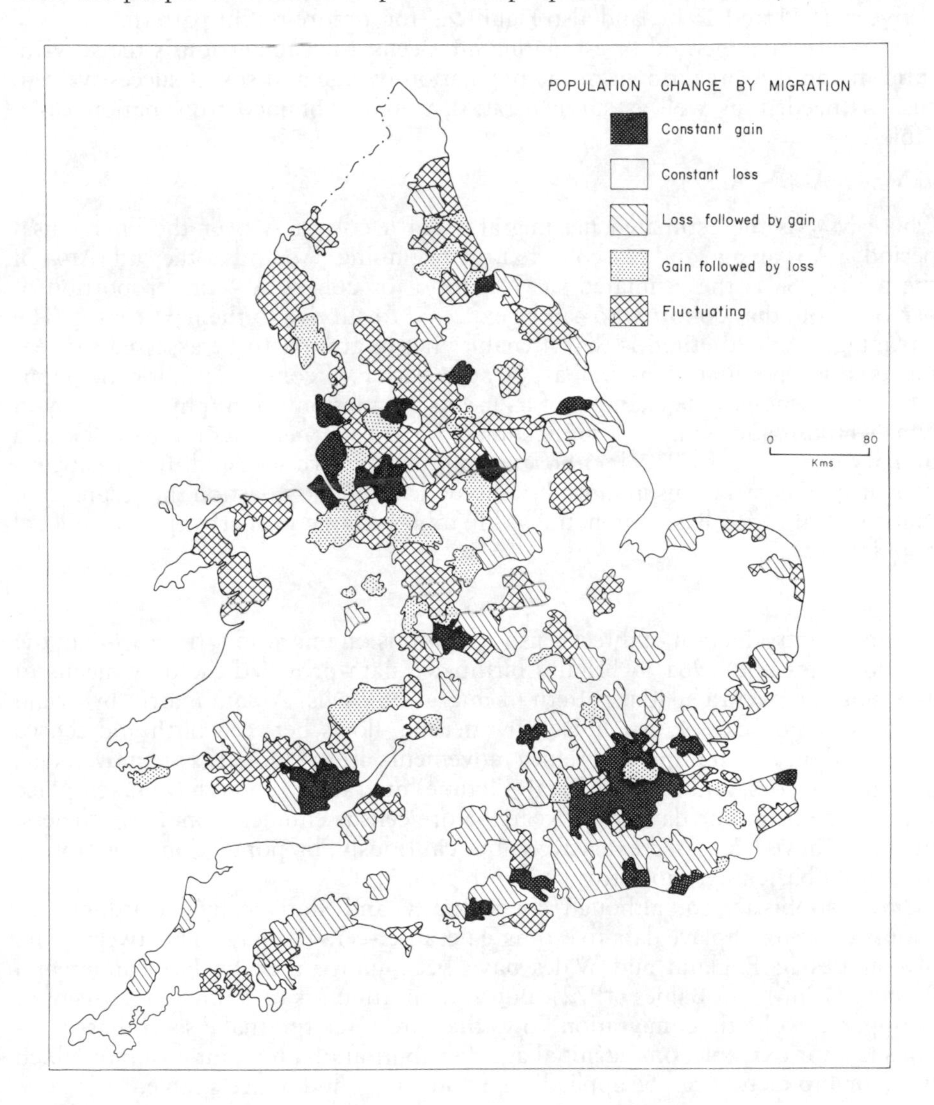

Figure 8.2
Migration trends, 1851–1911, by registration districts. Five types of trend are derived from the sequence of six net inter-censal migrations
Source: Lawton (1968), Figure 6.

cancelling each other out in aggregate terms, although not always in characteristics like age, sex and occupation; for example, a small net migration change in the whole population may conceal a large net loss among the younger working population and a considerable net gain among the retired. Nevertheless, in the absence of more detailed data, net migration estimates do provide valuable insights into the nature of population growth and decline, and can give some indication of the comparative 'attractiveness' of areas at particular periods. Consider, for example, Figure 8.2, one of a series of maps used by Lawton (1968) to demonstrate the spatial dimension of population growth in England and Wales between 1851 and 1911, and also Figure 9.4 for a more recent pattern.

An alternative method of estimating inter-censal net migration is the survival ratio method. A breakdown of the population by age and sex at successive censuses is needed, as well as survival ratios, usually obtained from national life-tables:

$$M_{A_n} = A_{t+n} - (A_t . S_A)$$

where M_{A_n} is the estimated net migration of a cohort A over the inter-censal period n; A is an age and sex cohort at the beginning (A_t) and at the end (A_{t+n}) of the period; S_A is the estimated survival ratio for cohort A – the proportion of persons from that cohort who can be expected to survive to the next census. The advantage of the method is that it enables net migration to be assessed for segments of the population as well as, by aggregation of cohort data, for the population as a whole. One example of its use in population geography is the way in which House and Knight (1965) estimated 1951–61 net migration by sex and quinary age group for local authorities in north-eastern England. But an awareness of spatial variations in mortality should lead one to question the assumption that survival ratios based on national life-tables can be properly applied to local populations.

Birthplace data

Before the introduction of direct census questions on migration (from 1940 in the United States and 1961 in Britain) birthplace data provided the only means of assessing the overall spatial pattern of migration flows. A data matrix by origin and destination can be assembled for 'lifetime' flows between birth and census enumeration. But multiple and return movements are excluded, as are movements of non-survivors, and the individual 'lifetime' migrations could have taken place at any time from one day to 100 years before census enumeration. Nevertheless, such data have been used extensively, but cautiously, by population geographers to identify basic migration streams (e.g. Figure 9.1).

More sophisticated, although conceptually and statistically hazardous, attempts to use birthplace data to estimate net inter-censal migration between pairs of counties in England and Wales have been undertaken by Friedlander and Roshier (1966) and Baines (1972). But in their attempts to estimate inter-censal, as opposed to lifetime, migration flows they are forced to make several assumptions (as, for example, on the initial age distribution of a migrant group to which survivorship factors can be applied) that some may find unacceptable.

Enumerators' returns

The manuscript returns of census enumerators can yield a rich harvest of migration data to researchers prepared to engage in the laborious task of data extraction and collation (Lawton, 1978), although confidentiality restrictions normally prevent access to more recent censuses – in the case of Britain to those later than 1891. It is, therefore, the censuses of the nineteenth century that have seen most research activity on migration at a detailed local scale. One approach is represented by the way in which Pooley (1977), using a 10 per cent systematic sample of households in Liverpool in 1871, was able to display the different distributions in the city of persons born in Ireland, Scotland, Wales and the rest of England; clear relationships could be drawn with the city's social topography.

Another application has been the use of birthplace information from households with children to detect the presence of stepped migration. For example, Gwynne and Sill (1976) note the birthplace of children in some 200 families

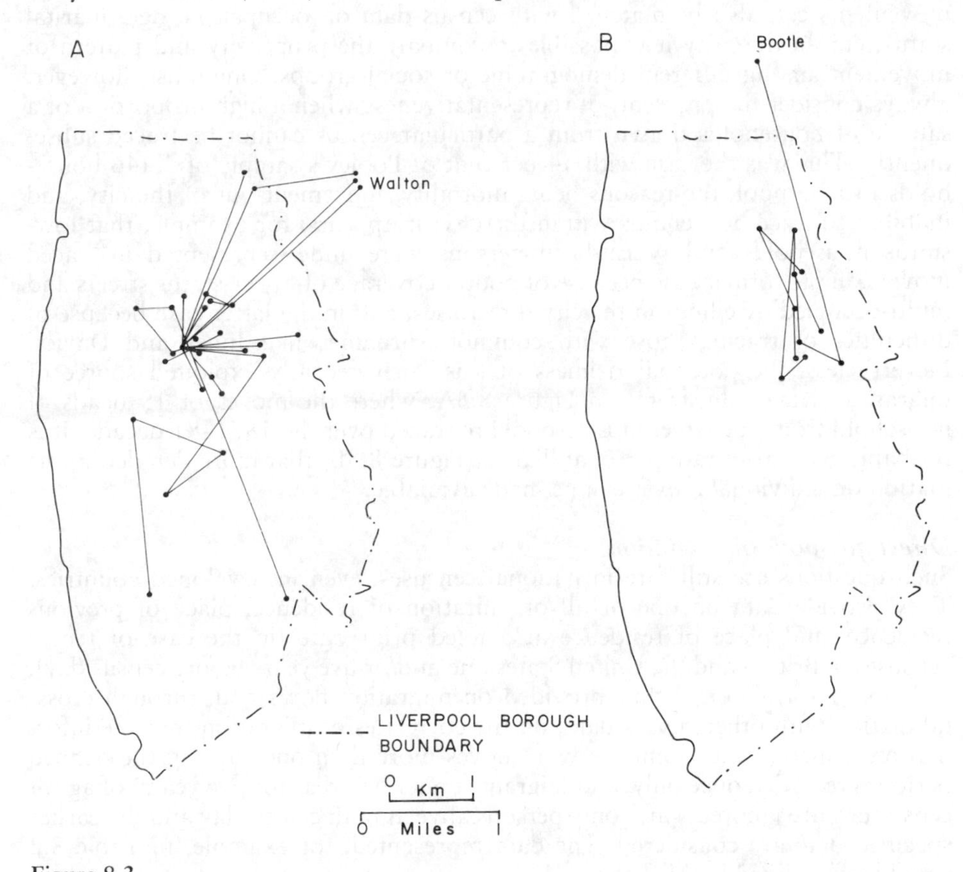

Figure 8.3
A: Individual moves from Virgil Street, Liverpool, 1871–81
Although some evidence is provided of suburban movement from an inner-city working-class area, the major feature is the short distance over which most moves ocurred, 78 per cent being under 1.6 km
B: Residential mobility of David Brindley, Liverpool, 1882–90, as recorded in his diary. This further illustrates the closely circumscribed area within which working-class people moved in Victorian cities. The diary records Brindley's community attachment to that part of Liverpool within which he moved, largely, it seems, in association with life-cycle stages. All his homes were rented, facilitating frequent movement
Source: Pooley (1979), Figure 5.

resident in Middlesbrough in 1861 where the father had been born in Wales; clear evidence emerges of stepped migration among adults from South Wales to Middlesbrough via other metallurgical districts in England.

More ambitious have been attempts to monitor residential mobility by using the techniques of historic record linkage (Wrigley, 1973) to trace individuals between successive censuses by means of city directories, rate books and the like. North American studies include those on Boston (Knights, 1971) and Hamilton (Katz, 1976), while comparable work in Britain embraces Leicester (Pritchard, 1976), Huddersfield (Dennis, 1977) and Liverpool (Lawton and Pooley, 1976; Pooley, 1979). Not only is it possible to trace the spatial and temporal pattern of residential movement, at least within the community under study, but individual movements can also be matched with census data on occupation, age, marital status, etc. In this way it is possible to compare the propensity and pattern of movement among different demographic or social groups. One must, however, always consider the problems of representativeness when a high proportion of a sample of households drawn from a particular census cannot be traced subsequently. This was the case with 54 per cent of Pooley's sample of 2,446 households in Liverpool, the reasons being mortality, movement out of the city, and inability to trace households within the city; it appears, for example, that low-status households and Welsh-born persons were under-represented in traced movers, in the former case because of poorer coverage of lower-status streets and multi-occupied dwellings in the city directories, and in the latter case because of difficulties of tracing those with common surnames like Jones and Davies. Nevertheless, the potential richness of this fairly recently exploited source of migration data is illustrated in Figure 8.3A, where the moves of 19 heads of household from one street in Liverpool are traced over the 1871–81 decade. It is probably only from rare personal diaries (Figure 8.3B) that more detailed information on individual moves can be made available.

Direct questions on migration

Such questions are still rare in national censuses, even in developed countries. They provide data on one or all of: duration of residence, place of previous residence, and place of residence at a fixed prior date (in the case of recent censuses in Britain and the United States one and/or five years before census day). Valuable information is thus provided on migration flows and, through cross-tabulation with other census data, on the correlates of migration, but the information is incomplete. Someone who moves more than once during the defined period is recorded once only, and migrants under one year (or five years) of age at census day are ignored, since only persons alive at both census day and the earlier specified date are considered. The data represented, for example, in Table 8.2 should be qualified in this light.

Longitudinal study

In a very few national census systems it has been possible to trace individuals at successive censuses. In Britain, the Longitudinal Study of the Office of Population Censuses and Surveys has matched the census details of some 500,000 individuals at the 1971 and 1981 censuses. The use of these data in social medicine research has been considered in Chapter 4, but there are also applications in migration research, well demonstrated by the work of Fielding (1989) on the inter-censal relocation of individuals in both class structure and geographical space.

A few analysts have set up their own monitoring systems of migration-type flows into and out of communities (e.g. Chapman, 1975, in the Solomon Islands) but inevitably these are confined to small areas and limited periods of time.

Partial registers

In most countries there are quasi-registers of population which provide direct or indirect data on migration for considerable segments of the population. Such registers include electoral lists, social or health insurance registers, local authority housing lists, public utility files, tax registers, school rolls and employee records, although not all are freely accessible to researchers. Two examples illustrate their use in migration studies.

In Britain the National Health Service covers virtually the whole population, and whenever anyone changes his or her doctor, this is recorded, together with any associated change of address. If such a movement transcends the boundary of an area health authority (in 1989 there were 98 family practitioner committees in England and Wales and 15 area health boards in Scotland), a record is kept on a central register. In the 1980s this became the major source of official data on

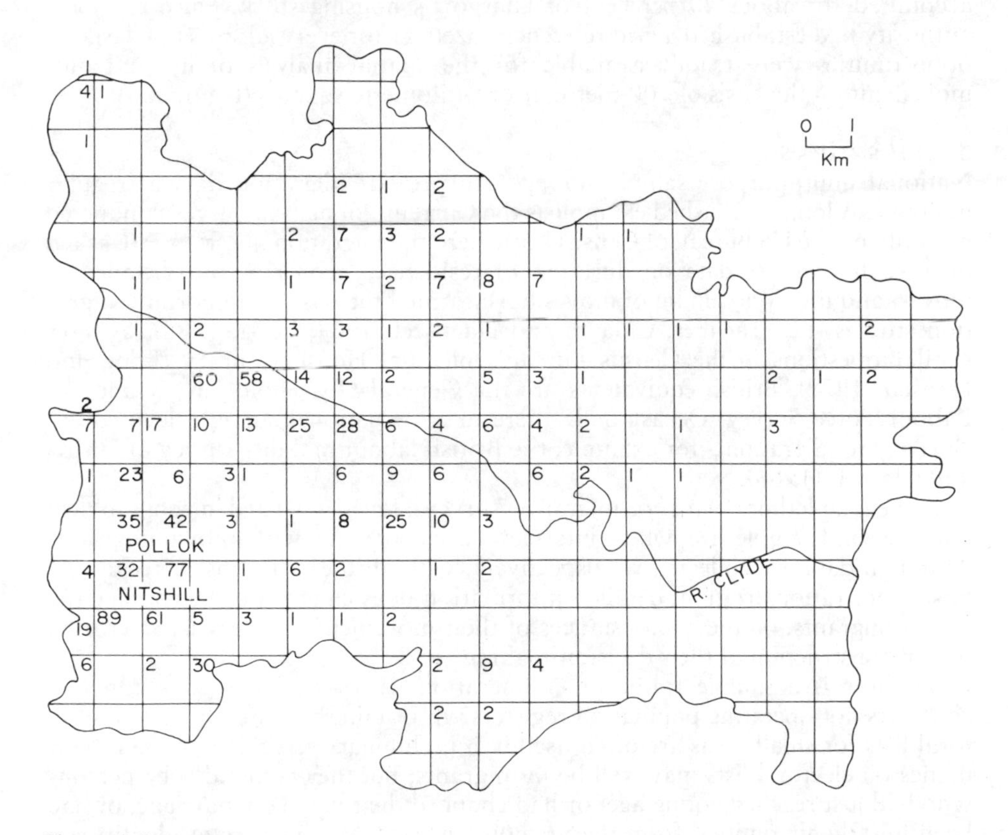

Figure 8.4
Migration field within Glasgow of in-migration households to the Pollock and Nitshill public housing estate during 1974. The number of households originating in each square kilometre is shown
Source: Forbes, Lamont and Robertson (1979), Figure 31.

internal migration in England and Wales, although it had long been used extensively in Scotland (Hollingsworth, 1970; H. Jones, 1970). It is more difficult, but not impossible (e.g. Rees and Rees, 1977), to access registers held at area level to monitor intra-area flows. Comprehensive as the NHS registers are, they do not cover the small numbers of *exclusively* private patients, and more important, they can identify only those changes of address which accompany a change of doctor, so that short-distance moves will be considerably underenumerated. Researchers must also consider problems posed by the variable time-lag between migration and registration with a new doctor.

A study of intra-urban migration in Glasgow (Forbes, Lamont and Robertson, 1979) illustrates the use of housing registers in migration studies. About 60 per cent of the city's housing is rented within the public sector, where centralized housing files provide information, including previous address, on all tenants. This information was assembled for all movers into public housing in 1974. Similar information was collected for moving owner-occupiers, in this case from the Register of Sasines (National Land Register, Scotland). The only moving households excluded from the study were those in the private rented sector, which then accounted for about 20 per cent of Glasgow's housing stock. Since the local authority had established a grid reference gazetteer for every address in Glasgow, opportunities were readily available for the spatial analysis of in- and out-movement on the basis of 100-metre or one-kilometre squares (Figure 8.4).

Social surveys

National multi-purpose sample surveys can provide migration data at regular intervals. A long-established example is the Current Population Survey conducted monthly by the US Bureau of Census but where the migration questions are asked once each year. Its data on duration of residence, frequency and distance of moves, and life-cycle timing of moves have formed the basis of important migration studies (e.g. Taeuber, Chiazze and Haenszel, 1968; Long, 1988), as have similar questions in the Census Bureau's biennial Housing Survey (Long and Hansen, 1979). British equivalents are the General Household Survey and the Labour Force Survey. Occasionally there may be national surveys devoted exclusively to migration – for example, the British labour mobility survey of Harris and Clausen (1966).

There is a plethora of *ad hoc* migration surveys, fairly restricted in geographical coverage and sample size, where interviews are conducted with known migrants. Although data are collected retrospectively, with all the problems of recall and post-event rationalization, detailed information does emerge on the characteristics of migrants, on the circumstances of their movement and, less satisfactorily, on a reconstruction of their decision-making.

But how is a sample frame or enumeration list of migrants constituted in countries not operating population registers? In Britain the annually revised electoral lists for small areas are often used in a preliminary screening process. New names on electoral lists may well be in-migrants; but they could also be persons who had just reached voting age, or had changed their name on marriage, or had been mistakenly omitted from the previous register. Much easier to identify are very particular migrant groups: coalminers (R. Taylor, 1969), farmers (Nalson, 1968; Sublett, 1975), managers (House, 1968), university graduates (Johnston, 1989), new town residents (H. Jones, 1976), and employees of relocated firms (Mann, 1973).

Macro-analytical models

A significant division exists between, on the one hand, migration models derived from social physics which interpret aggregate behaviour as the outcome of impersonal macroscopic laws and, on the other hand, micro-analytical perspectives which examine individual migrant behaviour as the expression of decision-making which need not be economically or spatially rational.

A common theme of macro-analytical migration models is the search for regularities which are capable of mathematical expression. The models generally have an ecological basis in that migration is measured between areas and is thought to reflect the comparative attraction powers of potential origins and destinations; explanation is thus based on the environmental and community context of migration flows. Yet the emphasis is on universal explanation through principles which are independent of specific situations or embrace all of them.

The classical gravity model

The first formal statement of the critical role of population size and distance in fashioning migration patterns was made by Ravenstein (1885) in his 'laws' or generalizations of migration which he derived inductively from an analysis, using census birthplace data, of inter-county movements within Britain in the nineteenth century:

1. . . . the great body of our migrants only proceed a short distance and . . . there takes place consequently a universal shifting or displacement of the population, which produces 'currents of migration' setting in the direction of the great centres of commerce and industry which absorb the migrants . . .
2. It is the natural outcome of this movement of migration, limited in range, but universal throughout the country, that the process of absorption would go on in the following manner:

   The inhabitants of the country immediately surrounding a town of rapid growth, flock into it; the gaps thus left in the rural population are filled up by migrants from more remote districts, until the attractive force of one of our rapidly growing cities makes its influence felt, step by step, to the most remote corner of the kingdom. Migrants enumerated in a certain centre of absorption will consequently grow less with the distance proportionately to the native population which furnishes them . . .
3. The process of dispersion is the inverse of that of absorption, and exhibits similar features.
4. Each main current of migration produces a compensating countercurrent.
5. Migrants proceeding long distances generally go by preference to one of the great centres of commere and industry.

(Ravenstein, 1885, pp. 198–9)

In the 1930s and 1940s the roles of population size and distance in determining interaction of all sorts became formalized in what have become known as 'gravity models'. They are based on Newton's Law of Universal Gravitation, which states that two bodies in the universe attract each other in proportion to the product of their masses and inversely as the square of the distance between them. Of course, people are not molecules, but they may be regarded as predictable in their aggregate behaviour on the basis of mathematical probability.

Zipf's P/D hypothesis (1949) is representative of the classic gravity model of social interaction pioneered also by W. J. Reilly, E. C. Young, J. Q. Stewart and W. Warntz (Carrothers, 1956). Zipf regarded the movement of goods, information and

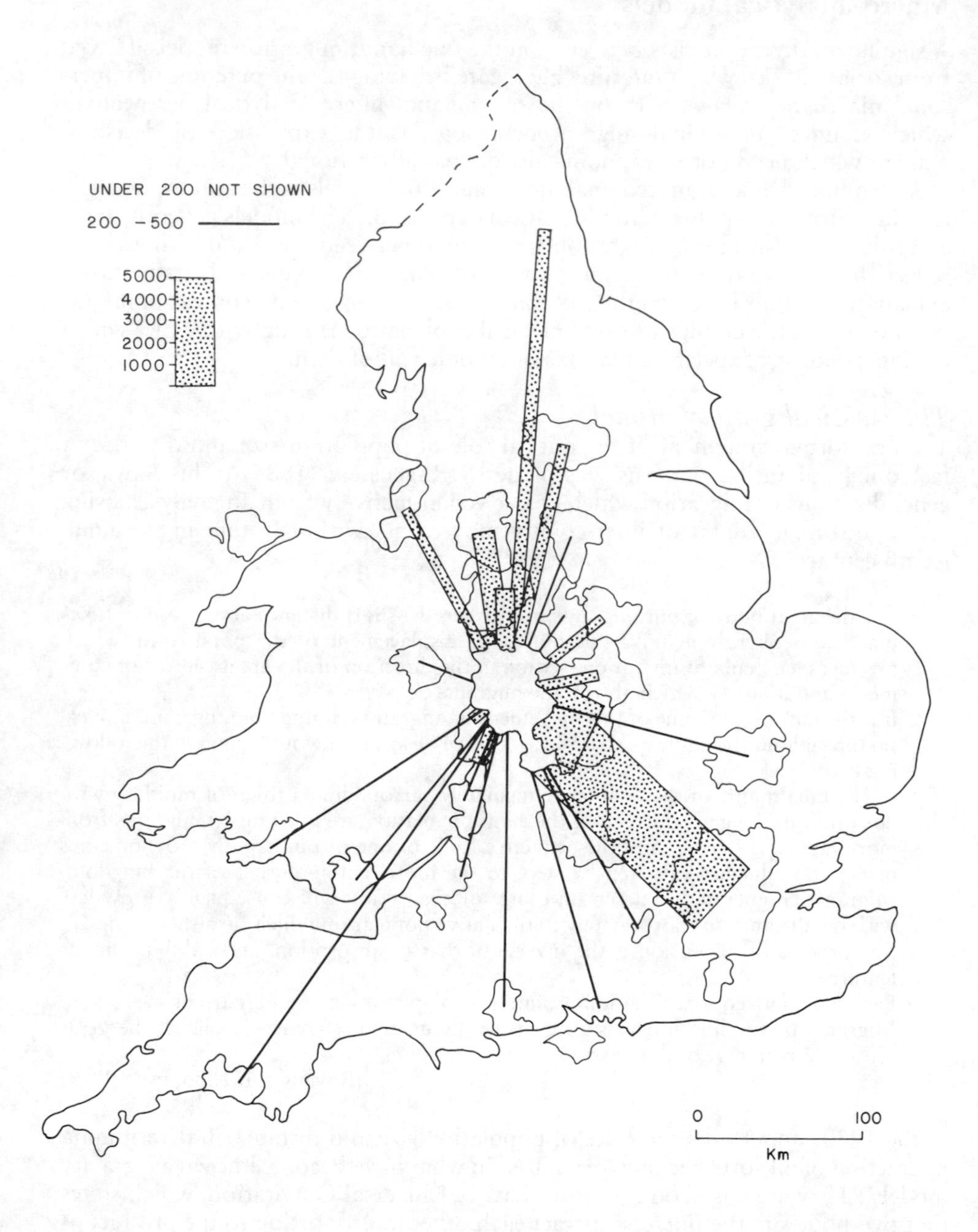

Figure 8.5
In-migration to Birmingham Standard Metropolitan Labour Area from other SMLAs, 1965–6, recorded by 1966 census. Boundaries are shown only for those SMLAs contributing over 200 migrants
Source: Johnson, Salt and Wood (1974), Figure 4.4.

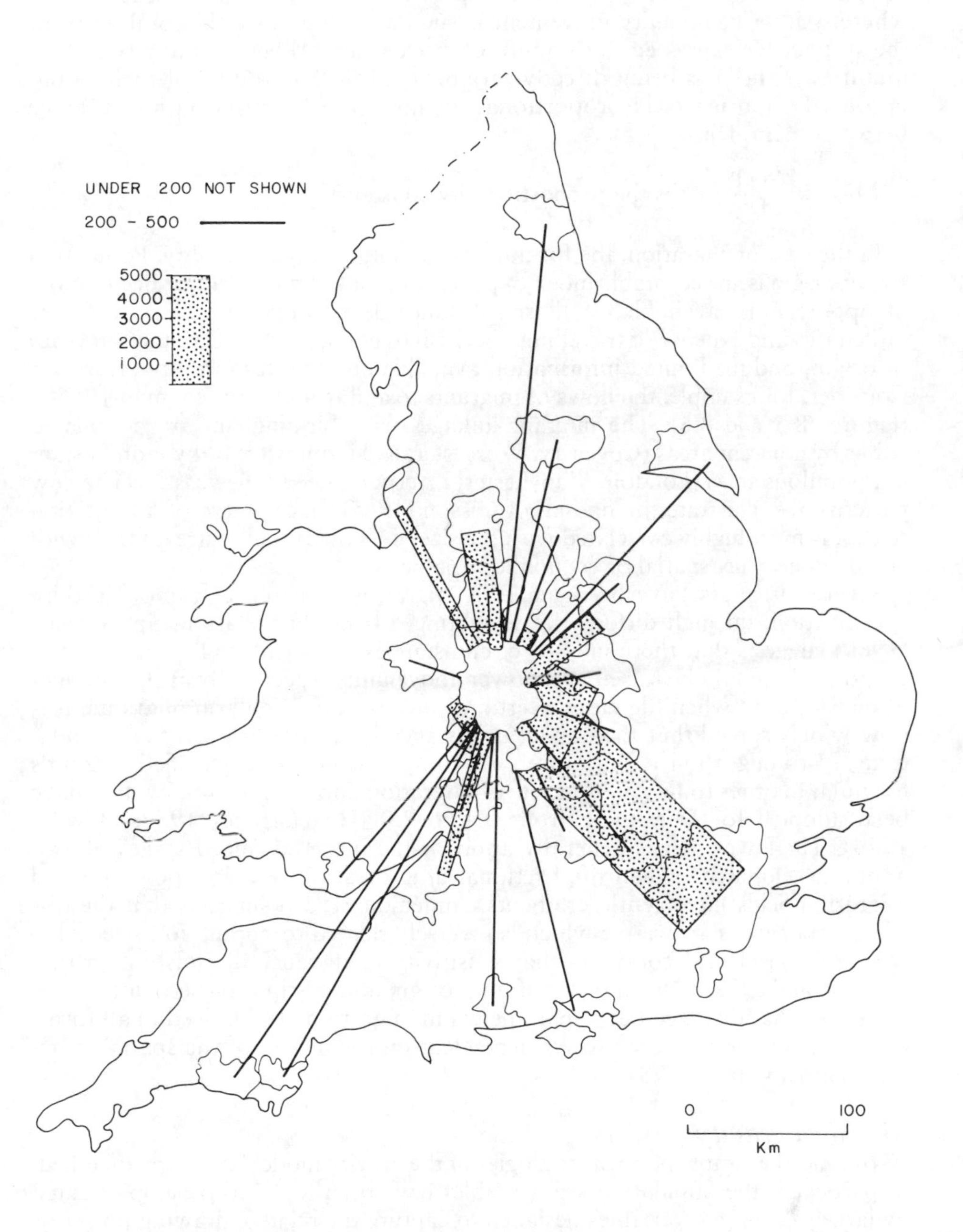

Figure 8.6
Out-migration from Birmingham Standard Metropolitan Labour Area from other SMLAs, 1965–6, recorded by 1966 census. Boundaries are shown only for those SMLAs receiving over 200 migrants
Source: Johnson, Salt and Wood (1974), Figure 4.5.

people within the social system as an expression of his 'principle of least effort', whereby inter-community movement is such as to minimize the total work of the system. He expressed the amount of movement (M) between any two communities (i and j) as being directly proportional to the product of their populations (P) and inversely proportional to the shortest transportation distance between them (D):

$$M_{ij} = k \frac{P_i \cdot P_j}{D_{ij}} \quad \text{(k is the proportionality constant)}$$

In the case of migration, the formula has an intuitive reasonability. Population size at origin is an acceptable index of people anxious to move and, at destination, of opportunities available. Similarly, distance deters migration because of the difficulties and expense of travelling, the wish to maintain social contacts at place of origin, and the limited information available on long-distance opportunities. Consider, for example, the flows of migrants to and from Birmingham in 1965–6 (Figures 8.5 and 8.6). The migrant linkages with Birmingham are dominated either by adjacent areas (Coventry, Worcester, Kidderminster) or by more distant but populous areas (London, Manchester, Liverpool, Leeds, Newcastle). The flow patterns also illustrate the finding of Olsson (1965) that the size of a migration stream is more highly correlated with the size of its own counterstream than with any demographic, spatial or economic variables.

Several attempts have been made in migration research at a more precise specification, through differential weighting, of the P/D relationship. Stewart (1960) suggests that there may be circumstances in which the P component in the formula should be raised to a power above unity, specifically at those stages of development when big cities exert a disproportionate pull on migrants. It is now widely agreed that the impact of distance is not uniform, so that its relationship to migration is only rarely the simple inverse one specified in Zipf's formula. In order to fit different sets of migration data, exponential values have been adopted for D ranging from 0.4 to 3.0 (Hagerstrand, 1957; Haynes, 1974). The lower values generally apply at the more advanced stages of economic development, when the frictional drag of distance is less powerful and migration fields have gentle gradients. Anderson (1955) suggests that the distance exponent is a variable which is inversely related to population size. Likewise, Olsson (1965) considers that sensitivity to distance diminishes with an increase in the hierarchical order of both origin and destination settlement. It is clear then that distance-decay parameters in migration (and indeed in all forms of spatial interaction) are always dependent on the nature of the spatial structure (Fotheringham, 1981).

Modified gravity models

With time, the simple physical analogies of the gravity model have been qualified. In particular, the population size variables have been replaced or supplemented by largely economic variables designed to capture the relative drawing power or comparative advantage of areas (Lowry, 1966). A typical example is the expanded gravity model used by Rogers (1967) in his 'successful' analysis of intercounty migration in California ($r^2 = 0.92$).

$$M_{ij} = k \left[\frac{U_i}{U_j} \cdot \frac{WS_j}{WS_i} \cdot \frac{LF_i \cdot LF_j}{D_{ij}} \right]$$

where M_{ij} = number of migrants from i to j
U = civilian unemployment rate
LF = labour force eligibles (i.e. of working age)
WS = per capita income
D_{ij} = highway distance between i and j.

Another variation (Hagerstrand, 1957) adopts as a functional measure of population size the number of personal contacts between areas of origin and destination at the beginning of the migration period, thereby recognizing the critical importance of friends and relatives in lubricating migration flows through the provision of information, initial housing and the like. A common proxy for number of personal contacts is the amount of previous migration between particular areas, but this raises the charge that a good fit in the model may simply reflect constancy in migration-inducing conditions, rather than any functional role of 'the friends and relatives multiplier'.

Intervening opportunity models

An American social psychologist, Stouffer, introduced his influential 'intervening opportunities' hypothesis in 1940. He argued that linear distance was less important a determinant of migration patterns than the nature of space; that distance should be regarded in socio-economic rather than geometric terms; and that because migration is costly, socially as well as financially, a mobile person will cease to move on encountering an appropriate opportunity.

His basic hypothesis was that 'the number of persons going a given distance is directly proportional to the number of opportunities at that distance and inversely proportional to the number of intervening opportunities' (Stouffer, 1940, p. 846). This may be expressed in a formula – using the version of Strodbeck (1949).

$$y = k \frac{\Delta x}{x}$$

where y is the expected number of migrants from a place to a particular concentric zone or distance band around that place, Δx is the number of opportunities within this band, and x is the number of opportunities intervening between origin and midway into the band in question.

An operational example is given in Table 8.3, based on detailed data from the Swedish population register. The formula is used to distribute between zones the known total number of out-migrants (801) from one parish. The measure of opportunities, in-migration from all origins during the period under study, corresponds with measures adopted in several other tests of Stouffer's formula (Bright and Thomas, 1941; Isbell, 1944; Strodbeck, 1949). The actual opportunities to prospective migrants could be housing, employment or environmental; they are the kind that have attracted migrants and are measured by them. This is a neat way of avoiding a direct specification of opportunities, which would never be comprehensive, but it does pose conceptual and technical problems. Circularity is present when migration from the particular origin under study (Vittsjö in Table 8.3) comprises part of the measured opportunities in surrounding bands (Table 8.3 does, however, indicate the very limited double-counting involved in this instance). Another problem is that in-migration represents only opportunities or vacancies filled, so that in areas of economic buoyancy and employment expansion it will invariably underestimate opportunities available.

Table 8.3 Stouffer's intervening opportunity theory applied to out-migration from the parish of Vittsjö, Sweden 1946–50

Zone (measured by distance of parish centres from Vittsjö)	Total in-migration per zone Δx	intervening x	$\frac{\Delta x}{x}$	Expected number or migrants from Vittsjö $k\frac{\Delta x}{x}$	Observed number of migrants from Vittsjö O
1. Adjacent parishes	4,104	2,052	2.00	251.5	167
2. 10–20 km	1,636	4,922	0.33	41.5	33
3. 20–30 km	24,156	17,818	1.36	171.0	203
4. 30–40 km	19,160	39,476	0.49	61.6	43
5. 40–50 km	35,596	66,854	0.53	66.7	58
6. 50–60 km	48,549	108,927	0.45	56.6	69
7. 60–70 km	82,141	174,272	0.47	59.1	96
8. 70–80 km	55,719	243,202	0.23	28.9	41
9. 80–90 km	26,849	284,486	0.09	11.3	11
10. 90–100 km	158,803	377,312	0.42	52.8	80
(The proportionality constant $k = \frac{\Sigma O}{\Sigma\frac{\Delta x}{x}}$)			Σ6.37	Σ801.0	Σ801

An example can illustrate how intervening in-migrations (opportunities) are calculated. For zone 3, the number is derived by summing Δx (zone 1), Δx (zone 2) and ½ Δx (zone 3).
Source: Hagerstrand (1957), table 5.3.

In 1960 Stouffer refined his 'intervening opportunities' model. He had been concerned with the operational inflexibility of the original model in that it could cope only with migration flows from a given centre to surrounding distance bands. He also came to realize that the take-up of opportunities in place B by inhabitants of place A through migration is inversely proportional not only to the opportunities intervening between A and B but also, as a further recognition of the variability of space, to the number of competing migrants from elsewhere. His refined formula takes the following form (using the version of Galle and Taeuber, 1966):

$$y = k\,\frac{X_O \cdot X_I}{X_B \cdot X_C}$$

where, during a particular period

y = the number of migrants from city 1 to city 2
X_O = all out-migrants from city 1
X_I = opportunities in city 2, measured by total in-migrants
X_B = opportunities intervening between cities 1 and 2, measured by total in-migrants to a circle having as its diameter the distance from city 2 to city 1 (Figure 8.7)
X_C = migrants potentially competing for opportunities in city 2, measured by total out-migrants from all cities within a circle having as its centre city 2 and as its radius the distance from city 2 to city 1 (Figure 8.7).

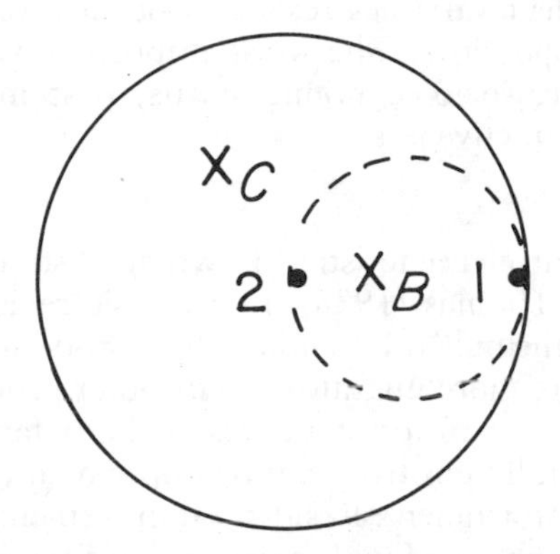

Figure 8.7
Schematic representation of the areas (circles) relevant to intervening opportunities (X_B) and competing migrants (X_C) for a migration flow between city 1 and city 2 according to Stouffer's refined 'intervening opportunities' formula.

Exponents for the terms in the formula can be determined empirically to improve the flexibility and goodness of fit of the model.

Application of the formula to 116 inter-city migration streams in the United States, for 1935–40 (Stouffer, 1960) and 1955–60 (Galle and Taeuber, 1966), produces r^2 'explanations' of more than 90 per cent of the migration variance. Although the major contribution is made by the $X_O \cdot X_I$ size effect, both intervening opportunities and competing migrants make independent and roughly equal contributions to the model's predictive utility. Another test of the model has been by Jansen and King (1968) on inter-county migration in Belgium. Here the formula performed well only *within* the French and Flemish linguistic regions, emphasizing important cultural constraints on migration.

Multiple regression models

Resource to such models dominated aggregate migration analyses in the 1960s and 1970s. In the basic single-equation model the dependent variable is some measure of migration (gross or net, absolute or population-standardized) for an area, while the hypothesized independent (predictor) variables comprise a selection of demographic, social and economic variables.

The flexibility of the multiple regression model is its great attraction. Several of the variables from gravity and intervening opportunity models may be adopted in multiple regression analyses (indeed, the Lowry–Rogers modified gravity model was operationalized in its original form in this way). But the range of possible independent variables extends well beyond these to embrace demographic, occupational, environmental and other measures. The flexibility is such that different sets of independent variables may be used for different sets of migration data; a variable offering a high degree of 'explanation' in one application may not be adopted in another. Such flexibility does have a cost, in that reliance is often placed more on *ad hoc* empiricism than on plausible theory: 'The proliferation of aggregative multiple regression analyses has had more the character of intelligent treatment of data than the testing of theoretically derived hypotheses' (Margolis, 1977, p. 41).

The choice of independent variables rests a good deal on an appreciation of differential migration propensities. The most important variables determining these propensities are age, socio-economic status, past migration experience, migrant stock and areal attractiveness.

Age
This is the most important characteristic known to distinguish migrants from non-migrants. Indeed, D. Thomas (1938), in a classic review of migration differentials, concluded that the only differential which stood up to several contexts was that young adults were more migratory than other groups. This reflects the demands generated in the expansionist phases of both family and career life-cycles, as well as the generally positive relationship which exists between a person's age and community attachment. Residential movement is clearly associated with leaving the parental home and with marriage and family growth, while the more speculative forms of labour migration more readily attract those without major family commitments and without the seniority, job security and pension rights derived from extensive employment in a single firm; older workers may also experience discrimination against them in hiring policies.

Such is the regularity in the age pattern of migrants in developed countries that Rogers (1984) has been able to represent it by means of a mathematical function known as the model migration schedule.

Figure 8.8 provides a typical example of age-specific migration propensities, characterized by the high mobility of young adults, some of whom will be accompanied by young dependent children (the reluctance to disrupt *secondary* school education is a deterrent to family migration). Figure 8.8 also indicates the characteristic small increase in migration propensities for largely amenity reasons at retirement (Rogers, 1988), the earlier peak for females reflecting the characteristic age-gap between husbands and wives. A further increase in migration among the over-75s is a consequence of growing dependency on care provided by relatives and institutions.

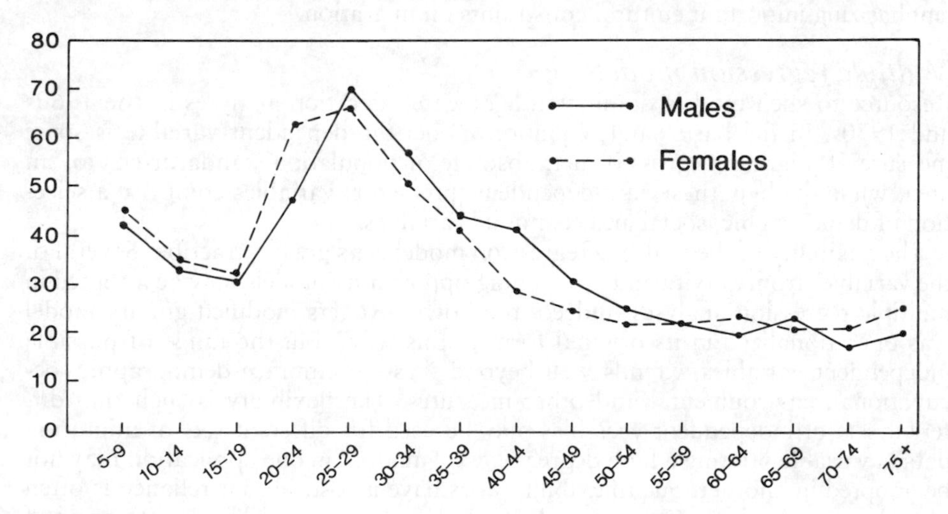

Figure 8.8
Percentage of the Australian population which moved between 1976 and 1981, by age and sex
Source: Hugo (1986), Figure 5.13.

Socio-economic status
In societies like that of nineteenth-century Britain it appears that the most mobile elements in the population have been the lower-status social groups unencumbered by possessions and with ready access to rented housing and to lodging opportunities (Redford, 1926; Pooley, 1979). But in contemporary developed countries there is overwhelming evidence that the more skilled, better educated and higher-income groups have the greatest propensity to migrate, especially over longer distances (Johnson, Salt and Wood, 1974; Long, 1988; Fielding, 1989). The higher-status groups have the ability and training to obtain and analyse many information sources on jobs and housing over a wide geographical area. Although the labour market for their jobs is territorially extensive, it is often a small one in terms of actual jobs available, so that changing an employer or even one's position within a company will often entail mobility of the 'spiralist' type:

> The progressive ascent of the specialists of different skills through a series of higher positions in one or more hierarchical structures, and the concomitant residential mobility through a number of communities at one or more steps during the ascent, forms a characteristic combination of social and spatial mobility.
>
> (Watson, 1964, p. 147)

Even although higher-status workers are well able to bear the financial costs of moving, they are often reimbursed by employers, and they have ready access to private housing through their creditworthiness with funding institutions.

Not all the higher-status groups have high rates of spatial mobility. Ladinsky (1967) shows that the self-employed professions – doctors, dentists, lawyers, architects, accountants, etc. – require appreciable investments in capital equipment and the building-up over several years of clienteles, so that mobility is restricted. In contrast, there is a migratory élite of white-collar workers comprising high-level managerial and technical staff encouraged to move among the branches of large, multi-locational government and private organizations (McKay and Whitelaw, 1977; Johnson and Salt, 1980; Salt, 1984).

At the other end of the migration propensity spectrum, unskilled workers scan much more restricted horizons for job opportunities. For them there is no equivalent of *The Times* or the *Guardian*, which provide nationwide coverage in Britain of job opportunities in a wide range of professional occupations; nor do they have specialized journals like those in teaching, law and medicine with extensive coverage of vacancies. Because their kinds of jobs are widely available spatially, there is often no requirement or incentive for the unskilled to make long-distance moves. Their lower incomes and lower mobility allowances make the financial costs of migration a significant deterrent, while their access to housing in a new area is restricted by the difficulties faced by incomers, as opposed to long-term residents, in acquiring rented public or private housing. They are also less socialized than the midle classes in adapting to new environments.

Past migration experience
It is now widely recognized that migration potential is functionally related to migration experience, in that in-migrants to a community are more likely to become out-migrants than are long-term residents (Morrison, 1967; Land 1969; Clark and Huff, 1977).

Migration is thus not unlike sinning – once done, it is easier to do again! The theoretical justification is that personal and community bonds deter migration and that these bonds gain strength with increasing duration of residence (Zimmer, 1955; Toney, 1976).

Migrant stock

Hagerstrand (1957), Nelson (1959) and Greenwood (1970) are among those who have incorporated measures of personal contacts between areas in migration models which give high levels of explanation. Their necessarily indirect measures are based on past migration or birthplace data. There is, however, a fundamental difficulty in including such a 'migrant stock' measure as an independent variable in regression analyses: both past and present migration are likely to have been influenced by socio-economic conditions which are fairly constant over time. In other words, one of the independent variables, past migration, is likely to be influenced strongly by several of the others; multi-collinearity of a particularly severe form could be present. Thus Vedder and Galloway (1970, p. 480) state that they 'know of no satisfactory means of quantifying friends and relatives in a fashion which would not result in a tautology being created in our [regression] analysis'.

Areal attractiveness

Multiple regression models often include measures not only, as above, of the population's propensity to migrate, but also of the environmental attractiveness of areas. Greenwood (1970) and Gober-Meyers (1978) include a temperature variable in their analyses of inter-state migration in the United States, on the grounds that, other things being equal, warm winters will attract migrants, especially retirees, to states like Florida, Arizona, New Mexico and California. About 6 per cent of households moving between American states in the 12 months preceding the 1979–81 Housing Surveys indicated a change of climate as their main reason for moving (Long, 1988). More commonly adopted in migration regression analyses are measures of economic performance and opportunities like per capita income and unemployment rate (Weeden, 1973; McNabb, 1979). It should be noted that the invariably positive correlation found both temporally and spatially between rates of unemployment and net out-migration is not due to high mobility among the unemployed themselves (they are, in fact, relatively immobile), but to the way in which poor employment prospects discourage in-migration and encourage the younger, better educated and trained workers to leave.

Analytical problems

The use of migration data in multiple regression analyses does pose some distinctive conceptual and operational problems. Since migration is a flow responding to variables operating at some earlier period, it is often desirable to lag some of the variables (Greenwood, 1970), although this is often not possible with census data. A striking example can be drawn from a time-series regression of net migration from Scotland to England and Wales, 1954–68, on relative unemployment in the two countries. The regression accounted for only 26 per cent of total variation in migration with no lags applied, but for 85 per cent with a one-year lag (Scottish Economic Planning Board, 1970).

Another problem concerns the choice of areal units, particularly if one wishes to exclude short-distance residential movement from the analyses. A partial solution has been to compile relevant data sets on the basis of city regions (Fielding, 1971) or geographical labour markets (Flowerdew and Salt, 1979; Kennett, 1983) defined by journey-to-work criteria; 72 per cent of all moves within Britain during 1980–1 were within such areas (Owen and Green, 1989).

There has also been a growing awareness that most multiple regression studies of migration regard inter-regional migration purely as a response to socio-economic conditions, while ignoring the important promotional role of migration in socio-economic change. A conceptually sounder approach than the single-equation regression model is the use of a system of simultaneous equations in which migration and socio-economic change are regarded as interdependent. In an analysis of inter-state migration and economic growth in the United States, Gober-Meyers (1978) adopts a study model comprising a set of three equations simultaneously determined; the three dependent variables are net migration, growth in per capita income and change in population composition. Other examples of this approach are provided by Greenwood (1981) and Cadwallader (1985).

Probabilistic models

The previous groups of models are characterized by a deterministic approach, involving a specification of the relationship between migration and its explanatory variables. They are generally ill-suited to forecasting because of the difficulty of predicting values for the independent variables. Probabilistic models of migration, on the other hand, are specifically designed for forecasting migration flows, but only under restrictive assumptions. Such models generally apply Markov Chain methods to a matrix representation of a migration system in which known inter-area flows are represented by transition probabilities. Assuming, somewhat unrealistically, a constancy of such probabilities over time, future patterns of migration may be predicted (Rogers, 1966; Joseph, 1975). Woods (1979, p. 189) sums up well: 'Markov Chain models are essentially of a descriptive and at best exploratory nature. Their function is not to interpret reality, but to let it change along pre-determined lines.'

Micro-analytical perspectives and models

The models discussed in the previous section often give adequate statistical explanation for the volume, distance and direction of migration flows by submitting aggregate data to macroscopic analysis. But since they are seeking empirical regularities in aggregate data, they tend to be descriptive, rather than truly explanatory, models. What they explain are patterns rather than processes, so that the motives of migrants can only be inferred.

As a reaction to the impersonal, mechanistic nature of such models, growing attention has been paid to the decision-making process among individual movers and occasionally, for control purposes, of stayers. A migration pattern is, after all, a composite expression of the aspirations, needs and perceptions of real persons; and mobility behaviour is one of several means by which individuals can seek well-being or utility maximization. The focus, therefore, of micro-analytical migration models is on the behaviour of individuals rather than on the characteristics of places and populations, and there is a ready recognition that prospective migrants may perceive and respond to environments with varying degrees of rationality. Compared with macro-analytical models, the behavioural migration models deal with people rather than places, and processes rather than patterns, while their more loosely framed and conceptual nature makes them more flexible, but less precise. Having said this, there is little analytical value in considering every household to be unique in the way it responds to environmental stimuli. Analytical progress can be achieved only through some degree of generalization, although 'the question arises as to what level of disaggregation is acceptable for the assumptions of within-group behavioural homogeneity' (Clark and Moore, 1978, p. 4).

Some micro-modelling landmarks

The first detailed study of the migration process was that of Rossi (1955) on intra-urban residential mobility in Philadelphia. He concluded that the major process involved was the way in which families adjusted their housing to the needs which arise at specific stages of the life-cycle – a finding generalized more recently by Clark and Onaka (1983). Table 8.4 is an attempt to model the relationships between urban structure, life-cycle stage and residential mobility for a particular sector of population, although complexities are posed by modern changes in household and family composition (Figure 8.9, Hall, 1986, and Kiernan, 1989). In particular, there is the important migration role of household formation and dissolution; about one-quarter of US households that relocated in 1972–3 had a change of head of household during the year, involving separation, divorce, death or new household formation (Quigley and Weinberg, 1977).

Germani (1965) regards migration as the outcome of objective factors working within a normative and psychological context. His objective factors are the characteristics of potential places of origin and destination, and the nature of the contact between them. The normative context embraces the norms, beliefs and values of the society at origin, and the psychological context takes into account the attitudes and expectations of specific individuals.

Lee (1966) sees the migration process as involving sets of perceived factors relating to origin and destination, and a set of intervening obstacles (Figure 8.10). The sets of factors and obstacles will vary among individuals in relation to life-cycle, socio-economic and personality characteristics. This is a refinement of the simple 'push–pull' view of migration, in which people are seen as being driven out of some areas by low wages, unemployment and the like and pulled to other areas by their superior opportunities.

A further refinement comes with the adoption of micro-economic perspectives in the so-called human capital model of migration (Sjaastad, 1962; DaVanzo, 1980). This attempts to measure the largely economic costs and benefits of

Table 8.4 Housing needs associated with different stages of the life-cycle for urban, middle-class, American families in the private housing market

Life-cycle stage	Housing needs/aspirations
1. Pre-child	Cheap, central city apartment
2. Child-bearing	Renting of single family dwelling close to apartment zone
3. Child-rearing	Ownership of relatively new suburban home
4. Child-launching	As in 3, or move to higher-status area
5. Post-child	Residential stability
6. Later life	Institution/apartment/live with children

Source: Abu-Lughod, Foley and Winnick (1960) and Short (1978).

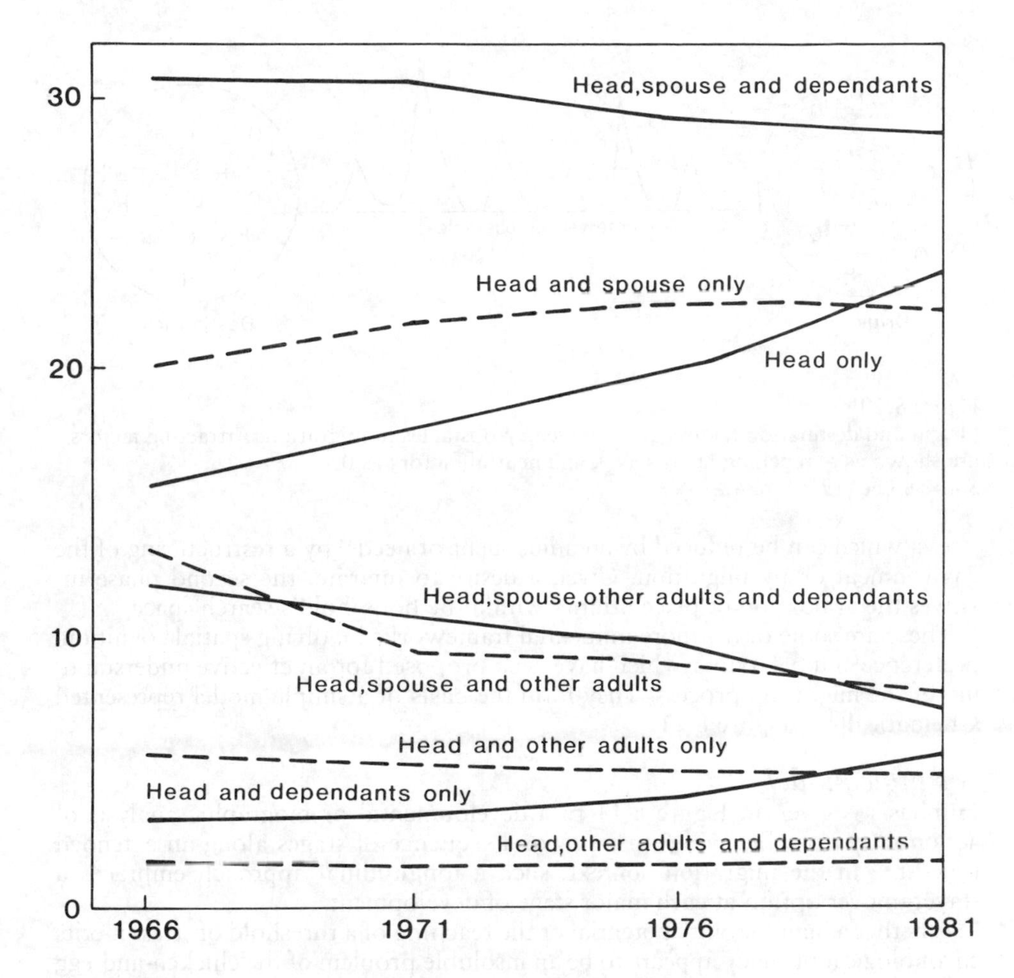

Figure 8.9
Percentage of all Australian families in particular family types, 1966–81
Source: Hugo (1986), Figure 7.4.

migrating between particular locations, allowing for the expected period over which the likely net benefits of any migration investment will be enjoyed. Given the trends in married female employment in recent decades, it is essential that such models recognize the growing dominance of two-worker families, particularly since most evidence suggests that such families migrate less than one-worker families (Sandell, 1977; Mincer, 1978).

Wolpert (1966) conceives of migration behaviour as embracing, above all, the evaluation of place utilities. These measure a person's level of satisfaction with respect to particular locations within that limited portion of the environment – one's action space – that is relevant to decision behaviour. He shows that the evolution of a person's action space over time is associated with a complex of institutional and social forces.

Brown and Moore (1970) suggest that the migration process comprises two major phases. The first involves the development of a state of dissatisfaction or

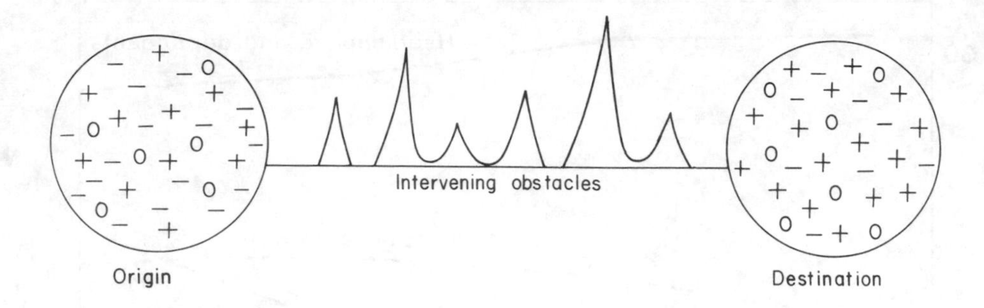

Figure 8.10
Origin and destination factors and intervening obstacles in migration. Attracting factors are shown as +, repelling factors as –, and neutral factors as 0
Source: Lee (1966), p. 48.

stress, which can be reduced by an adjustment of needs, by a restructuring of the environment or by migration. Given a desire to migrate, the second phase involves the evaluation of place utilities within the household's search space.

These are some of the more influential frameworks involving spatial cognition, preference and behaviour which have been proposed for an effective understanding of the migration process. They form the basis of a simple model represented schematically in Figure 8.11.

A simple model

Emphasis is given in Figure 8.11 to a developmental or biographic analysis of action to explain behaviour in terms of a sequence of stages along an extended time-line. In the migration context, such a longitudinal approach embraces a stayer–mover option at each major stage of development.

Whether a high mobility potential or the reaching of a threshold of stress merits chronological primacy appears to be an insoluble problem of the chicken-and-egg type. They interact so intimately that stress is not simply a function of objective characteristics like inadequate housing, unsatisfactory employment or environmental decay; it also reflects the ability to consider escape from such conditions

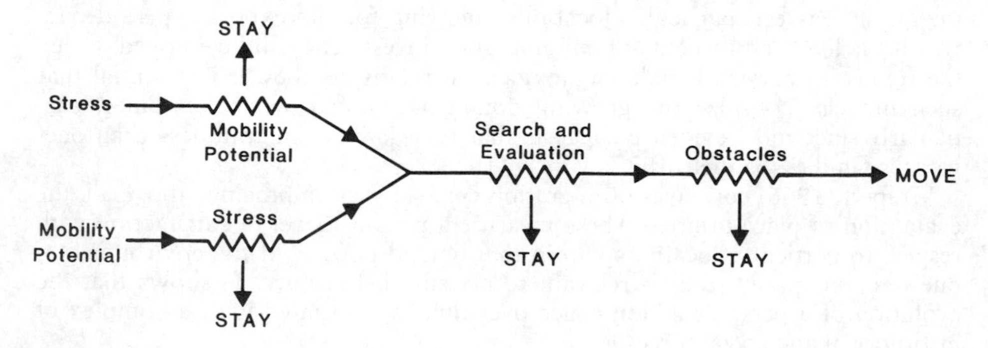

Figure 8.11
A longitudinal representation of the migration process

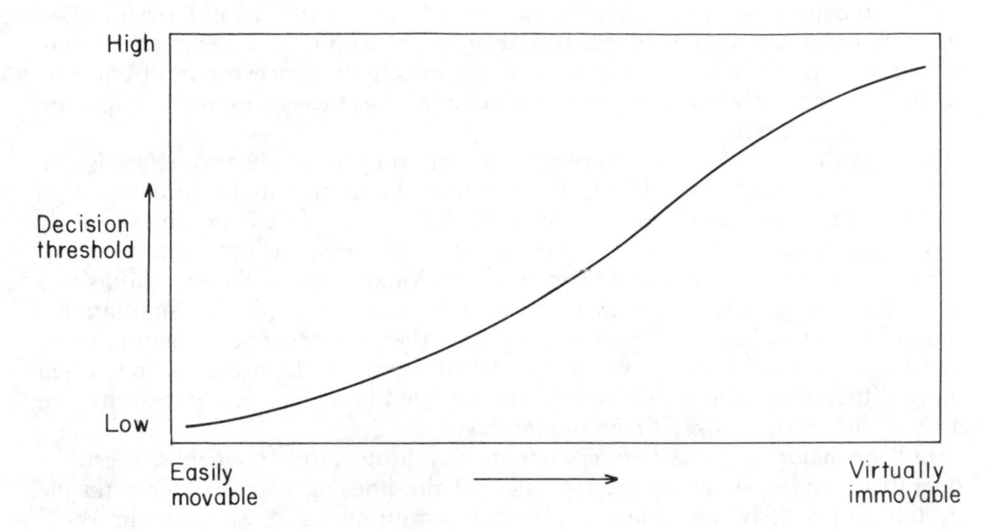

Figure 8.12
A continuum of household-moving potential
Source: Short (1978), Figure 5.

by migration. Persons with low migration potential (determined largely by age, socio-economic status, etc.) could well rationalize the same features which others find intolerable.

It is also possible to conceive of mobility intentions being pursued by people of high mobility potential – the so-called chronic movers – quite independently of any significant level of stress. Morrison (1973) suggests that the population may be regarded as a continuum of migratory potential. At one end are the highly mobile households with low decision thresholds (needing very little stress to make them move), while at the other end are the virtually immobile households with high decision thresholds (Figure 8.12). In a study of out-migration from several mining villages in Durham, R. Taylor (1969, 1979) distinguishes 'a sense of dislocation' among would-be migrants from 'a sense of belonging' among stayers.

Given a decision to contemplate migration seriously, the next step is to search for and evaluate relocation possibilities. Silk (1971) has identified a sequence in basic search behaviour: goal specification, procedure selection, and information gathering. Geographers are particularly interested in the spatial parameters of information gathering, termed 'awareness space' by Brown and Moore (1970). Such space can be divided into 'activity space', where information is obtained by direct observation on a regular, often day-to-day basis, and 'indirect contact space' where information is derived from more general forms of communication like relatives, acquaintances and mass media. For many low-status groups the 'indirect contact space' may be highly constrained, in contrast to groups like university teachers, who travel extensively, attend conferences, consult a wide range of colleagues and follow several professional publications. Another division in information gathering is between the use of formal channels such as newspapers, estate agents and manpower agencies, and of informal contacts like friends, relatives and workmates.

Search behaviour rarely operates optimally. Herbert (1973) and Barrett (1974) in their empirical studies of residential mobility within Swansea and Toronto both report that searching is not a thorough, systematic process: not infrequently it consists of a fairly casual consideration of a few vacancies within a familiar and fairly restricted area.

If an evaluation of the place utilities thrown up by search behaviour leads to a desire to move, there are still several obstacles to be surmounted before migration can take place. Some of these, like physical distance and family ties, may already have operated at earlier stages in the migration process, but others now emerge for the first time. There are sometimes institutional barriers like the quotas and stringent selection procedures that have been increasingly applied to international migration in this century; likewise, there were the poor relief Settlement Laws of eighteenth-century England which led Adam Smith to declare that 'it is often more difficult for a poor man to pass the artificial boundary of a parish than an arm of the sea or a ridge of high mountains'.

But the major obstacles to internal migration in the modern world concern the operation, and particularly the rigidities, of the housing market. In Britain the decline of privately rented housing, with its traditional ease of entry and rapid turnover, to less than 10 per cent of the housing stock has clearly restricted mobility. Owner-occupiers, too, unless they are élite employees eligible for intra-company relocation subsidies, have been inhibited from moving from economically depressed parts of northern England in pursuit of superior job opportunities in the economically buoyant South East by the great regional gulf in house prices. At an intra-urban scale there is also the 'gate-keeping' role of estate agents and building societies as they direct certain types of clients towards or away from particular parts of a city (Williams, 1978).

Among lower income groups, the role of public rented housing (28 per cent of Britain's housing stock in 1985) has had a particularly inhibiting effect on migration. In the allocation of tenancies, incomers are often discriminated against by a length of residence qualification, and it was not until 1980 that central government put pressure on local authorities to follow the example of the new towns and ease such restrictions in the interests of labour mobility and business efficiency. Yet the areas (in the South East) with the major current demand for both working-class labour and housing are the very areas where public sector housing has historically been poorly developed and is currently most subject to conversion to owner-occupation under the 'right to buy' legislation of the 1980s (Dunn, Forrest and Murie, 1987; Forrest, 1987). The inability, therefore, of job seekers from the North to find affordable family housing in the South has led to increasing numbers of manual workers sharing rented accommodation during the working week and returning, in an uncongenial form of circular migration, to their families in the North at weekends (Hogarth and Daniel, 1987). Furthermore, there are often restrictions on movement *within* the public housing sector of a local authority, as P. Taylor (1979) has highlighted in a study of 'difficult to let, difficult to live in, and difficult to get out of' estates in Killingworth.

All this suggests that 'individual household decisions [on residential mobility] are made in a decision-making environment, hierarchically structured by the central and local state, financial institutions and estate agents' (Short, 1978, p. 434). Similarly, Cadwallader (1986) emphasizes that supply-side constraints should always be incorporated to qualify the apparent consumer-preference sovereignty implicit in most models of migration decision-making.

Operationalizing behavioural models

A migration model like Figure 8.11 can be operationalized by a series of survey questions covering both objective background variables and attitudinal – evaluative – procedural variables relevant at each phase (e.g. H. Jones, 1976; Pryor, 1976). On the basis of survey data and analysis, the preliminary model can be verified, modified or rejected. Some have regarded migration as being well suited to this type of analysis. It possesses 'a definite locus in time, involving a fairly uniform sequence of acts. The structural orderliness of residential shifts helps movers to reconstruct the sequence of events and feelings which made up the act of moving' (Rossi, 1955, p. 457).

Nevertheless there are several dimensions of the migration process that pose difficulties to analysts. The recurrent problem, as in all behavioural studies, is the acquisition of good quality data. Most commonly, data are collected from cross-sectional, retrospective interviews with migrants who are successful in the sense that they have at least remained at their destination long enough to be detected by researchers. There tends to be little information on those who contemplate migration but eventually stay, on those who move but return quickly, or on the really chronic movers. Information from retrospective surveys is also subject to error from recall and from post-event rationalization. This certainly applies to reconstructions of decision-making, and even to more objective 'life history' information.

The alternative strategy of monitoring actual and potential migration situations of individuals as they evolve over an extensive time period by means of longitudinal study is arduous and hazardous. There are serious methodological problems in asking people about something they have not done, and there are also 'problems of reaching sufficient depth in an interview situation: the topic is such as to demand a long interview – probably repeated at several time intervals – and such goodwill in response may be difficult to obtain in a situation where point and purpose is not obvious to the layman' (Herbert, 1973, p. 44).

The one thing that should be obvious from this review is that a simple survey question – 'Why did you migrate?' – cannot, by itself, be expected to elicit meaningful answers. This is because several factors are involved in any move, because migrants find difficulty in recalling and articulating their thinking at the time of migration, and most important, because the question can be answered at several levels of generalization. Reference, therefore, in migration surveys to 'employment reasons', 'housing reasons', 'environmental reasons' and the like are very crude summaries of complex behavioural situations. They have greatest credibility when a funnelling approach is adopted in interviews, in which the total migration context is explored before the respondent is requested to sum up by indicating the most important reasons for his or her move.

A systems approach

The strategies and methodologies devised to explain migration are highly varied, and it has proved particularly difficult to integrate the macro- and micro-analytical perspectives into a successful grand model. One widely acknowledged attempt has been Mabogunje's conceptualization of the problem within the framework of general systems theory. He considers rural–urban migration, with particular reference to Africa, as 'a circular, interdependent, progressively complex, and self-modifying system in which the effect of changes in one part can be traced through the whole of the system' (Mabogunje, 1970, p. 16). In his specification of the complex of interacting elements, particular emphasis is given to the social, economic, political and technological environments. While such emphasis is salutary at a conceptual level, there are, nevertheless, enormous operational problems in specifying, and certainly in measuring, such all-embracing contexts. Such an approach parallels the more explicitly structuralist explanations of migration, addressed in the next chapter, in which the *fundamental* causes of migration are seen to lie in the political economy of societies.

9

THE POLITICAL ECONOMY OF INTERNAL MIGRATION

The macro- and micro-analytical models of migration reviewed in the previous chapter have increasingly become subject to the criticism or, more accurately, the qualification that they provide only partial explanations of migration. They deal with the proximate or intermediate determinants, rather than the fundamental determinants which lie in the creation of spatially uneven development in the first place. These models also fail to address adequately the pivotal role that migration has played – as cause, effect and integral component – in socio-economic formation and transformation. Thus, there 'has been a general failure to ground migration work in any basic, comprehensive social theory . . . What we have instead is a series of *ad hoc* generalizations or, at best, middle-range theories that float within an intellectual limbo' (Zelinsky, 1983, p. 39).

An influential attempt to break out of this impasse was Zelinsky's (1971) model of the mobility transition (Figure 9.1) which he saw as paralleling (and interacting with) the demographic transition. It was based on:

> definite, patterned regularities in the growth of personal mobility through space–time during recent history, and these regularities comprise an essential component of the modernization process; . . . [in particular], a transition from a relatively sessile condition of severely limited physical and social mobility toward much higher rates of such movement always occurs as a community experiences the process of modernization.
>
> (Zelinsky, 1971, pp. 221–2)

This model demonstrates a population geographer's use of space–time awareness in the formulation of a valuable, pioneering overview of some important links between development theory and demography. Explanation is essentially in terms of Rostow's stages-of-growth and modernization theory, on the basis of a linear progression from what Zelinsky calls pre-modern traditional society to super-advanced society. As modernization proceeds, various forms of migration and circulation ebb and flow, with different types of movement succeeding one another as the dominant wave.

As a broad description of events, the mobility transition has been found generally acceptable, provided it is modified for regional cultural conditions. But, as

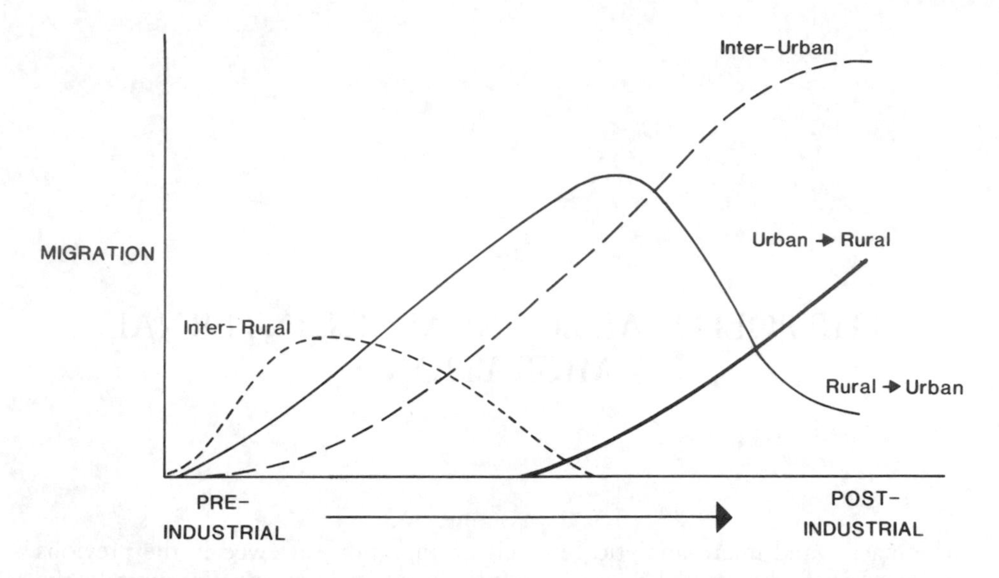

Figure 9.1
A schematic representation and extension of the major migration elements in the Mobility Transition Model (Zelinsky, 1971); counterstreams to the dominant flows are not shown

with the demographic transition, its explanatory basis has often been regarded as inadequate. A more powerful driving force than rather vaguely specified modernization is the role of temporal changes in the dominant mode of production and the spatial interaction or articulation between coincident modes of production. In particular,

> the major spatial rearrangements of population, which all countries have experienced at some time throughout the past 200 years, are associated with the establishment and expansion of the capitalist mode of production.
>
> (Forbes, 1984, p. 157)

Nowhere, in fact, are the spatial inequalities that give rise to migration more pronounced than in societies undergoing capitalist intrusion and transformation.

Marxist-based interpretations of migration evolution have become particularly prominent in the contexts of movements within Third World countries (Amin, 1974; Omvedt, 1980; Shrestha, 1988) and of international labour migration, particularly from southern Europe, North Africa and Turkey to north-western Europe (Castles and Kosack, 1973; Nikolinakos, 1975; Seers, Schaffer and Kiljunen, 1979). But the application of such perspectives to the explanation of internal migration evolution in today's developed countries has been much more restricted. Two examples are Rees and Rees (1981) for South Wales and H. Jones (1986) for Scotland. The latter case study will now be highlighted since Scotland demonstrates in a particularly vivid but also representative way how migration from the eighteenth century to today has responded sequentially to the emergence, maturing and current restructuring of capitalism and to all that entails in changing social relations of production (the social arrangements by which labour is mobilized and deployed) and changing spatial divisions of labour. Migration is thus interpreted more as a socio-economic process than merely a demographic component or an aggregation of individual movers with individual motives.

Migration and capitalism: the Scottish experience

A traditional viewpoint, based more on theoretical deduction than empirical verification, has been that until well into the eighteenth century territorial mobility in Scotland was restricted severely by the subsistence nature of the agricultural economy, the strength of local social ties, poor communications, sharply circumscribed awareness-space and minimal disposable incomes. However, recent research on sources like parish registers, estate records and movement certificates of the Church has been uncovering a pattern of substantial population turnover in local communities in pre-industrial Scotland, as in north-western Europe generally (Pryce, 1982; Clark and Souden, 1987), dominated by the movement of unmarried farm servants on short-term contracts. Although there was some long-distance migration in Scotland associated with particular groups, notably vagrants, apprentices, students and clergy, the great bulk of migration was over very short distances, characteristically less than half a day's walk. It needed the spur of emerging spatial inequalities through the economic transformations associated with early capitalism in the eighteenth and early nineteenth centuries to promote significant longer-distance movement expressed in the rising levels of inter-rural and then rural–urban migration that comprise the early part of the mobility transition model (Figure 9.1).

The Lowlands

Capitalist relations in agriculture emerged when Scottish landlords, following the lead of their English counterparts, responded to the profit opportunities offered by the combination of good arable land and the new productive methods of the Agricultural Revolution. Accordingly, they replaced their many, small unproductive tenants by a few, innovative, high-rent-paying tenants on new, enlarged, enclosed farms.

The traditional view of the demographic impact of such restructuring was that it led to a rural labour surplus, which then flooded on to the labour market on the basis that the new farms used paid labour more efficiently than the small customary tenants had formerly used the labour of their families and subtenants. One expression was that harvesting needs were now met by itinerant bands of Highlanders and Irishmen. But the modern consensus view is that the new husbandry created at least as many jobs as it shed, particularly in building dykes, in planting and maintaining hedges, in ditching, draining, marling and liming, and in the labour-intensive needs of turnip husbandry.

The agricultural improvements in Lowland Scotland are unlikely, therefore, to have stimulated extensive out-migration and rural depopulation in any simple cause-and-effect manner. They did, however, involve a considerable spatial rearrangement of the rural population at a local scale. Moreover, the new pattern of rural social relations (capitalist landlord, tenant farmer and landless labourer) loosened the land bond for the growing section of rural population that no longer controlled its means of production and livelihood, enabling it to provide the necessary labour pool for, initially, localized proto-industrialization and, subsequently, major industrialization in central Scotland.

Proto-industrialization emerged widely in rural areas of seventeenth- and eighteenth-century Europe during the lengthy transition to industrial capitalism (Mendels, 1972). Its basis was the domestic cottage industry of small-scale commodity production organized by urban merchant capital seeking out rural labour surpluses. Historical demographers have been particularly interested in the fertility-enhancing role of proto-industrialization (see Chapter 5), but there were also important implications for migration. Thus early nineteenth-century Scotland saw the speculative establishment by landowners of planned villages to attract both labour and merchant capital for domestic textile industry. Examination of the origins of the new village inhabitants (Lockhart, 1982) shows that the overwhelming majority (tradesmen and artisans as well as handloom weavers) generally travelled less than ten miles, further confirming that migration in Lowland Scotland at this time was dominated by appreciable redistribution of rural population. There was only a very modest representation of the frontierward movement that figures prominently in the early stages of Zelinsky's model. In Scotland this took the form of reclamation and settlement of marginal land, especially in the North East where landlords consciously established smallholding tenants on such land as agents of reclamation.

The Highlands and Islands

This remote, north-western part of the country experienced a remarkably abrupt impact of capitalist influences on a primitive clan-based mode of production – with revolutionary migration consequences.

Until the second half of the eighteenth century there was little participation by the Highlands in the emergent capitalism being experienced by the rest of Britain, essentially because bonds of kinship and mutual obligation in clan society precluded the development of impersonal money relationships. Rapid changes occurred, however, in response to two factors. First, a considerable increase in southern demand for Highland primary products (black cattle, sheep and kelp) exposed the inability of the traditional mode of production to respond effectively to the forces of market demand. Second, the Jacobite rebellions had demonstrated that traditional Highland society posed an unacceptable military threat to the British state. Therefore, after the 1745 uprising, commercial and political interests dictated a rapid dismantling of the clan system and, in particular, the transformation of patriarchal clan chiefs into profit-seeking landlords integrated into the national economic system.

The migration implications of these revolutionary changes arise from the transformation of the agrarian system and its associated settlement pattern that was required by landlords for successful large-scale production of wool and kelp (a form of alkali used in soap- and glass-making derived from the summer cutting, drying and burning of seaweed under the marketing control of landlords). In what Marx described as 'the last great process of expropriation of the agricultural population from the soil' (and therefore of primitive accumulation of capital) much of the better land was cleared of small tenants to provide sheltered feeding for the new large sheep farms rented profitably to southern flock-masters in response to the surging industrial demand for wool. In the first two decades of the nineteenth century much of the 'cleared' population in the Western Highlands and Islands was transferred internally within estates to coastal sites, often exposed and barren, and provided with tenanted plots of land. So small were the plots that tenants were effectively forced to gain most of their livelihood through fishing, distilling or kelping. This then was the origin of crofting, seen by landlords as a profitable means of dealing with a cleared population (Hunter, 1976).

Where kelping could not meet crofters' cash needs generated by an emerging money economy, temporary migration was a common response (Devine, 1979). Initially most migrants were seasonal harvesters, working in the grain and turnip fields of the newly enclosed farms of the Scottish Lowlands, although some Highland harvesters found their way to eastern England (Redford, 1926). Migration for periods of some years was associated with the Highland regiments, the navy, the Greenland fisheries and the Hudson Bay Company. More seasonal was movement to the building industry in growing urban centres in Lowland Scotland (Marx's 'light cavalry' of capitalism) and to the fishing, gutting and packing industries of north-east Scotland as they expanded in the nineteenth century in a commercial capitalist form which undermined traditional forms of fishing on the west coast and elsewhere. The crofting system, therefore, could be regarded as providing, for parts of the emerging capitalist system, a marginalized, seasonal labour reserve, much of whose costs of subsistence were provided outside the capitalist system, in much the same way as in temporary or circulatory migration from Third World countrysides to cities today. Urban capital was not, therefore, bearing the real costs of labour reproduction.

It was in landlords' interests, initially, to discourage overseas emigration. Thus the interests of this dominant class prevailed in the 1803 Passenger Act, which effectively doubled transatlantic fares. But the power structure of the state changed as the entrenched interests of the landed oligarchy gave way to those of the industrial bourgeoisie and its allied urban professional classes. Accordingly, the British government reduced and then eliminated tariffs and taxes on kelp substitutes, undermining the profitability of kelping. Similarly, the 1845 Poor Law Act required Highland landlords for the first time to be assessed formally as ratepayers. Consequently there was a dramatic change in landlord attitudes towards local population growth. Far from population retention, plot division and population growth being encouraged, landlords from the 1820s attempted vigorously to shed what they now regarded as a massively surplus tenant population (and thereby create even more acreage for profitable sheep farming), largely through exodus promoted by the very emigration agents that they had formerly despised. As in the Third World in the 1960s, overpopulation, rather than structural deficiencies, became adopted by the establishment as the basic cause of the 'Highland problem', providing another example of blaming the victim in circumstances when high population growth was more the *result* of poverty than its cause.

It is clear then that very special forms of migration were associated with the emergence of the crofting system as a unique form of petty commodity production developing between pre-capitalist and capitalist modes of production.

Industrial capitalism

This phase was dependent on the prior penetration of capitalist relations in agriculture, which had provided an important mechanism for capital accumulation, the freeing of labour from the land, and a food surplus to support population growth generally and an industrial proletariat specifically.

An important feature of the capitalist demographic regime is the massive, urbanward, inter-regional and sometimes international migrations required to provide labour in those economic sectors and geographical locations where capital finds its

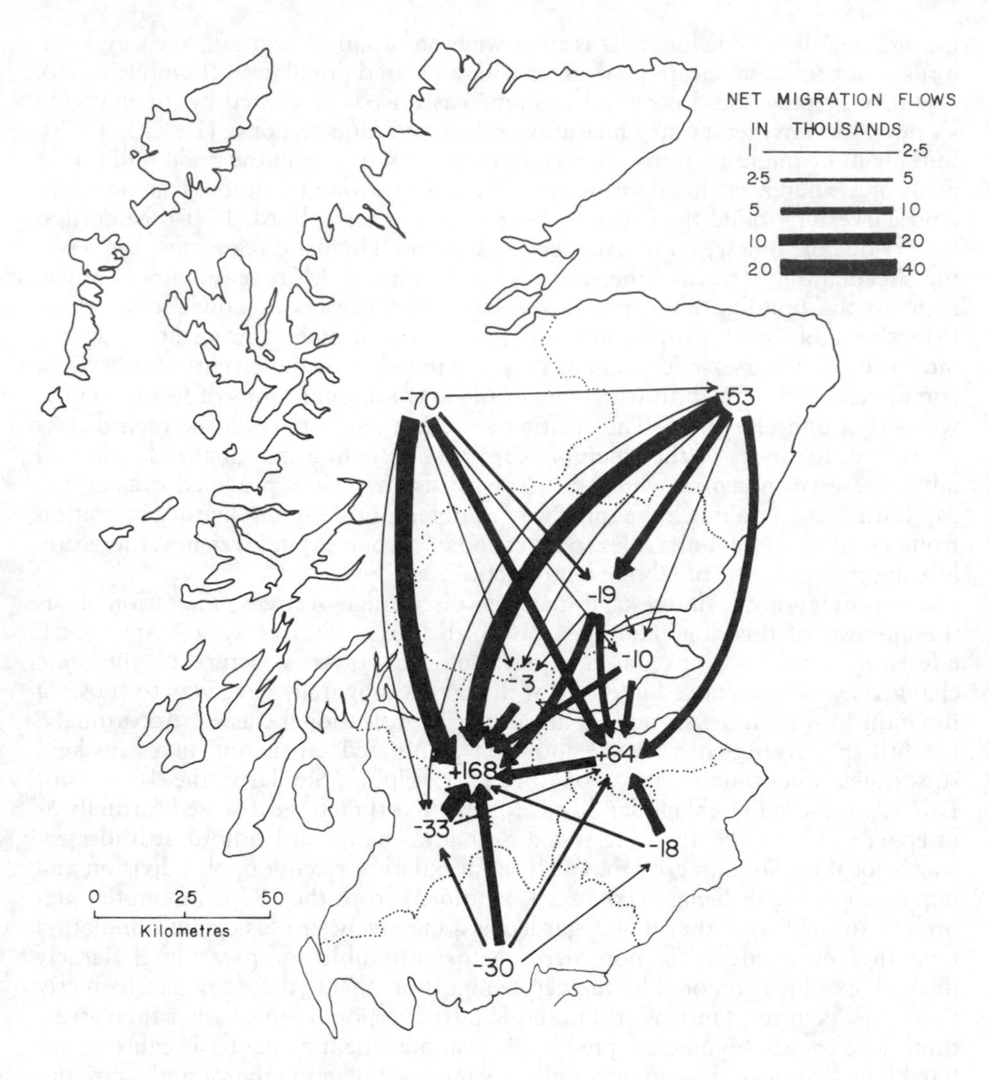

Figure 9.2
Net 'lifetime' migration between Scottish regions on the basis of place of birth and place of enumeration data at the 1901 census. The figure shown for each region is the overall balance in thousands. The pattern is dominated by inflows to the manufacturing, mining and urban belt of central Scotland
Source: Simplified from Osborne (1958), Figure 5.

highest returns. But the system, in terms of labour mobility, did take some time to evolve. Thus, when the cotton industry led the way in factory-based production in the late eighteenth century, its need for water-power sites meant that the new spinning mills were often located in thinly settled districts where the capitalist system was too immature to provide ready pools of accessible labour. Indeed, the history of the early Scottish spinning mills is one of a constant search for workers, at a time when there was no readily available reserve army of labour, and the provision of housing and basic services in what were virtually company towns.

Three sources of labour that Robert Owen tapped at New Lanark were Highlanders (sometimes lured from emigrant ships), pauper children from the workhouses of Glasgow and Edinburgh, and the Irish.

Institutional developments had a role to play in promoting the labour mobility needed by industrial capitalism. Thus the Scots Poor Law Act of 1845 ensured that the unemployed had no legal right to relief, while workhouses were thin on the ground and 'settlement' or entitlement to relief could be acquired only after five years' residence in a parish without receiving parochial relief. Thus the Poor Law intensified the migration impact of the cyclical downturns or slumps thought to be inherent in capitalism, particularly given the narrow sectoral base of the Scottish industrial economy and the occurrence of depressions within the context of a growing reserve army of labour.

Figure 9.2 shows how the spatial pattern of capital accumulation, concentrated in the textile, mining, metallurgical, shipbuilding and heavy engineering industries of the central belt, stimulated centripetal, inter-regional, migration flows which supplemented considerable local urbanward movement. Not only was urban employment expanding rapidly, but there was also the undermining of rural craft and cottage industries by the entry of capitalist mass-produced goods on the new rail and road systems. But there has been little research on the *specific* processes in labour recruitment and housing provision used by capital to shift labour around. One important factor was the nature of the rented housing market (Harvey, 1985), which did allow workers to adjust their housing and lodging readily to the booms and slumps in employment and wages that have been a feature of unregulated industrial capitalism.

Another integral part of capitalism is the class polarization between bourgeoisie and proletariat. Each class developed its own strategy of family formation, with the bourgeoisie in the late nineteenth century beginning to invest in the future human capital of its offspring by reducing fertility. Similarly, class polarization is evident in the spatial pattern of residential differentiation and intra-urban migration being established in nineteenth-century Scottish cities (Elliot and McCrone, 1980), in particular the beginnings of suburbanward flight by the bourgeoisie and allied middle classes. This produced a new social topography in cities, with the poor now on the inside and the rich on the outside – something that Engels described vividly but accurately for Manchester in *The Condition of the Working Class in England* long before Burgess and the Chicago school of human ecologists.

Restructuring of capital

The First World War brought to an end over a century of British capitalist expansion. Britain now had to face the increasing challenge of other national capitals – a challenge too great for older industrial regions like central Scotland, where capital responded to overcapacity in export-reliant, capital-goods industries and to labour militancy by mergers, closures, redundancies and capital flight overseas. The capitalist state increasingly intervened to staunch the falling profitability of capital and to preserve the stability of capitalism's social system by market regulation, selective nationalization, regional development policy and the welfare state. Nevertheless, the continuing severity of the economic structural problems of central Scotland, accentuated recently in light industry by several foreign-owned companies rationalizing their European operations, has been the basis of appreciable net out-migration to the rest of Britain and overseas. By the

1960s and 1970s the dominant manufacturing region, west central Scotland, was providing easily the highest number of Scottish emigrants overseas (on a per capita as well as absolute basis) and the Highlands and Islands and the North East the lowest (H. Jones, 1979).

As far as redistribution of population *within* Scotland is concerned, two forces intimately linked to late capitalism are dominant: oil developments and counter-urbanization. Oil has become the most important natural resource necessary for the prosperity of modern capitalist societies and their most powerful multinational corporations, providing the incentive to seek out new, sometimes expensive, but strategically secure energy resources on those peripheries of developed countries where geological conditions are favourable. The exploitation of North Sea oil and gas resources from the early 1970s has attracted appreciable capital and therefore labour to north-east Scotland. In terms of labour relations and work practices, safety record and labour recruitment (Philip, Taylor and Hutton, 1982), the oil industry in Scotland has demonstrated many of the classic features of unbridled, exploitative, nineteenth-century capitalism – hence the radical theatre's response in *The Cheviot, the Stag and the Black, Black Oil*. One outstanding form of labour recruitment has been the harnessing of an industrial reserve army in depressed west central Scotland for offshore work and short-term

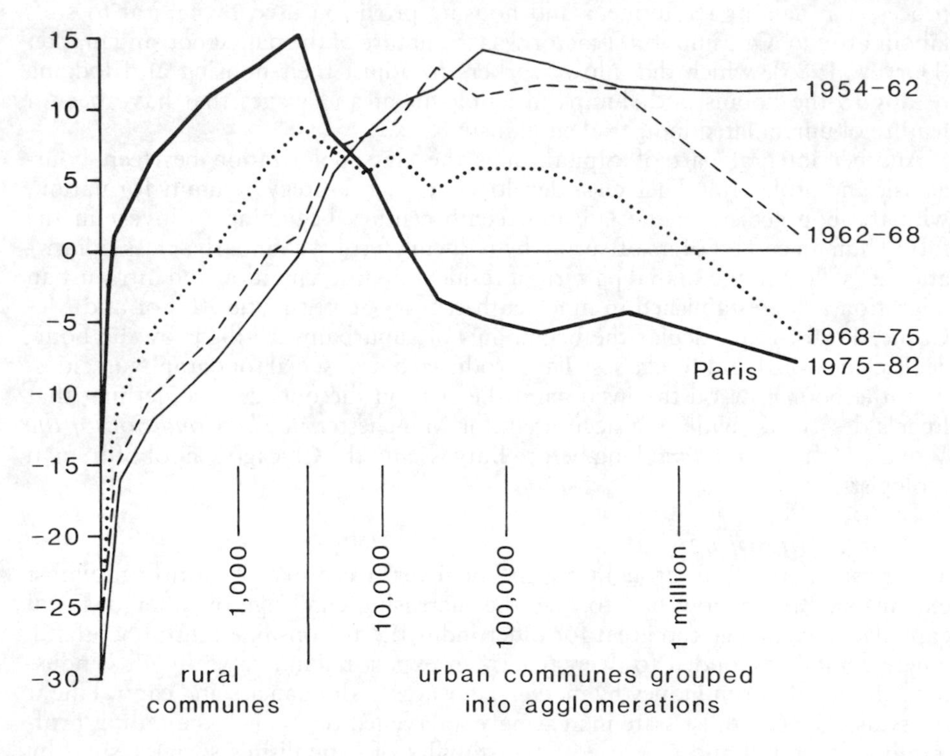

Figure 9.3
Average annual net migration rates (per 1,000 population) by settlement size category, France, 1954–82
Source: Fielding (1986), Figure 8.4.

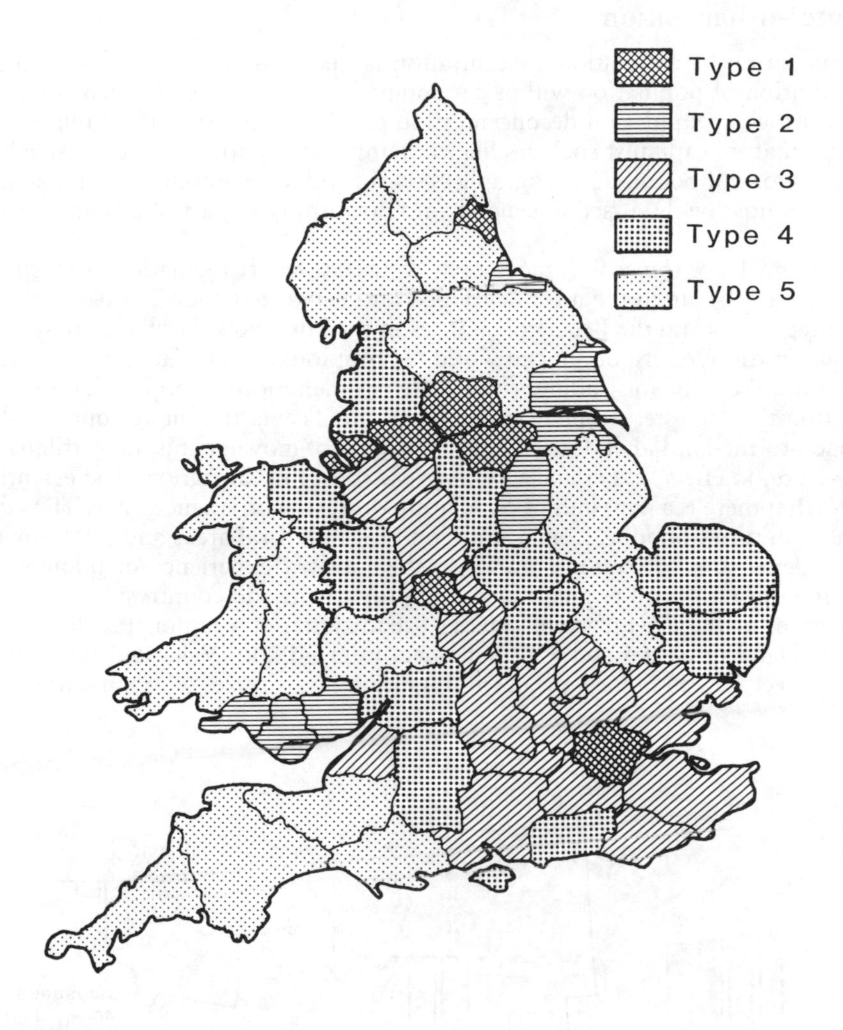

Figure 9.4
Net migration experience of counties in England and Wales in the 1961–6, 1966–71, 1971–4, 1974–7 and 1977–80 periods
Type 1 Net out-migration in four or five periods
Type 2 Net out-migration in three periods
Type 3 Net in-migration high in 1960s, lower in 1970s
Type 4 Net in-migration in at least three periods, but no overall trend
Type 5 Significant improvement in migration balance in 1970s, after net out-migration or low net in-migration in 1960s
Source: Champion (1983), Table 8.3.

construction work at fabrication yards and landfall terminals. Workers have been accommodated in camps, lodgings and, on occasion, redundant liners, returning to their permanent homes during periodic 'leave' in what sociologists have termed 'the intermittent husband syndrome'.

Counter-urbanization

If urbanization or population concentration is taken to characterize the spatial redistribution of population within developing capitalist societies, then counter-urbanization or population deconcentration can be regarded as the hallmark of modern, mature capitalist societies like contemporary Scotland. So extensive has been the work of population geographers on this relatively modern phenomenon that it will now be discussed in general, and not simply as part of a Scottish case study.

The trend back towards a more even distribution of population over space seems to have begun in the most developed and urbanized countries like Britain, the United States and the Benelux countries in the late 1960s but has now spread throughout the Western developed world (Champion, 1989). The term 'counter-urbanization' can be misleading, since there is much more to population deconcentration than the stereotype of urban drop-outs fleeing to remote countrysides in a back-to-the-land idyll. Indeed, the great mass of movement is inter-urban, so that we are, in effect, experiencing a new pattern of urbanization. The essential point is that there is a new shift of population *down* the urban hierarchy. Thus the 'population turnaround' is from a situation in which rural areas and small towns were experiencing net out-migration, to the modern experience of population losses in large cities and population gains in the towns and countryside.

Evidence for the emergence of counter-urbanization at a national scale is provided in Figures 9.3, 9.4 and 9.5 for France, England and Wales and the United States respectively. In France, the net in-migration into all sizes of urban settlement

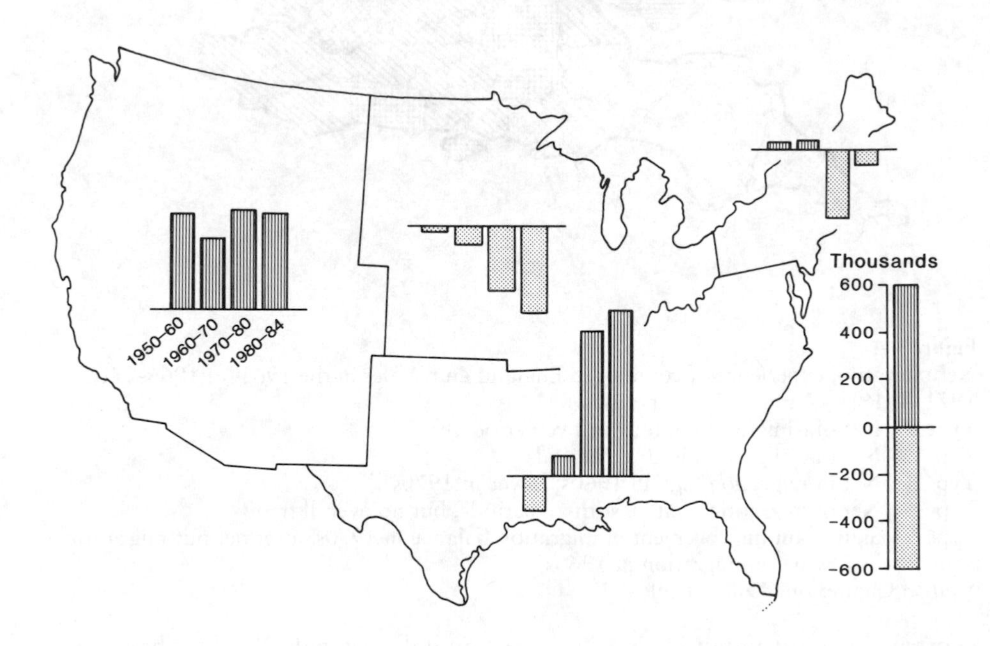

Figure 9.5
Average annual net migration for major US regions, 1950–84
Source: Long (1988), Figure 5.1.

in the 1950s and 1960s had given way to net out-migration in Paris in 1968–75 and in all but the smallest towns in 1975–82. Meanwhile, the rural communes had switched almost entirely from traditional net outflow of population to net inflow.

In England and Wales net outflow from all major conurbations has been a consistent feature from 1961 to 1980. Some other industrial areas like South Wales and Teesside have also experienced significant periods of net out-migration. The suburban and commuting zones around the major conurbations, particularly London, show continuous net in-migration, with rates being higher in the 1960s than the 1970s. On the other hand, the outlying, more rural regions experienced a conversion to, or increases in, net in-migration in the 1970s.

In the United States the long-established net inflow to the industrialized and urbanized North East is seen to have continued in the 1950s and 1960s, *entirely* due to the continued movement of blacks from the South (Long, 1988, p. 144). But counter-urbanization is evident in the net outflow from the region in the 1970s and early 1980s, as indeed from the Midwest and particularly from its eastern industrialized section. The South has been transformed from a rural reservoir of out-migrants to the primary destination – even more than the West – for inter-regional migrants.

In discussing the components and causes of counter-urbanization some commentators have focused on country-specific factors, but such has been the universality of the phenomenon that real explanation can be achieved only in relation to developments common to all Western industrial societies. Three fundamental determinants can be recognized: metropolitan expansion, economic restructuring and changing life-style preferences.

Metropolitan expansion

The spillover into surrounding countrysides of largely middle-class metropolitan populations seeking superior residential and environmental conditions is a long-established feature of urban ecology. In countries like Britain it has been complemented by the state's decanting of population from congested, nineteenth-century, inner-city housing to public sector housing in new towns and in those established towns having overspill agreements with London, Glasgow and the like. More recently, developments in transport have permitted an extension of the commuting field of more affluent city workers.

There is some evidence for a new trend from the 1980s of demographic renaissance in some of the metropolitan centres (Champion and Congdon, 1988). Contributory factors include environmental improvements in the inner city and a switch of housing policy towards the rehabilitation and upgrading of existing stock. There has also been the spectacular growth of financial and business services in some of the more prestigious metropolitan centres like London, Paris, Milan and Boston (Harrison and Kluver, 1989). In such cities many of the new young middle class have turned their back on the suburbanward movement sought by earlier generations. To them an inner-city residence, whether in gentrified, formerly working-class neighbourhoods or in land-use conversions like the London Docklands, offers easier access to work, entertainment and their fellows (Short, 1989). They conceive of suburbs as for families, where the needs of children are the dominant concern, whereas in archetypal 'yuppie' households the wants of adults are paramount. It is important to realize, however, that such inner-city social trends and population regrowth are unlikely to occur extensively, if at all, in the economically and environmentally disadvantaged, nineteenth-century manufacturing cities.

Economic restructuring

Since relocation of production is the basis of inter-regional migration, counter-urbanization must be underpinned by the changing space-economy associated with modern capitalism (Fielding, 1986). In response to the declining profitability of capital in many developed countries from the mid-1960s, there has been appreciable reorganization of productive capacities and labour processes.

There has been an important shift in relative attractiveness for industrial capital investment and therefore job creation/retention from metropolitan centres to small and medium-sized settlements. Goverment regional development policy encouraged this in the 1960s and early 1970s in countries like Britain where continued metropolitanization was thought at the time to prejudice public welfare, but more fundamental and universal is the appreciation that metropolitan concentration no longer provides compelling external economies for capital. 'The time-eliminating properties of long-distance communication and the space-spanning capacities of the new communication technologies are combining to concoct a solvent that has dissolved the agglomeration advantages of the industrial metropolis' (Berry, 1980, p. 18). This has permitted exploitation of advantages that outlying regions like the American South often offer to capital in land, taxes and the costs, turnover, flexibility and docility of labour. Myrdal's cumulative causation model (see Chapter 11) is thus being undermined by the emergence of a new spatial division of labour (Massey, 1984; Bradbury, 1985), based on the dispersal of routine production.

Initially, much of the dispersal took the form of corporate restructuring by large companies in the form of 'Fordist' branch-plant mass production. But as such production became subject to competition from newly industrializing countries in the Third World, the major industrial, and therefore population, generation during the 1980s has been in small, often high-tech firms producing new products for specialized markets by means of new work relations. A key locational factor is the quality of life attractions of smaller settlements in prestige environments to entrepreneurs and skilled personnel (Keeble and Wever, 1986; Aydolot and Keeble, 1988). But at a national level such growth is dwarfed by the de-industrialization of the major cities; in Britain, for example, the conurbations lost almost two million manufacturing jobs between 1971 and 1981.

The restructuring of office employment and consequent population shifts have been much less dramatic. Core regions like South East England continue to consolidate their advantages, but within them there has been some dispersal of the more routine activities.

Changing life-style and residential preferences

These seem to underlie the demographically least important and perhaps least sustainable element of counter-urbanization, but ironically the one that is perhaps the most widely discussed. This is the turnaround of many of the most rural and peripheral regions like the Scottish Highlands, northern Norway, the Massif Central and northern New England from their traditional pattern of depopulation to one of novel growth through net in-migration (Brown and Wardwell, 1980; Perry, Dean and Brown, 1986; H. Jones *et al.*, 1986).

There is a long history of anti-urbanism and idealization of pastoral escapism in Western intellectual tradition, so that we need to consider why there seems to have been a very real recent shift in residential preferences in an anti-urban, pro-ruralist direction (Dillman, 1979). Three factors may be cited, all related to important features of modern capitalist economies and societies:

1. An important precondition allowing more people to 'live out their preferences' is that more tolerable 'creature comfort' conditions in remote rural areas have been achieved through the widespread provision by the state of such facilities as electrification, piped water supplies, all-weather roads and network television. This can be achieved only in wealthy nations with surpluses to support populations in peripheral areas, despite apparent inefficiencies in so doing.
2. The expansion of private affluence in many developed countries, especially in the 1960s and early 1970s, has given more people, at least in the higher income groups, the financial means to indulge their residential preferences. An important financial enabling factor has been the capital raised by migrants selling expensive housing in metropolitan areas and purchasing cheaper property in remote rural areas. A not uncommon sequence is from holidaymaker to second-home owner to settler.
3. The trade-off that people make between material enrichment on the one hand and quality of life and environmental considerations on the other seems to shift significantly towards the latter at high levels of economic development. Bell (1976) interprets this as an expression of a fundamental contradiction in late capitalist society between a social structure predicated on the values of routine production and efficiency and a culture that emphasizes self-expression and fulfilment; while the Marxist notion of increasing alienation as a specific weakness of capitalism can be adapted to explain dissatisfaction with the work structures, consumption patterns and normative life-styles associated with metropolitan economies.

> Higher crime rates, traffic congestion and the hassle of commuting, and the day-to-day aggravation of having to deal with surly waitresses and cabbies, and the suffocating crush of unsmiling humanity have discounted to some degree the opportunities for upward social mobility and cultural advantages of the great Northern metropolitan centres.
>
> (Biggar, 1979, p. 26)

Also relevant has been the great surge of interest in environmental issues.

It is wrong to think of the new rural incomers mainly as affluent retireds. There are many economically active incomers who have certainly had to accept a lower material standard of living, but who find some 'enabling' income in a variety and often ingenuity of ways (small-scale farming, craftworking, tourism, and self-employment sometimes linked to metropolitan demand).

Because it is a novel trend, it is easy to exaggerate the role of people-led, as opposed to employment-led, migration to the periphery – hence the infectious hyperbole of terms such as population turnaround, rural revival and green wave. It is salutary to note, therefore, that many parts of rural peripheries continue to experience depopulation, particularly among the younger and better educated. For example, in the Scottish Highlands and Islands about half of the rural enumeration districts continued to experience depopulation between 1971 and 1981, while in southern Scotland the proportion exceeds three-quarters. In rural

Scotland the demographic impact of long-distance in-migration has been concentrated largely in areas like Mull, Skye and Wester Ross that have obvious scenic attractions and romantic imagery. A particularly contentious issue there, as elsewhere, is the interaction of incomers and locals, particularly in the housing market and in local power structures.

The future of counter-urbanization?

There have been several attempts in the 1980s (e.g. Rogerson, 1987) to demonstrate what are thought to be significant new trends in counter-urbanization, notably a reduction in overall mobility and a slowing down, or even reversal, of the net flow from urban to rural areas. However, the identification of such trends from country to country has lacked consistency, and no convincing relationship has been identified with the deepening business depression from the mid-1970s to the early 1980s and the subsequent recovery. In reality, any 'new' trends are merely modest changes in small net balances between large gross flows.

At high levels of economic development, spatial convergence in socio-economic conditions ensures that there is 'no simple unidirectional movement of people but rather a complex pattern of gross flows, which merely add up to a prevailing tendency towards deconcentration' (Champion, Fielding and Keeble, 1989, p. 80). Consequently, 'the orientation of migration research may shift from the dichotomous metropolitan–non-metropolitan or urban–rural focus of the past to more complex, but more analytically useful, categorizations of origin and destination' (Long, 1988, p. 206).

Rural-to-urban movements in less developed countries

These comprise the dominant form of territorial mobility during Zelinsky's early transitional stage of the mobility transition (Figure 9.1), associated with a growing rate of natural increase, the beginnings of economic transformation, the provision of modern communications and a general rise in material aspirations.

Scale of urbanization

UN estimates suggest that between 1950 and 1980 the urban population of less developed countries (excluding China) increased by almost 600 million, comprising almost 40 per cent of the *global* (excluding China) population increase. In 1975, for the first time in world history, a majority of the world's urban population was located in the less developed countries, and by 2000 the proportion is expected to reach two-thirds. The *rate* of urbanization (i.e. the increase in the urban proportion of population) is remarkably similar to that in the earlier European and North American urban transition. Thus, in the quarter-century from 1950 to 1975, the urban proportion grew from 17 per cent to 28 per cent in less developed countries, compared with from 17 per cent to 26 per cent in today's developed countries between 1875 and 1900. But, because of higher rates of natural increase, the actual growth of the urban population is much higher in today's less developed countries – 188 per cent between 1950 and 1975, compared with 100 per cent in developed countries between 1875 and 1900 (Preston, 1979). Within the Third World, urbanization levels are much higher in Latin America (68 per cent of the population in 1985) than in Africa and Asia (both 30 per cent).

Models of dualism and dependency

Urbanward migration as a spatial and sectoral redeployment of human resources can now be considered in relation to general theories of development.

In the 1950s and 1960s the modernization perspective dominated development literature. This was based on a model of dualism, or the coexistence within a national society of two largely separate sectors or systems: the traditional sector and an expanding modern sector. The traditional sector was seen as subsistence-based, of low productivity and technologically backward, in contrast to the productive, innovative modern sector which concentrates industrial development, modern government services and material advancement generally.

Rural-to-urban migration was thus regarded as a positive agent in beneficial structural change. Opportunities in the modern sector, initially present in areas of plantation agriculture and mining created by European investment, became increasingly concentrated in large urban centres, particularly the primate cities which have commandeered the great bulk of modern manufacturing investment and centralized government services. In the early days of modern sector development there is evidence of forced labour transfer, but a more potent and durable cause of population movement has been the cash needs, particularly for taxes, school fees and rising material aspirations, generated by the development of a money economy. The need for urbanward movement was enhanced by what was widely thought to be conditions of excessive fertility, over-population and environmental vulnerability in crisis zones like the Sahel and north-eastern Brazil.

Many of the value judgements underpinning the classic, essentially Eurocentric, dual economy model, in particular the implication that desirable development is generated within and diffused from an expanding modern sector, have been attacked in modern Marxist analyses of labour migration. It is argued that the realities of the world political economy are such that traditional, community-based modes of production in the Third World are transformed by Western capitalist penetration, operating through its colonial metropolitan beach-heads, into conditions of dependency and *under*development (Frank, 1969), one of the strongest manifestations of which is labour migration.

In such analyses the effective functioning of the traditional domestic mode of production is seen as being undermined or 'maimed' (Cliffe, 1978, p. 329) by several external dislocating forces. First, capitalization of agriculture, particularly when associated with rapid population growth, results in land concentration and landlessness. Second, capitalist mass-production and marketing of consumer durables destroy domestic craft production. Third, colonial and neo-colonial state policies discriminate against the traditional mode in taxation, in investment allocation and in the pricing of rural products: 'everywhere it is government by the city, from the city, for the city' (Gugler, 1986, p. 202). Under such conditions of dislocation and underdevelopment, migration becomes a channel through which the surplus value of migrants' labour is driven or drawn from the periphery into the capitalist core. It is argued that capitalism is thus subsidized by its labour force being 'fed and bred in the domestic sector' (Meillassoux, 1981, p. 95), to which it returns for sustenance – regularly in the guise of circulation and permanently as return migration.

More sophisticated structuralist arguments (e.g. de Janvry, 1981) have qualified this simple model of capitalist penetration and dissolution of the traditional economy. More emphasis is placed on the varying functional conjuncture or articulation of the two modes of production under varying historical and geographical conditions.

> The interests of the capitalist class involved, as well as those of the peasantry, vary over time and across regions. The 'maintenance' or destruction of peasant production reflects the opportunities available in a particular period and the outcome of the struggles between these groups in various regional contexts.
>
> (Collins, 1988, p. 18)

Clearly, then, there will be different forms and levels of migration associated with different interactions between modes. But beyond dispute is that 'internal migration is a manifestation of the progressive and cumulative incorporation of provincial areas into the dominant national urban economy' (Roberts, 1978, p. 98). Nevertheless, there are three important ways in which rural-to-urban migration in the Third World differs from its earlier counterpart in developed countries: primate city dominance, circulation and urban underemployment.

Primate city dominance

In most Third World countries a metropolis-periphery model is the basis of spatial inequalities, and the overwhelming target of migration is the national metropolis, where the concentration of modern sector activity and the focusing of transportation systems have created hyper-urbanization of striking proportions. Resources for both investment and consumption are disproportionately allocated to the capital city because that is where the power-base of the government lies. Bangkok is perhaps the most primate city of any country in the world. Containing just over half of Thailand's urban population, its population in 1960 was already 26 times that of the second largest city, but by 1980 this had risen to over 45 times. Another example is provided in Tunisia, where the major net migration flows within the country between 1969 and 1975 can be seen to concentrate heavily on the city of Tunis (Figure 9.6); net flows to other large settlements like Sousse and Sfax have been small.

Circulation

This term denotes a variety of movements, usually short term and repetitive, in which people leave their places of residence for varying periods but eventually return to them. The dominant movements link rural and urban areas, so that appreciable sections of the population have a dual dependence on city and village. Such bi-local life-styles were not uncommon in Europe during its transition to capitalism, as illustrated by the movements between the Scottish Highlands and urban central Scotland discussed earlier. But the resort to circulatory movement rather than permanent migration does appear to be much greater in the contemporary Third World.

It is impossible to monitor the scale of circulation because censuses, which are modelled on Western practice, rarely employ questions and definitions appropriate to this flexible form of migration. Since most circulators regard their usual residence as their rural family home, censuses tend to over-enumerate *de facto* rural populations and under-enumerate urban populations. Most information on circulation has to be obtained from intensive small-scale studies, in which population geographers have been well to the fore (Prothero and Chapman, 1985).

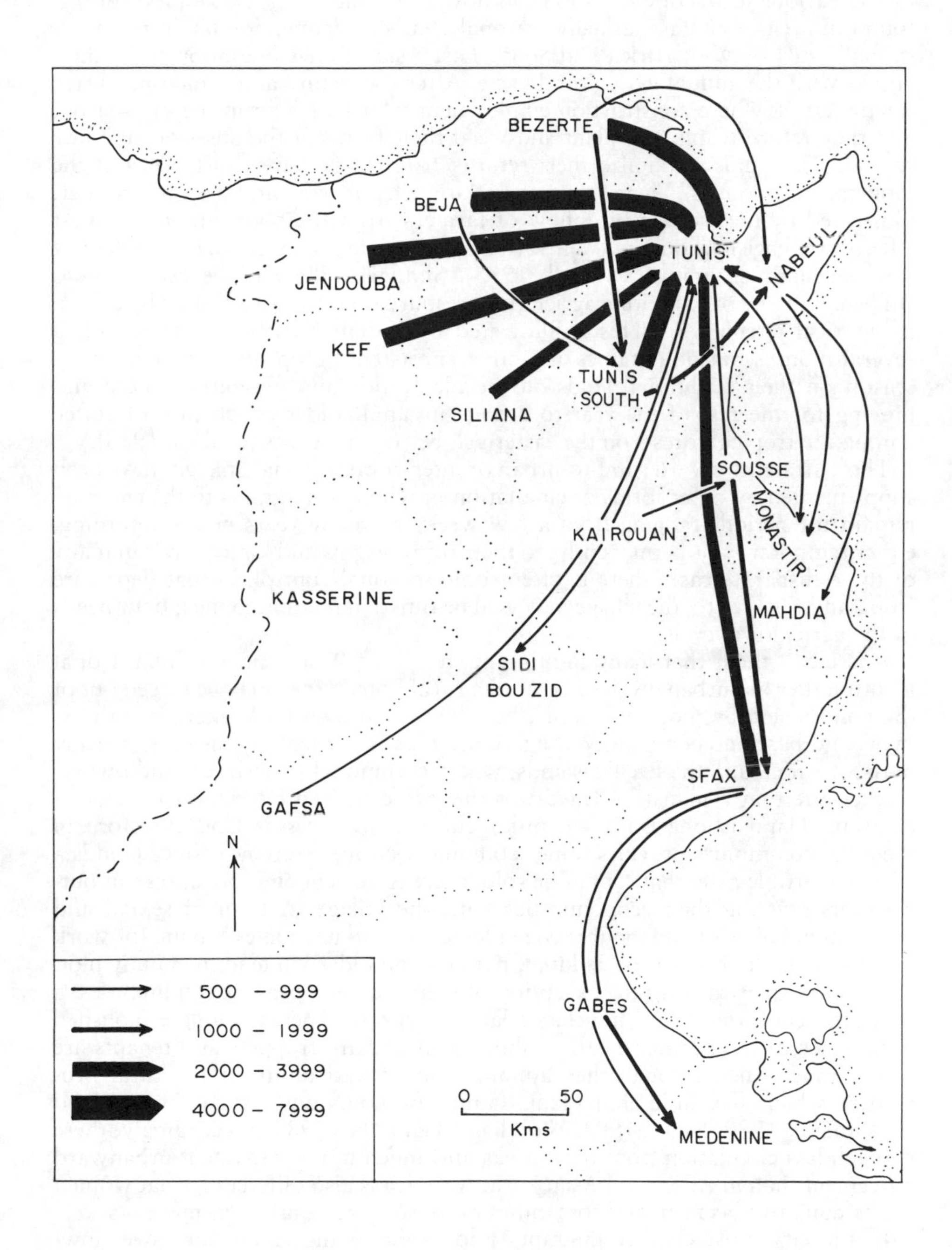

Figure 9.6
Net migration between administrative divisions of Tunisia, 1969–75
Source: Findlay and Findlay (1980), Figure 1.

The various forms of circulation can now be considered. The simplest form is found in areas where a markedly seasonal climatic regime, notably in the monsoonal lands of West Africa and South East Asia, allows migration to be integrated with the annual agricultural cycle. After the main harvest migrants leave for the cities, where construction employment is at its height in the dry season, and they return to plough, plant and weed their farms at the onset of the rainy season. They thus maximize their returns from both the subsistence and the monetary economies. There is also circulation from one rural area to another, illustrated by the wet-season inflow of farmers from the poor interior of West Africa to engage in groundnut cultivation in The Gambia under what is virtually a form of share-cropping (Swindell, 1985). Similarly, there is the centuries-old Andean strategy of exploiting agricultural resources at diverse altitudes by a form of 'vertical circulation'. This is illustrated by Indian peasant groups spending several months raising potatoes and other subsistence crops during the growing season on their traditional lands on the bleak altiplano of southern Peru and moving for the rest of the year to engage in smallholder production of coffee among the tropical forests on the eastern slopes of the Andes (Collins, 1988).

These straightforward rural-to-urban or inter-rural seasonal linkages have been supplemented by other forms of circulation in which in-migrants to the city may remain for periods ranging from a few weeks to many years before returning, either temporarily or permanently, to their rural origins and kinfolk. But in many of these disparate cases there is often some seasonal control, in that departure from, and re-entry to, the village may well be timed in relation to the labour needs of the agricultural cycle.

It is clear, then, that many in-migrants to Third World cities exhibit a dual identity; they are urban dwellers loyal to a rural home. They may be forced out of the countryside by monetary needs, but they are drawn back intermittently by their land base and community obligations. This is particularly the case in many parts of Africa and the Pacific islands, where communal patterns of land ownership ensure that a man who leaves for the city can keep his claim to a plot of communal land as long as his wife or kin cultivate it. Access to land, therefore, in the natal community provides some economic security. Moreover, since land is a non-convertible asset that cannot be sold, there is no economic advantage in out-migrants severing their economic ties with the village. A form of sexual and generational division of labour is engendered, with young males leaving for work in the city while their wives, children, parents and older kin tend the family plot.

In Latin America, with the exception of some Andean peasant communities, a very different form of land and class relations prevails. There is a long-established dominance of large feudal estates, whose peons, share-croppers and tenants are being transformed through the capitalization of agriculture into a rural proletariat which has little individual, family or community stake in the land (Buksmann, 1980; Balan, 1983; Wood and Carvalho, 1988). Accordingly, there is much less circulation from rural areas and much more permanent urbanward movement than in Africa and Asia. Latin America is also different in that women are as numerous as men, and sometimes more so, among urban in-migrants.

In the city most circular migrants find work in the labour-intensive, low-productivity, informal or unregulated sector – the so-called bazaar economy – embracing activities like street vending, barbering, shoe shining, dishwashing and pedicab driving. This is partly because formal sector employment demands certain levels of education and skills as well as a commitment to regular working

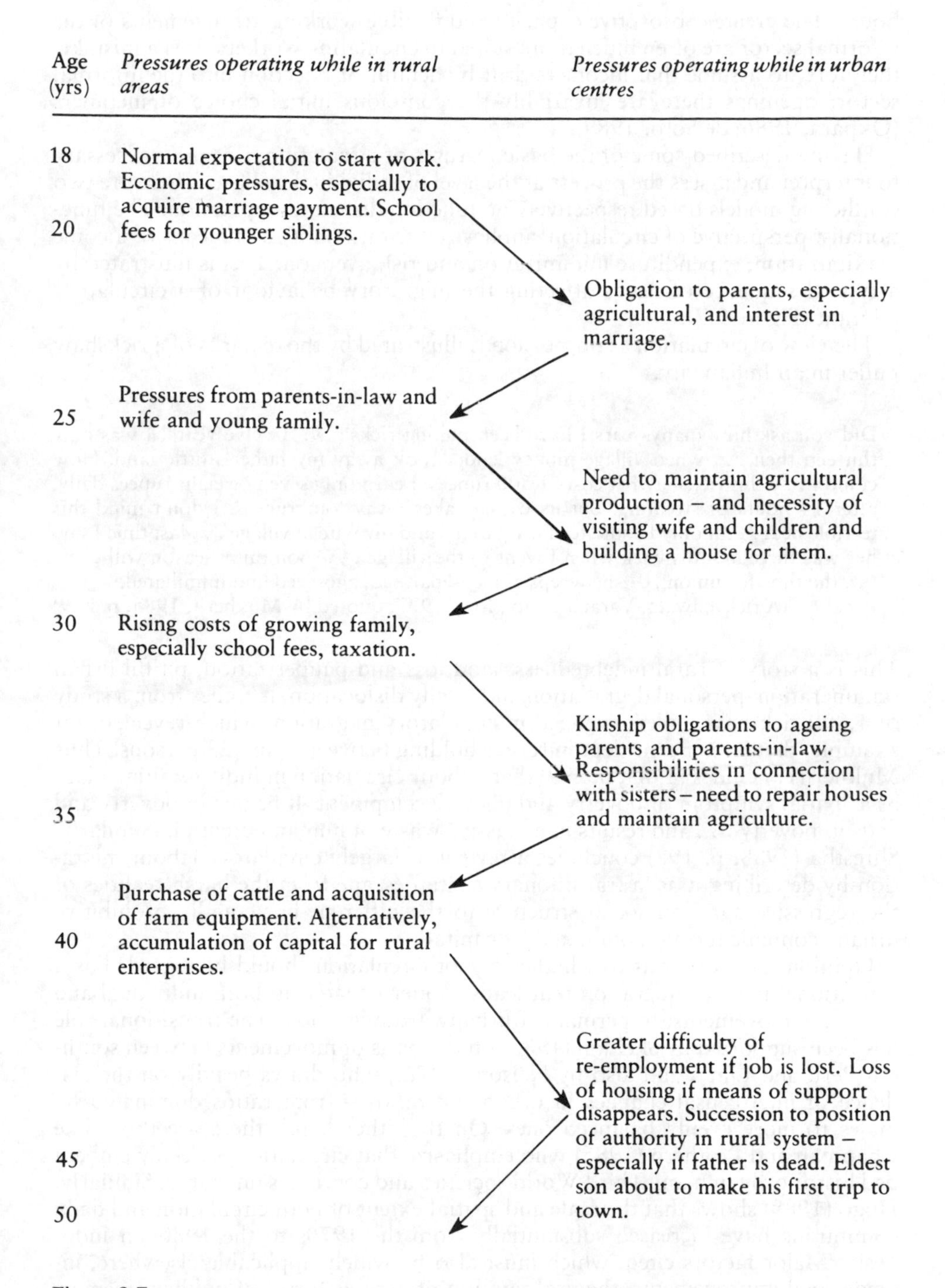

Figure 9.7
Representation of a migrant worker's career
Source: Mitchell (1969), Figure 3. By permission of Cambridge University Press.

hours. The greater absorptive capacity and flexible working arrangements of the informal sector are often much more suited to circulating workers. It is a mistake, therefore, to assume that incomers drift by default or rejection into the informal sector; openings there are invariably the conscious initial choice of incomers (Okpara, 1986; de Soto, 1989).

Having described some of the basic features of circulation, it is now necessary to interpret and assess the process at the level of individual welfare. There are two conflicting models based respectively on functionalism and oppression. The functionalist perspective of circulation emphasizes its rationality as a means of income maximization, expenditure minimization and risk aversion. This is illustrated by a typical sequence of needs affecting the migratory behaviour of a circulatory migrant (Figure 9.7).

The view of circulation as oppression is illustrated by the remarks of a rickshaw puller in an Indian city.

> Did you ask how many years I have been pulling rickshaw? Twelve years! I was only thirteen then . . . when village money lender took away my father's little land. How could I buy this rickshaw? It costs 1,500 rupees. I earn only seven or eight rupees daily, after 14 hours of sweating. But its owner takes away four rupees. I don't mind this terrible heat . . . if only I could feed my parents and my wife in village . . . last time I saw her was three months back when I went to the village . . . soon rainy season will come . . . the time for union . . . she weeps alone, separated, agonized and humiliated.
>
> (A rickshawala, Varanasi City, May 1977, quoted by Mukherji, 1985, p. 279)

This is a story of rural indebtedness, land loss and pauperization, pitiful urban remuneration, personal degradation and family dislocation. It comes from a study of 1,000 rickshaw pullers engaged in circulatory migration, which revealed, for example, that 90 per cent slept in rooms holding between 8 and 23 persons. Thus Mukherji concludes (1985, p. 297) that 'labour circulation in India is taking place as a spatial symptom of poverty and underdevelopment. It begins in poverty and ends in poverty . . . and results in a massive waste of human potential.' Similarly, Shrestha (1988, p. 198) concludes a review of largely circulatory labour migration by describing it as 'a transitionary spatial escape from the harsh realities of the regressive agrarian social structure to the different but equally prohibitive urban economic terrain dominated by capital'.

Opinions also differ as to whether or not circulation should be regarded as a transitional form of migration that leads sooner or later, in both individual and aggregate movements, to permanent urbanward migration. The transitional role has been supported by Skeldon (1985) on the basis of movements between southern Peru and Lima, and also by Nelson (1976), who draws heavily on the evidence of inter-censal changes in urban sex ratios – from ratios dominated by males to more evenly balanced ones. On the other hand, there are those like Chapman and Prothero (1985) who emphasize that circulation is deeply embedded historically in many Third World societies and continues unabated. Similarly, Hugo (1988) shows that the scale and spatial extent of both circulation and daily commuting have increased substantially from the 1970s to the 1980s in Indonesia. Major factors cited, which must also be widely applicable elsewhere, include road improvements, the proliferation of various forms of public transport, and the impact of mass media in expanding the horizons of village-dwellers. Hugo quotes a 1984 World Bank estimate that at least 25 per cent of rural households in Java had at least one family member working for part of the year in an urban area.

Urban underemployment

Third World cities have not replicated Western experience in the nineteenth century when appreciable growth of manufacturing employment was the leading edge of urbanization and labour absorption. In Third World cities the mushrooming numbers of 'marginal' urban dwellers survive through the 'trickling down' of modern sector wages into the ever-expanding but 'involuting' informal sector. Urban capitalist development

> has gone beyond its intended objective as a result of its own dynamism . . . the 'rural-exodus' has become uncontrolled, uncontrollable and explosive . . . what is 'functional' at one stage becomes 'dysfunctional' at another.
>
> (Amin, 1974, p. 98)

> An increasingly greater number of workers perform a decreasingly smaller amount of work, and poverty is shared among them. Urban growth occurs without development.
>
> (Mukherji, 1985, p. 293)

> A significant proportion of immigrant workforces [in Indonesian cities] is occupied in low productivity petty commodity production, contributing primarily to their own subsistence and reproduction rather than economic development through the expansion of productive enterprises.
>
> (Forbes, 1984, p. 157)

Incomers to the urban informal sector may well be worse off, in real terms, than if they had stayed in the rural economy. But *some* in-migrants are widely known to have been highly successful, so that incomers are prepared to play an urban employment lottery game.

> There is a considerable element of chance in this game because much of the hiring is haphazard. And the game is very serious: rural income is forgone, costs are incurred in migration, severe hardship is experienced in urban employment. But new migrants keep joining in the 'gold rush' prospecting for urban employment.
>
> (Gugler, 1976, p. 192)

One influential economic model suggests that migrants will continue to move cityward as long as their *expected* income is greater than their rural income. The expected urban income may be defined as the product of a modern sector income and the probability of gaining a job in that sector (Harris and Todaro, 1970).

One effect of urban labour surpluses has been to modify the role of labour circulation. Migrants still maintain strong ties with their rural origins, but they are more reluctant to make periodic returns of long duration, because of the difficulty of regaining urban employment. This does not necessarily imply a transition to permanent family migration, since modern transportation developments permit a greater spatial range for both commuting and periodic visiting.

The political economy perspective: the way forward

Few would now dispute the value of structuralist perspectives in shifting migration studies away from their traditional emphasis on particularist, atheoretical studies focusing on the essentially voluntary, opportunistic behaviour of individuals. On the other hand, there has been criticism of the coarseness, rigidity and total determination of Marxist structuralist explanations which sometimes ride roughshod over complex and regionally diverse situations. This is why some migration analysts (e.g. Forbes, 1984; Gilbert and Kleinpenning, 1986) are attracted to Giddens's (1979) concept of 'structuration', which deals with the complexity of interaction between structural forces and human agency. It is particularly important to see how different individuals and groups respond in their migration behaviour to what are undoubtedly common structural causes of social and spatial inequality. What we certainly need are theorized case studies of migration – in other words, empirical investigations that are theoretically well informed.

10

INTERNATIONAL MIGRATION

The distinction between internal and international migration reflects the compartmentalization of the modern world into sovereign states and its effect on the recording of territorial mobility. Clearly, many features of migration are common to internal and international movements – the roles of spatial inequalities, distance decay, selectivity, decision-making, self-perpetuation, counterstreams and the like – but the extent of national government intervention and control does add a distinctive dimension to international migration. This dimension has become increasingly dominant in the modern world, as this spatio-temporal study of the evolution of international migration will demonstrate. Another important distinction is the greater change in socio-cultural environment involved in moving from one country to another, with all the implications, spatial as well as social, for migrant adaptation and assimilation.

Origins

At early periods, when national territories were only loosely defined, the distinction between internal and international migration is meaningless. A more appropriate distinction is that between short- and long-distance movement. Davis (1974) argues that it was the rise of urban civilizations in the Near East and Mediterranean regions that initiated a significant level of long-distance migration through the technological inequalities between territorial groups. The civilized core regions drew peasants and artisans from their immediate hinterlands, and also so-called barbarians from the periphery and beyond. Movements of the latter were of two types. First, there was the slave labour captured in successful military campaigns like those of Imperial Rome in North Africa, Gaul, Britain and the Rhine frontier. At the height of the Roman Empire between a quarter and a third of Rome's population are thought to have been slaves. It is hardly surprising that a proposal to make slaves wear distinctive dress in Rome was rejected on the grounds that it would be too dangerous to let them see how numerous they were! Plinny the Younger is known to have possessed 500 slaves between his town house and country estates, and any Roman who could muster less than eight slaves was of very little account (Cunliffe, 1978). The majority of slaves were employed in arduous or menial occupations like pyramid building in Egypt, silver mining in Greece and domestic service in Rome, although some skilled trades were also a slave preserve, like the Greek tutors of wealthy Roman households.

Second, there were the periodic surges of horse-riding warrior nomads from the throbbing pulse of interior Asia into its wealthier and more civilized fringes. A wave-like pattern of tribes pushing their neighbours before them can often be identified, with the outer ripples finding expression in such conquering invasions as the Hittites into Anatolia, the Scythians into Egypt, the Aryans into the Indus cities, the Huns, Goths and Vandals into the Roman Empire, the Seljuk Turks into Persia, the Mongols into China and Asia Minor, and the Ottoman Turks into the Balkans. Not dissimilar movements were those of the Angles, Saxons, Jutes and Vikings into Britain.

European expansion overseas

By about the seventeenth century long-distance migrations throughout the world began to comprise a single network organized in the interests of the politically and economically dominant states of north-western Europe (Wallerstein, 1979). The voyages of discovery and subsequent reconnaissance had opened European eyes to the possibilities of exploiting the New World, but the manner of exploitation, and hence the associated migration patterns, varied in relation to the particular physical and cultural environment encountered and to the particular European origin of the colonizers. Two environments stand out in terms of the amount of international migration generated.

The first type, comprising sparsely peopled, tropical and subtropical coastlands, was quickly exploited, because of maritime accessibility and the potential provided by warm, humid climates for the production of exotic crops. Thus, in the coastlands of the Caribbean and of the Americas from Virginia to Brazil, Europeans organized the commercial production of cotton, sugar, tobacco, coffee, tea, spices, indigo and rice. In the absence of any significant indigenous population, large quantities of cheap labour had to be imported. Initially, Europeans were used under an indenture system, whereby an individual would contract to work for an employer abroad for a fixed number of years in exchange for his passage and subsistence, and at the end of his term he could become a free settler. But since Europeans were poorly equipped to work in such climates and since the numbers coming forward under a basically oppressive recruitment system were always limited, plantation managers quickly resorted to slave labour from Africa, the first cargo of slaves arriving in Virginia, for example, in 1619. The number of slaves leaving Africa has been estimated as 11–12 million, with a mortality toll during the voyages of about one-tenth to one-quarter (Curtin, 1969).

With the abolition of slavery during the nineteenth century, British and Dutch colonialists substituted a semi-slave trade in indentured labour from the huge population and poverty reservoirs of Asia, especially India and China, where it was not difficult to recruit men willing to tolerate a period of hard labour abroad at a low wage; often the only alternative was starvation. While there were appreciable flows to Guyana and Trinidad, the major destinations were the new zones of plantation agriculture in Malaya, Sumatra, Ceylon, Fiji, Hawaii, Natal, East Africa and Mauritius. Coolie labour was also imported extensively for tin mining in Malaya and Sumatra and for railway construction in East Africa and California. Not all of

Table 10.1 Estimated number of ethnic Chinese in selected Asian countries about 1970

	Number ('000)	% of total population
Thailand	4,930	14
Indonesia	4,800	4
Malaysia	3,600	34
South Vietnam	2,660	15
Singapore	2.200	98
Cambodia	2,000	30
Burma	500	2

Source: Bouvier (1977), Table 1. By permission of the Population Reference Bureau, Inc.

these international movements within and from Asia were confined to indentured labourers, and, in addition, many labourers remained at their destinations on the expiry of their contracts, often to enter and eventually dominate many sectors of commerce. Hence, by the time that barriers were placed on immigrant contract labour early in this century, because of growing tensions at destination between immigrant and native workers and because of government concern at origin about notorious abuses of the system, major external pools of ethnic Chinese and Indians had been created, whose growth by natural increase provides substantial minorities in many countries today (Table 10.1).

The second type of New World environment to attract European investment and settlement was the sparsely peopled, temperate zone grassland and deciduous woodland, environmentally well suited to established European technologies, agricultural practices and settlement patterns. The huge outpouring from Europe to such environments in the Americas, South Africa, Australia and New Zealand constitutes arguably the most important migratory movement in human history, embracing some 55–65 million emigrants from Europe between 1820 and 1930, or about one-fifth of Europe's population at the beginning of the period. But for as much as three centuries after discovery, these lands had received a mere trickle of settlers. The earliest colonial powers, Spain and Portugal, had consciously discouraged permanent migration as they sought to plunder their possessions for luxuries and precious metals, while the Dutch, French and British encouraged only as many settlers as seemed necessary to consolidate their territorial claims. Demographic pressures within Europe were rarely acute before the nineteenth century, and 'few people were so poor or so persecuted that they wanted to transfer to a wild area to live under subsistence conditions and battle savages' (Davis, 1974, p. 97). Such distant lands were considered fit for military and commercial adventurers, deported criminals, paupers, political and religious dissidents, but hardly for the decent ordinary citizen. Hence by 1800 the white population of the United States had barely exceeded 4 million, and as late as 1840 there were less than 200,000 Europeans in Australia and a mere 2,000 in New Zealand.

What then were the factors that opened the emigrant floodgates in the decades following the Napoleonic Wars? These can be readily appreciated by reference to the simple push-pull and intervening obstacles conceptualization of migration (Figure 8.10).

Push factors

Demographic pressures built up in eighteenth- and nineteenth-century Europe as mortality rates fell with the onset of modernization. The resultant dislocation of agricultural and social systems involved severe rural congestion, farm subdivision and landlessness at a time when industrial growth was often insufficient to absorb surplus rural population; in fact, in the early stages of industrialization, mechanization often led to severe unemployment among artisans. Rural employment problems were exacerbated where transformation of farming structures involved land consolidation and tenant eviction, as in the British enclosure movement generally and the notorious Highland clearances specifically.

The role of emigration as a safety-valve to European population growth can be illustrated by the total of 750,000 emigrants from Norway and 1,100,000 from Sweden between 1840 and 1914, equivalent to 40 per cent and 25 per cent respectively of each country's natural increase during that period (Grigg, 1980). But the most spectacular, although atypical, example of blood-letting is Ireland. The population of the country had increased from 5 million in 1800 to almost 8.5 million by 1845, only to drop by nearly 2 million by 1851. Some 800,000 are thought to have died of disease and starvation, and a further million emigrated, during the fateful five years following the ravages of potato blight in 1846 (Cousens, 1960).

Pull factors

Complementing land hunger in Europe was the lure of virgin lands in the New World, by now tamed by the pioneers. Particularly attractive environmentally to agriculturists were the humus-rich grasslands and deciduous or mixed woodlands of the North American interior, the South American pampas, the South African veld and large parts of New South Wales, Victoria and New Zealand. These areas were sparsely peopled and barely exploited agriculturally by their indigenous populations, and an additional inducement in most, although not all, areas was a liberal government policy of public land distribution at little or no financial cost to *bona fide* settlers. For example, the Homestead Acts in the United States and Canada in 1862 and 1872 respectively offered to settlers, for a nominal fee of $10, title to a quarter-section (160 acres) of public land after five years' residence and farming.

Intervening obstacles

Major deterrents to overseas migration are often an unawareness of opportunities and the physical, financial and social hardships encountered in any long-distance movement; but both deterrents receded significantly in the nineteenth century. Knowledge of the new continents was enhanced by the development throughout Europe of pamphlets, newspapers and postal services, and once a significant movement had been initiated, information transmission through a friends-and-relatives network was assured.

Before the nineteenth century oceanic voyages were long and arduous, with the typical transatlantic crossing taking two months and a considerable mortality toll of passengers. Sailing vessels were improved early in the nineteenth century, but, more important, they were replaced by steamships from the middle of the century, with their greater space, speed and safety standards. Transport developments on land were equally important. Railways carried emigrants to the ports of embarkation – a vital consideration in continental Europe – and they also opened up the New World interiors.

Transatlantic shipping companies and American transcontinental railway companies lubricated intercontinental migration in another, more positive way. They drummed up business for themselves by their extensive promotional activities within Europe: they recruited for employers, they extended credit for fares and land purchase, and they widely propagandized the virtues of the new lands. A common poster was that of an American ploughman turning up piles of coined dollars, and the exaggerated claims of one railway company, the Northern Pacific, led to its territory being derided as 'the Banana Belt'.

One way in which obstacles to overseas migration can be eased is government promotion and assistance at origin, but there is little evidence of this in nineteenth-century Europe where *laissez-faire* policies held sway and where both business and military interests favoured the retention of large populations. In Britain the question of government assistance to emigrants was widely debated, and strongly advocated by parliamentary committees examining poverty in Scotland and Ireland, but proposals were always rejected, except in the case of particular groups like discharged soldiers. The proper role of the British government was thought to lie more in a general overseeing of emigration through the Emigration Commissioners, the Poor Law Commissioners and emigration agents in overseas ports, so that any financial assistance to British emigrants was left to parish authorities, philanthropic societies, landlords and emigration societies funded by public subscription. Such assisted emigrants never exceeded one-tenth of all British emigrants (Guillet, 1963).

Immigration into the United States

The United States has been the dominant destination in the overseas expansion of European population, attracting a gross immigration of some 35 million Europeans between 1820 and 1930 – almost two-thirds of European exodus at that time. The United States has proved more attractive than alternative destinations in its relative proximity to Europe, in its rich endowment of agricultural and mineral resources, and in the way its political and social institutions have promoted ready assimilation and social mobility among European immigrants. In this last respect, South American destinations were particularly disfavoured by the master-and-man relationship of aristocratic landowner and powerless tenant which was the legacy of Spanish and Portuguese colonial regimes. In Brazil and Argentina tenancy conditions were such that landlords were able to commandeer a good deal of the cash crop profits, as well as all of the increased land values consequent upon settlement expansion.

The era of unrestricted entry to the United States is traditionally divided into three periods. Up until 1820 immigration was slight and was largely from Britain. At the first federal census in 1790 persons of English descent accounted for 61 per cent of the population, with a further 18 per cent being Scottish and Irish descent. The only other significant European groups were the Germans in Pennsylvania, the Dutch in New York and the Swedes in Delaware (Thomlinson, 1965).

The period from the 1820s to the 1880s is known as the 'old migration'. Immigration increased rapidly as canals and railroads opened up the interior, bringing in their wake waves of pioneer settlers. The origins of immigrants widened, particularly to include appreciable numbers of Germans and Scandinavians (Table 10.2), but 95 per cent of all immigrants still came from northern and western Europe. Clearly it was not the poorest and most congested countries of Europe that led the way in emigration, and even within famine-ridden Ireland it was not the most destitute western regions that showed the highest regional rates of emigration (Figure 10.1). This indicates the importance

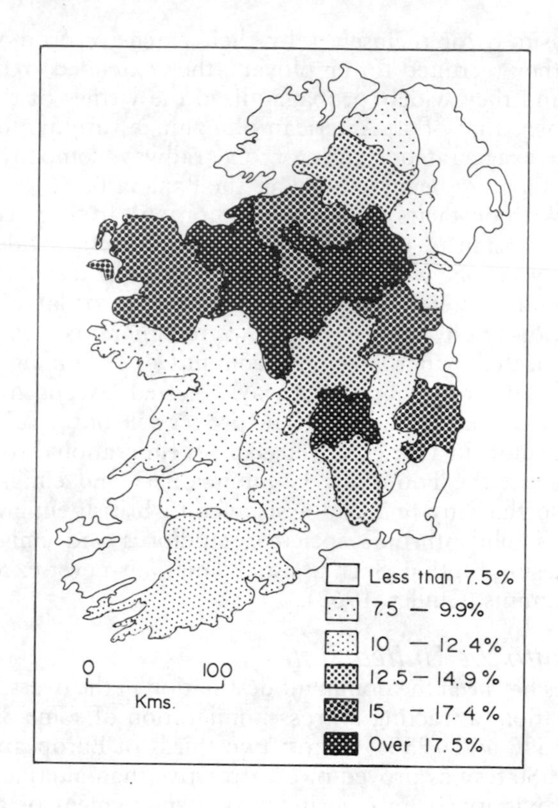

Figure 10.1
Emigration during the Irish famine, 1846–51, as a percentage of the 1841 population
Source: Cousens (1960), Figure 1.

of information provision, transportation facilities and some private capital in stimulating long-distance migration.

During the 1880s, the 'old' gave way to the 'new migration'. Immigration from much of north-western Europe subsided as fertility began to fall and as manufacturing employment expanded; indeed, Germany became a country of net immigration. The zone of heaviest emigration had now shifted to southern and eastern Europe, particularly Italy, Greece, Austria-Hungary, Russia and Poland (Table 10.2), where the spread of emigration fever was closely related to the progress of fundamental economic, social and demographic changes across Europe from the north-west to the south and east. Everywhere, incipient economic development, transportation improvements, better knowledge of opportunities abroad, falling mortality and accelerating population growth were followed by a surge of emigration.

The 'new migration' differed from the 'old' not only in being bigger and coming from a different set of European origins, but in being dominantly directed towards the cities. By the end of the century the spread of the agricultural frontier right across the United States meant that little good land was freely available, but manufacturing and commerical growth was creating major labour demands in rapidly expanding cities.

Table 10.2 Number of immigrants in thousands to the United States by selected origins, 1820–1930

	All countries	England and Wales and Scotland	Ireland	Germany	Scandinavia	Russia and Baltic States	Italy
1820–29	127	26	50	4	–	–	–
1830–39	537	73	170	124	2	–	1
1840–49	1,427	217	655	384	13	–	–
1850–59	2,814	444	1,029	975	25	–	8
1860–69	2,081	532	427	723	96	1	9
1870–79	2,741	577	422	750	208	34	45
1880–89	5,247	810	673	1,444	672	182	267
1890–99	3,693	328	404	578	391	449	602
1900–09	8,202	469	344	327	488	1,500	1,930
1910–19	6,346	371	166	174	238	1,106	1,229
1920–29	4,294	337	206	386	205	87	527

Source: US Bureau of the Census (1960), Series C88–114.

Urban ecological studies show that the influx of the 'new' immigrants to American cities usually took the form of spatial segregation, in which enclaves like Greektowns and Little Sicilies became established in the poorer neighbourhoods around city centres. Such segregation was largely by immigrant preference, as a means of cushioning the impact of an alien society, and in subsequent generations there were few barriers placed on dispersal. This is in marked contrast to the more recent ghettoization experience of blacks and Puerto Ricans, where the formative role has been played by external pressure rather than internal coherence (Morrill, 1965; Rose, 1971).

The smoothing of annual immigrant flows into decadal totals (Table 10.2) suggests a fairly regular rise in immigration from the low levels of the 1820s to a deluge in the early twentieth century. In fact, there were considerable annual fluctuations, involving many peaks and troughs (Figure 10.2). Economists have closely examined these fluctuations in relation to economic conditions generally, and business cycles specifically, on both sides of the Atlantic (Jerome, 1926; D. S. Thomas, 1941; B. Thomas, 1954, 1972). There is general agreement that before about 1860 the best correlation is an inverse one with harvest conditions at origin – poor harvests leading to appreciable migration, and vice versa. Later, with the growing maturity of the American economy, the dominant influence becomes the business cycle at destination, with cyclical upsurges in business prosperity in the United States attracting immigrants almost regardless of conditions in Europe. More ambitious, but controversial, is the view of B. Thomas (1954, 1972) that the economies on both sides of the Atlantic underwent complementary, interdependent, but opposite *long* swings in association with capital and population flows between them; to him, the pivotal role was played by investment transfer.

Towards the end of the nineteenth century there were signs of a new era in international migration with the imposition for the first time in the United States of barriers on immigration in response to charges of 'cheap labour' and 'unfair competition'. Immigrants under the 'old migration' had occasionally

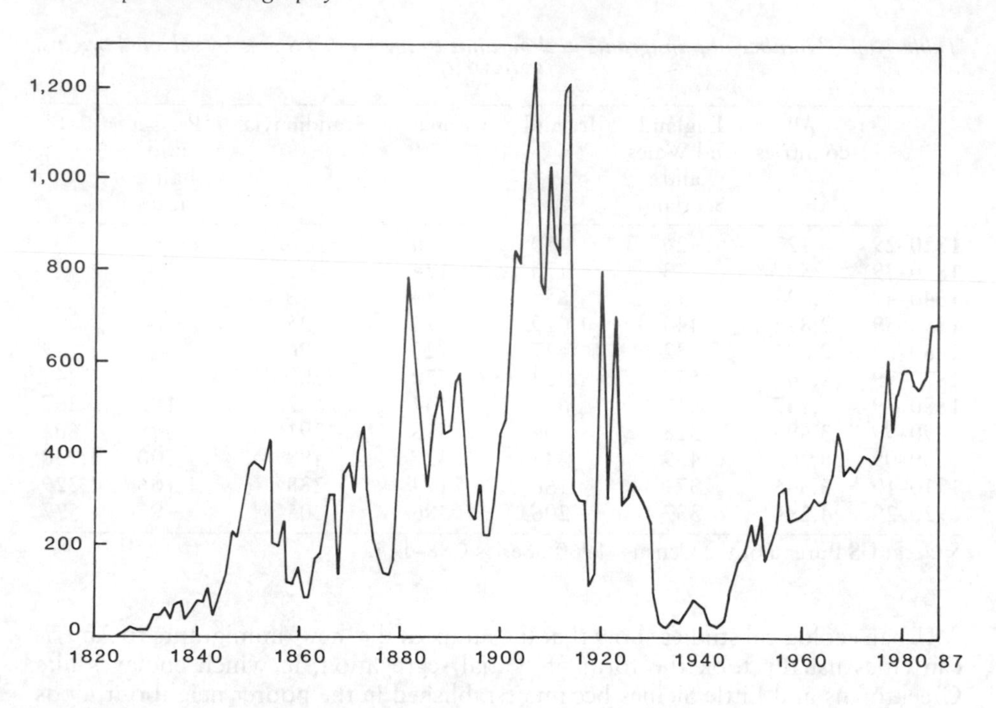

Figure 10.2
Legal immigration (thousands) to the United States, 1820–1987
Source: Bouvier and Gardner (1986), Figure 1. By permission of Population Reference Bureau, Inc.

met with hostility and prejudice, especially the destitute Irish, but it was generally accepted that immigration was necessary to develop the country's resources. It was the 'new migration' which caused a fundamental change in attitudes. Not only was the flood of immigrants much greater, but they were concentrated visibly in cities and comprised those nationalities which were thought to be more difficult to integrate – hence their branding as 'polacks', 'bohunks' and 'wops'. The first exclusions were against prostitutes, convicts, the mentally ill and persons likely to become public charges (1875 and 1882), the Chinese (1882), the Japanese (1907) and nationals of an Asiatic Barred Zone (1917). But the ultimate expression of selective immigration control was embodied in the Quota Act of 1924, which took full effect in 1929 and continued with little modification until the mid-1960s. This limited the annual immigration from origins outside the western hemisphere to 154,000, with quotas distributed among countries in the same ratio as these countries contributed to the national origin (often through several generations) of the United States population in 1920. This was flagrant ethnic discrimination, since countries like Italy and Poland, with only short but tumultuous migration links with the United States, were granted derisory annual quotas of 6,000 compared with 66,000 for Britain and 26,000 for Germany. The countries of north-western Europe never filled their quotas, and this, together with economic depression, actually led to a small net outflow of aliens from the United States during the 1930s.

Forced migrations and refugees

The First World War heralded the rebirth of forced migration, but unlike the earlier movements of slaves to the Mediterranean empires and to the colonial economies of the Americas, the force was now normally applied by the sending country rather than the receiving one, and in the cause of ethnic and ideological purity rather than economic gain.

> Two world wars, ignited by a nation obsessed with the separateness and solidarity of its own folk, were ironically ended by a legitimation of that obsession for nations in general. Under the Wilsonian banner of 'national self-determination' it was all peoples, not only the Germans, who could claim folk sovereignty. Carried to its extreme, this ideal, which justified the dismemberment of the defeated German, Turkish and Austro-Hungarian empires, encouraged every minority to seek a territory of its own and every colony to seek 'independence'.
>
> (K. Davis, 1974, p. 100)

Between 1950 and 1975 the number of independent countries tripled, and the wars and revolutions which accompanied these changes created millions of displaced persons and refugees. Refugees are defined by the United Nations as individuals who, owing to well-founded fear of persecution on the basis of race, religion, nationality or membership of a particular social group or political opinion, are outside their country of nationality and are unable, or owing to such fears, unwilling to avail themselves of the protection of that country.

During the late 1980s the Annual Reports of the UN High Commissioner for Refugees estimated current numbers of refugees at around 14–15 million, although the cumulative total during this century exceeds 100 million. There has been an endless and seemingly haphazard series of refugee movements rising and often disappearing in short periods of time and occurring under diverse historical and geographical circumstances, so that it is important to identify some common principles underlying the movements.

Proximate and fundamental causes of flight

Five proximate or intermediate causes can be identified, although they are far from mutually exclusive categories.

Wars of independence

Between the mid-1950s and mid-1970s most refugee movements resulted from indigenous populations struggling against colonial domination. Neighbouring territories were often supportive of the struggle and were therefore willing hosts, particularly given an expectation of ultimate victory and repatriation. Major examples included mass movements from the former Portuguese colonies in Angola, Mozambique and Guinea Bissau, and from the white settler state of Southern Rhodesia (now Zimbabwe). Even in the 1980s such movements remained important in Africa, evidenced in Figure 10.3 by flights from the domination of Namibia by South Africa and the annexation of Western Sahara by Morocco.

International conflicts

The two world wars have been responsible for most refugee movements originating in Europe. Major recent equivalents in the Third World have been the war

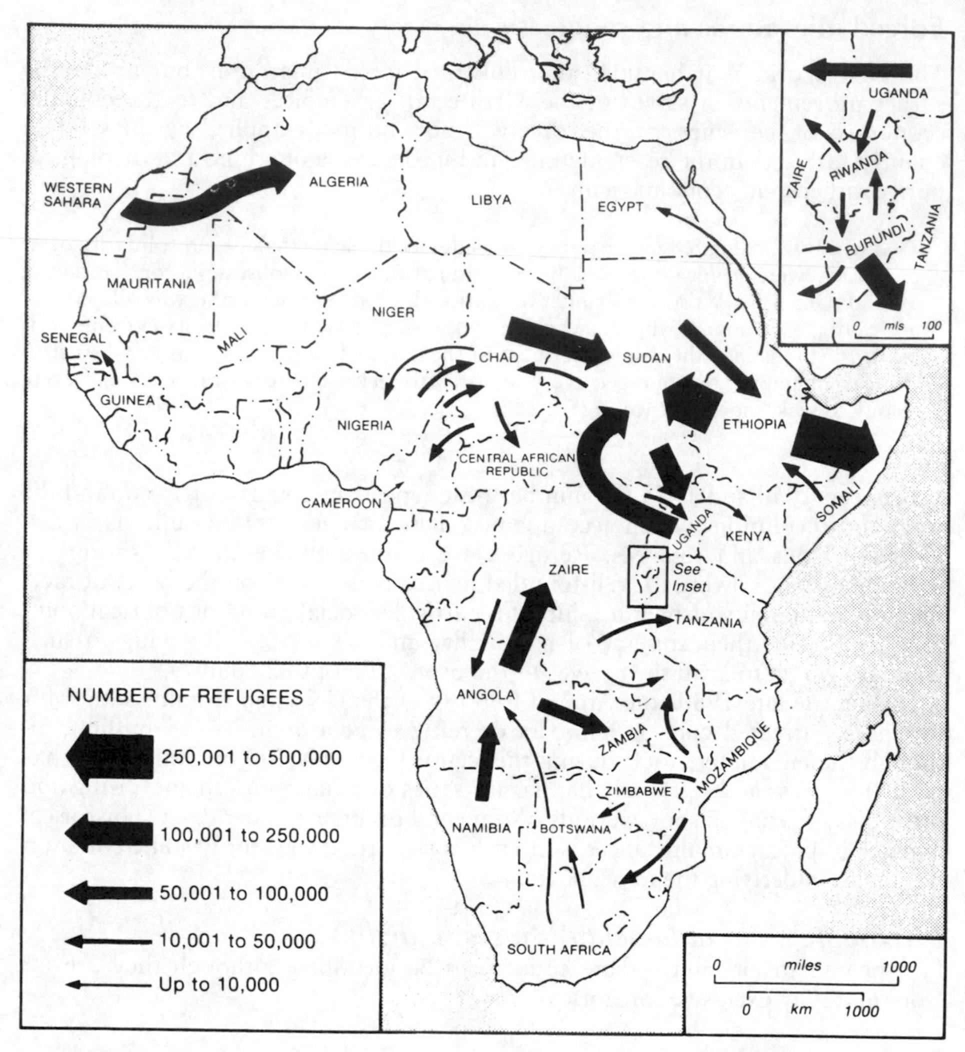

Figure 10.3
Refugee movements in Africa, 1985–6
Source: Kliot (1987), Figure 1. By permission of The Geographical Association.

between Ethiopia and Somalia from 1977, Vietnam's invasion of Cambodia in 1978, and the Russian occupation of Afghanistan between 1978 and 1989, which provided the largest single concentration of refugees in the mid-1980s – over 3 million Afghans, largely women and children, in Pakistan.

Internal revolutions and civil war
A first wave of refugees under these circumstances is caused by mass violence and dislocation, and a second wave by actual or anticipated oppression of the defeated group. Historic examples include the flights following the American, French, Russian and Chinese revolutions. Modern Third World cases involve: in Asia flights from successive regimes in Vietnam, Cambodia and Laos; in Latin

America the exodus of Chileans after the 1973 overthrow of Allende, and of Cubans and Nicaraguans following their revolutions in 1959 and 1979 respectively; and in Africa a series of prolonged civil wars, particularly in Ethiopia, Sudan, Chad, Uganda, Mozambique and Angola (Figure 10.3).

Ethnic conflict

The quest for 'ethnic purity' was a dominant theme in European refugee movements in the first half of this century. There was a renaissance of such movements in the 1980s as the Soviet Union and east European governments increasingly permitted, and in a few cases encouraged, ethnic minorities, notably Jews, ethnic Germans and ethnic Turks, to leave for 'historic homelands'.

In the Third World the formation of independent states has often been accompanied by ethnic violence and consequent expulsions, including, for example, many Tutsi from Rwanda and Hutu from Burundi. Ethnic conflict on a more prolonged basis has also led to the exodus of groups like Sri Lankan Tamils and southern Sudanese. There are occasionally explicit government policies of expulsion, particularly when a vulnerable and visible minority can be made the scapegoat for economic problems, as in the expulsion of Jews from Germany in the 1930s and of Asians from Uganda and Kenya in the early 1970s (Gould, 1982).

There are also *de facto* refugees *within* countries, like the Ibos in Nigeria in the 1960s, South African blacks removed forcibly from cities to newly created 'homelands', and the recent victims of growing inter-communal violence in the Soviet Republics of Armenia, Azerbajan, Georgia and Uzbekistan. In this last case, the modern easing of central state control has permitted the re-emergence of vigorous and sometimes persecutive nationalism.

Partition of states

The 1947 partition of the Indian sub-continent led to some 6–7 million Muslims leaving for Pakistan and about the same number of Hindus and Sikhs for India. Other examples are the divisions of Korea and Vietnam in the 1950s, the separation of Bangladesh from Pakistan in 1971 and the partition of Cyprus in 1974.

More fundamental explanations for the seemingly chaotic pattern of refugee movements over time and space must lie in the global political and economic conditions which determine the five proximate causes discussed above. Three major determining conditions can be identified:

1. Most newly independent Third World states are weak – an almost inevitable legacy of political and economic underdevelopment in colonial times. Many of their boundaries, arbitrarily imposed in the nineteenth century, ignore ethnic distributions. Moreover, within states, colonial policies often compounded ethnic tensions by the preferment of particular ethnic groups.
2. Internal strife is often fomented externally by rival neighbouring states and by super-powers furthering their geo-political interests (Barber, 1984; Zolberg, Suhrke and Aguayo, 1986). Such interests often encourage the establishment of threatening 'refugee-warrior' communities just outside the borders of states. Major examples in the late 1980s included the Palestinian camps, the SWAPO bases on the Namibia–Angola border, the Nicaraguan Contra bases in Honduras, and the Khmer Rouge camps on the Thai–Cambodian border. Another

type of political consideration is represented by the long-standing willingness of the United States government – for both humanitarian and propaganda purposes – to provide refuge for those fleeing communist countries.

3. Persistently poor economic conditions exacerbate the dislocating impact of civil strife, as do irregular environmental deteriorations like drought in the Ethiopia–Sudan–Somalia region. There is also the contentious problem of the extent to which refugees are really economic migrants, not so much fleeing persecution as seeking to improve their standard of living and using refugee status to circumvent immigration barriers. The debate on this issue in 1989 centred on the Vietnamese 'boat people', and also on Central Americans fleeing to the United States and Canada.

Resettlement of refugees

Three options are available: repatriation to the country of origin, resettlement in the country of first asylum, and resettlement in a third country.

Repatriation

Just as anti-colonial wars, revolutions, coups and ethnic conflict cause flight, so the granting of independence, peace, amnesty and prospects of stability can bring return. Voluntary repatriations of this type from neighbouring states have been common throughout Africa in recent decades, but there have also been some cases of forcible repatriation, like those involving Kenya and Uganda in the 1970s.

Resettlement in country of first refuge

Again, this has been an extensive response in Africa, not so much in government resettlement projects as in spontaneous settlement in border areas with ethnic affinities and available land. It is a sad fact, however, that the countries of first and, more often than not, only refuge are invariably among the world's poorest countries, and their peripheral areas of refugee settlement are often particularly disadvantaged. Refugee aid organizations and research studies focus on the needs of refugees rather than of their hosts, yet the case is overwhelming for additional development aid to poor countries hosting substantial refugee populations; but reality has rarely matched reason in this field.

The refugees' impact on local and often fragile economies has been discussed by Chambers (1986). He shows that the better-off locals generally gain, because additional cheap labour and enhanced consumer demand benefit their farms and businesses. On the other hand, the poor will be adversely affected by labour competition. A serious longer-term ecological impact concerns the growing competition for common property resources like grazing land, water, trees and fish in fragile environments overloaded with refugees like the North West Frontier Province of Pakistan and the Horn of Africa. Problems arising from the collection of firewood are a recurrent issue, and Christensen (1983) has reported that Afghan refugees in parts of Pakistan could gather fuel safely only at night.

Resettlement elsewhere

Resettlement in affluent Western countries is often the goal of refugees, but extensive resettlement there is prevented by fears in the West about the social and economic impact, particularly in recessionary periods. Thus in 1987, 67 per cent of applications for political asylum in France were rejected and 90 per cent of those in Italy (OECD, 1989), although an unknown proportion of unsuccessful applicants stay on illegally. Press reports suggest that only 10 per cent of the increasing numbers of Vietnamese 'boat people' arriving in Hong Kong in 1989 were being granted official refugee status.

The major acceptances of refugees occur when particular historical ties and responsibilities are recognized by Western governments. Thus Britain accepted the entry of Asian refugees from its former colonies of Uganda and Kenya in the early 1970s and would be likely, despite current posturing, to act as a refuge from any possible future oppression of the Hong Kong population by China. The West German government has looked sympathetically on refugees from East Germany, and now on freely moving migrants, particularly given the age structure, education levels and occupational skills of those taking advantage of the dismantling of the Iron Curtain in 1989. By far the largest resettlement in the West has been by refugees from Vietnam, Laos and Cambodia in the United States. From the fall of Saigon in 1975 to 1989, almost a million refugees from the region have been settled permanently in the United States. But such movements to the West are the exception rather than the rule. The great mass of refugees are stranded, often under conditions of great deprivation, in poor Third World countries of first refuge.

Modern permanent-settlement immigration

Apart from forced movements, immigration has become focused on the developed countries of the West, with Third World countries replacing Europe as the major supplier of emigrants. A broad distinction is usually made between the theoretically temporary, migrant-labour movements to north-western Europe, the Middle East and elsewhere, and permanent-settlement immigration with citizenship rights to the still uncrowded countries of North America and Oceania. But this distinction has become increasingly blurred as many of the intendedly temporary migrant-workers stay on and are joined by their families.

The major countries of traditional settlement immigration have all responded positively to Third World demands for less discriminatory immigration policies, so that formal ethnic quotas or preferences were discarded in the United States in 1965, Canada in 1962 and 1967, and Australia in 1973, in favour of preferences based on family reunion and labour market skills. Thus, in the United States immigrants from Asia rose from 7 per cent of the total inflow in 1965 to 46 per cent in 1985, when the Third World contribution as a whole was 84 per cent and Europe's only 11 per cent. Similarly, in Canada the share of immigrants from Asia rose from 12 per cent in 1966 to 38 per cent in 1987, while the European share dropped from 87 per cent to 24 per cent. One consequence has been the charge that scarce development skills, which are costly to produce, are being lost to the Third World by a highly selective 'brain drain' (Glaser, 1978). On the other hand, some Third World countries, notably India, exhibit a 'brain overflow' situation, in that university education has expanded well beyond the economy's absorptive capacity for graduates. There is also some criticism within immigration countries of large entries of physicians and other professionals from Third World countries on the ground that it allows inadequacies in the indigenous system of professional training to be perpetuated. In the United States, for example, there were more Filipino than black doctors in the 1970s – a sorry indictment of the country's medical education system (Bouvier, 1977).

Canada: a representative case study

In a world of growing population pressure, Canada has been one of a small handful of countries to maintain an expansionist immigration policy. Traditionally its governments, both Liberal and Conservative, have been convinced that the benefits of immigration generally outweigh the costs, at least for a huge, resource-rich country with one of the world's lowest population densities. A larger population has been actively sought as a means of achieving economies of scale – in resource development, market enhancement and public services provision – and also as a means of consolidating national sovereignty in a continent dominated demographically by the United States.

In recent decades the immigration issue has become more controversial, particularly in relation to the so-called Visible Minorities, so that major government efforts were made in the mid-1970s to encourage a national public debate on the policy options available, through the publication of a four-volume discussion document, the *Green Paper on Immigration*, and through the establishment of a parliamentary committee on immigration policy to receive briefs and hold public meetings throughout Canada. This debate culminated in the 1977 Immigration Act, which required the government to determine, in advance, annual immigration targets; these were not to be regarded as fixed quotas but rather as flexible targets for planning purposes. Continued consultation between government and representative opinion groups in the 1980s (Proudfoot, 1989) has confirmed the support for levels of immigration well above the estimated annual emigration of 75,000, largely because of concern about future population decline and adverse dependency ratios as fertility rates remain well below replacement level.

In Canada, as in the other permanent immigration countries, three broad groups of immigrants can be identified on the basis of selection criteria: relatives of Canadian residents, independent workers and refugees. Their respective proportions of all immigrants during the 1980s have been about 50 per cent, 30 per cent and 20 per cent. An important tension, very evident also in Australia, has been between the demands from ethnic minorities for a further extension and liberalization of family reunification categories and the demands from employers for occupationally skilled, and even unskilled, entrant workers. Another common feature of immigration policy has been the identification of, and preference accorded to, a so-called 'business class' of entrepreneurs and investors. An important element in this class in the late 1980s has been the Hong Kong Chinese, who have been particularly drawn to Vancouver by its buoyant economy and long-established Chinese community.

An overriding control on the level of Canadian immigration in recent decades, even for refugee and family categories, has been the state of the internal labour market. Figure 10.4 clearly reveals an inverse relationship between Canadian unemployment rates and immigration totals, a relationship dependent partly on prospective immigrants' assessment of the Canadian labour market, particularly through advice from friends and relatives within Canada, and partly on conscious regulation of inflow by immigration officials – the so-called 'tap on–tap off' policy. The one exceptional year when the relationship breaks down is 1957, when a dramatic surge in immigration reflected the Suez crisis and the Hungarian revolution.

The occupational composition among independent-worker immigrants has been related increasingly to labour market needs within Canada. In 1966 this relationship was formalized by the creation of a new government department,

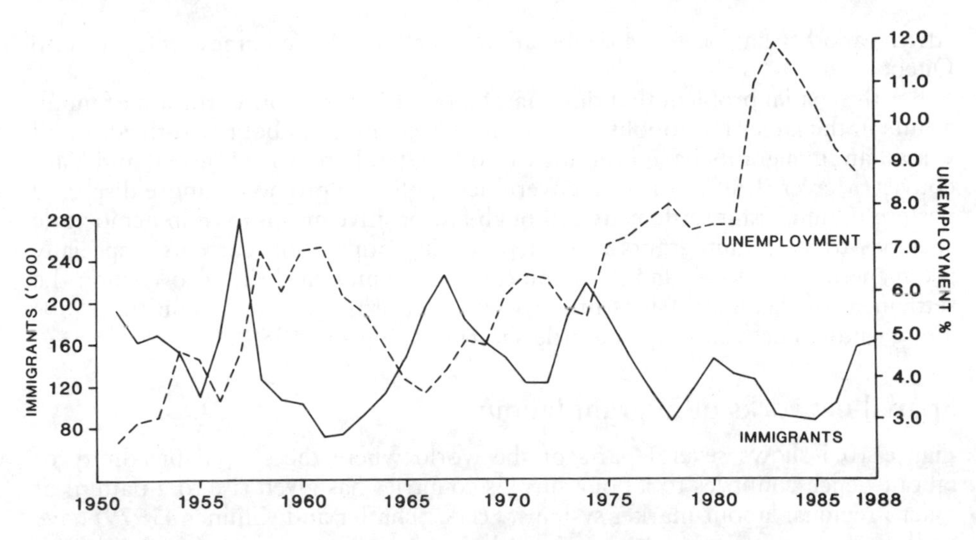

Figure 10.4
Annual immigration into Canada and annual average unemployment rate (number unemployed as a percentage of labour force), 1951–87

Manpower and Immigration (now Employment and Immigration), and in the same year a White Paper on Immigration set the tone of subsequent immigrant selection policy when it stated: 'We are not a country of virgin lands and forests waiting to be settled by anyone with a strong back and a venturesome spirit . . . Canada has become a highly complex industrialized and urbanized society. And such a society is becoming increasingly demanding of the quality of its work force.' Consequently, from 1967 there has been a points-assessment scheme to select, as objectively as possible, suitable workers. These selection procedures have operated such that managerial, professional and other highly trained occupations have increased markedly at the expense of unskilled occupations, although most immigrants can escape full occupational assessment through being a refugee or having Canadian relatives. An ancillary means of tuning immigration to labour demand has been the increasing issue in the 1980s of temporary employment authorizations, often for seasonal agricultural and domestic work by West Indians. The annual number of such authorizations usually exceeds that of permanent immigrants, and a problem of illegal over-staying certainly exists.

One important part of the immigration debate concerns the spatial distribution of immigrants within Canada. Clearly there is an influence over and above that of existing population distribution exerted by provincial variations in economic buoyancy. Thus Ontario, British Columbia and Alberta have the greatest immigrant attraction relative to population, while the other Prairie Provinces, Quebec and particularly the Maritimes are relatively unattractive. A particularly sensitive problem has been the falling proportion of immigrants entering Quebec, since the traditional safeguard of Quebec's demographic, cultural and linguistic position within the confederation, a high natural increase rate, has disappeared in recent decades, to the extent that its provincial fertility rates are now the lowest in the whole of Canada. One response has been provincial and federal efforts to identify and tap promising sources of French-speaking immigrants such as West

Africa; another has been the deliberate channelling of Vietnamese refugees into Quebec.

But the spatial problem that dominates all others is the concentration of immigrants in the larger metropolitan areas, to the extent that about two-thirds of all Canadian immigrants have been drawn to Greater Toronto, Montreal and Vancouver (Mercer, 1989). It is firm government policy to promote a more dispersed pattern of immigrant settlement, although few positive means exist in democratic societies to steer immigrants against prevailing population currents, especially given the importance of family reunification. Attempts in Britain to fashion the settlement of Ugandan Asians in the 1970s and of the Vietnamese in the 1980s along similar lines certainly had little success (Robinson, 1989).

Spatial networks of migrant labour

Figure 10.5 shows several parts of the world where the spatial proximity of labour-deficit countries to labour-surplus countries has given rise to a pattern of macro-regional labour market systems. Seers, Schaffer and Kiljunen (1979) have applied the core–periphery concept to these systems, since the more developed, capital-rich core countries have possessed the power to organize labour inflows from their less developed peripheries very much – at least initially – in their own interests.

Europe

International labour movements within Europe had already become significant by the early decades of this century. The two important destination countries were Germany, recruiting migrants from central and eastern Europe for its rapidly expanding manufacturing industries, and France, where labour needs were created by its falling fertility, ageing population and wartime manpower losses (Ogden, 1989).

Causes

The 1960s saw the development of a larger scale and wider network of international labour flows. Three major causes may be identified. First, there were the labour demands created by rapid economic growth in the more developed countries. Second, the demographic echo effect of low fertility in the 1930s and early 1940s was a sluggish growth rate in the 1960s in the indigenous labour forces of north-western Europe – about 0.5 per cent per annum. At the same time population pressure was high in the Mediterranean region, where in Spain, Portugal, Yugoslavia and Greece the expansion of the labour force was about 1.0 per cent per annum, and the share of the labour force in agriculture still as high as 25–40 per cent in 1970; in Turkey, Algeria, Morocco and Tunisia the pressure was even greater, with equivalent figures of over 2.0 per cent per annum and 45–65 per cent. Clearly there was a reserve of labour in the peripheral countries available, and often anxious, for recruitment. Third, because of social mobility aspirations, the indigenous workers and even the unemployed in north-western Europe were simply not prepared to undertake unpleasant, menial and sometimes dangerous occupations. Thus in 1979 the foreign proportion of the workforce in the Paris region was 23 per cent in catering, 34 per cent in automobile manufacture and 64 per cent in construction (Ogden, 1985b).

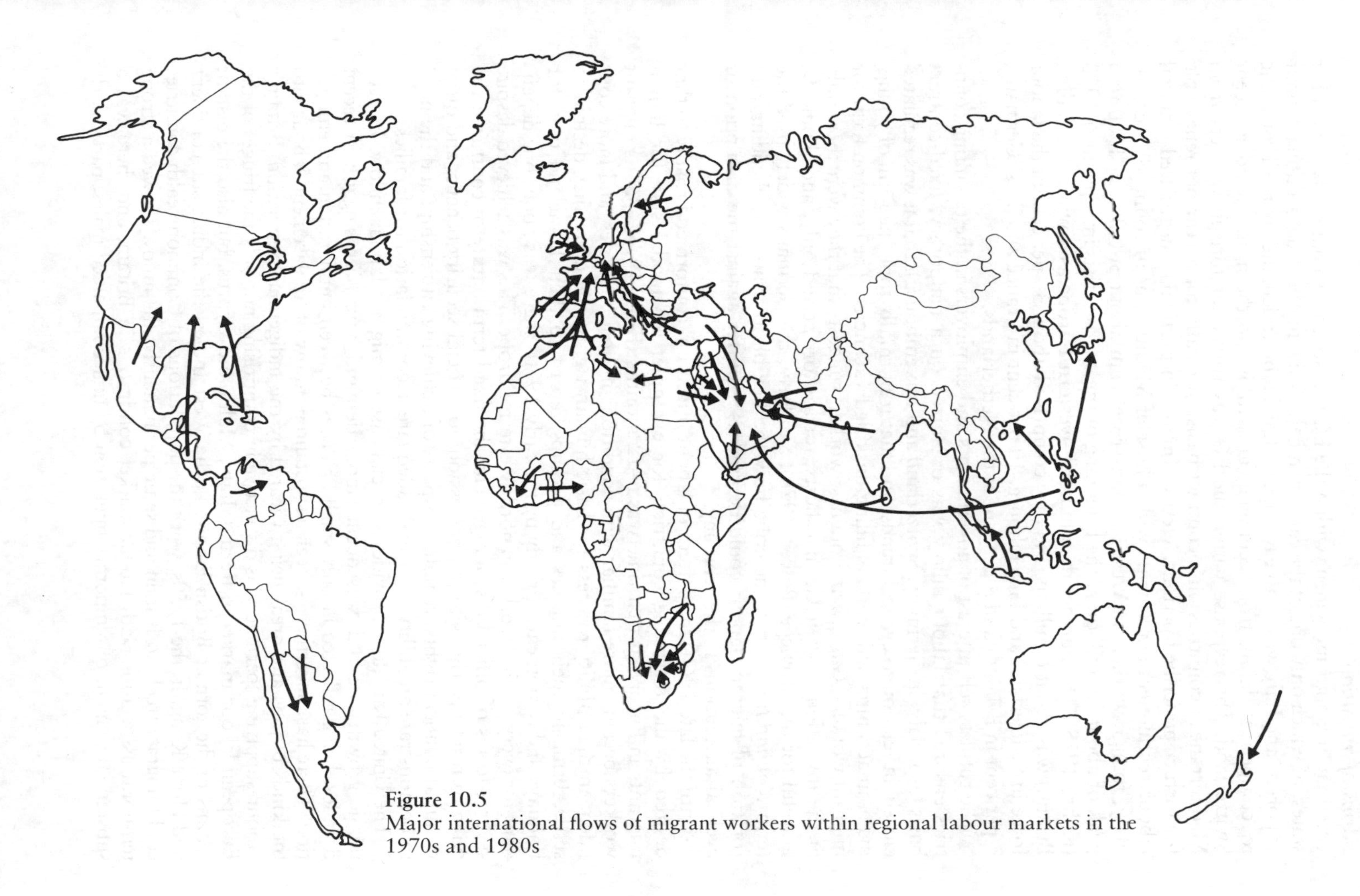

Figure 10.5
Major international flows of migrant workers within regional labour markets in the 1970s and 1980s

Forms of migration

Freedom of labour movement within the EC is restricted to nationals of member states, of which only one of the early members, Italy, has been a traditional labour exporter; and Italy was unable to cope with EC labour demands in the 1960s, due to competing claims from economic growth in northern Italy and adjacent Switzerland. The response within the EC was to recruit foreign workers from Mediterranean countries on a contract basis, normally for a year but renewable for further prescribed periods. West Germany, in particular, negotiated bilateral labour recruitment agreements with a host of Mediterranean countries. Accordingly, foreign workers in West Germany rose from about 300,000 in 1960 to a peak of about 2.6 million at the beginning of the energy crisis in September 1973 (Kuhn, 1978). The number then in north-western Europe as a whole (excluding Britain) was about 6 million, and they comprised about 30 per cent of the labour force in Switzerland and Luxembourg, 10 per cent in France and West Germany, 7 per cent in Belgium and 3 per cent in the Netherlands.

The ruthless self-interest of governments and employers in the recruiting countries ensured that labour inflow was organized on a temporary, fixed-contract basis (especially in Germany, Switzerland and Luxembourg), that it was regulated closely in relation to cyclical trends, and that it normally took the form of young, male, unaccompanied workers filling unskilled vacancies. The common term for migrant workers, *Gastarbeiter* or guest worker, suggests that they were there only at the invitation of their German hosts, for a short period only, and should be grateful for the privilege (Power, 1976). Similarly, the Germans widely used the term *Konjunkturpuffer* to describe how the flexible recruitment, shedding and rotation of migrant labour could be used as a buffer against the disruption of cyclical fluctuations in the economy.

Until the late 1960s the pattern of labour migration conformed broadly to that desired by the recruiting countries; for example, the modal stay for Italian migrants in Europe at that time was 6–12 months (King, 1976). But as migrant workers became more familiar with conditions in host countries, and more confident in their ability to press for basic human rights, the more they desired to settle, bring in their families and perhaps return to their homeland only for holidays and retirement. Already by the early 1970s there was about one dependant for every foreign worker in north-western Europe. This was achieved despite several forms of official discouragement: the fixed-term nature of contracts; the tying of residence rights to job possession; low social security benefits and often minimal political rights; remote prospects of achieving citizenship; and the need to demonstrate an ability to provide adequate accommodation for families.

The United Kingdom provides a special case of migrant labour inflow. Its economic growth rate and consequent labour demand have been sluggish in recent decades, being reflected in only small balances between immigration and emigration. There has, nevertheless, been a substantial inflow of 'replacement' labour into the kinds of jobs and areas being vacated by the indigenous population. Britain's labour-supplying 'periphery' is distinctive in comprising not only the fringes of the European labour market in Ireland, Italy, Malta and Cyprus, but also the distant realms of the former British Empire in the West Indies, the Indian sub-continent and Hong Kong. In the 1950s, when the populations of Commonwealth countries could migrate freely to Britain and settle permanently, the inflow showed a strong temporal association with labour market conditions in Britain, and there was a substantial element of temporary migration by unaccompanied males, particularly from India and Pakistan. But, as a response to growing public disquiet, the 1962 Commonwealth Immigrants Act and subsequent restrictive legislation transformed the nature of this immigration, so that the great majority of coloured immigrants have for some time been dependants seeking family reunion and permanent settlement.

Spatial factors

Distance has been a major factor determining the spatial pattern of international migration linkages within the European labour market (Table 10.3); hence the important flows from Spain and Portugal to France; from Italy to Switzerland; from Yugoslavia, Greece, Italy, Turkey and now East Germany to West Germany; from the Irish Republic to the United Kingdom; and from Finland to Sweden. In addition, traditional political connections are evident in the channelling of North African outflow into France.

Analysis of the regional origin of migrant workers within the countries of southern Europe (King, 1976) suggests that in countries at the initial stage of emigration, like Turkey and Yugoslavia in the 1960s, the outflow is concentrated in their more developed regions – Slovenia and Croatia, and the western provinces of Turkey; the role of opportunity awareness seems critical, as it was in fashioning the regional pattern of European migration to the New World in the nineteenth century. Where there is a longer experience of labour migration, outflow is heaviest in areas of least development and most population pressure – Portugal north of the Tagus, Galicia, Andalucia and much of the Meseta in Spain, Gozo in Malta, the Mezzogiorno in Italy, and the islands and northern mountainous regions of Greece.

The distribution of immigrants within host countries is best interpreted in relation to the immigrants' role as replacement population, attracted to job and often residential vacancies created by the upward social mobility and outward migration of the indigenous population (Peach, 1975). In Britain, the eight major conurbations, which contained 36 per cent of the country's 1981 population, all showed substantial population decline between 1971 and 1981; yet they

Table 10.3 Estimated stocks of total foreign population (with three leading origins specified) in selected countries, 1987 (1985 France), in thousands*

Origin	Belgium	France	West Germany	Sweden	Switzerland
Finland				131	
Italy	252		544		385
Portugal		751			
Spain					113
Turkey	76		1,481	22	
Yugoslavia			598	39	87
Algeria		821			
Morocco	126	516			
Total	853	3,752	4,630	401	979
As % of total population	8.6	6.8	7.6	4.8	15.0

* Excluding temporary residents and illegal immigrants.

Source: OECD (1989), Table A.2.

accounted for very high proportions of the 1981 census population born in India (63 per cent), Pakistan (69 per cent), Bangladesh (73 per cent) and the West Indies (77 per cent). Similarly, in France 60 per cent of foreigners are concentrated in the regions centred on Paris, Lyon and Marseille; in Paris one in six of the population is foreign (Ogden, 1985b).

Turn of the tide?

From the end of 1973 one host country after another imposed severe restrictions, and in many cases total bans, on further recruitment of foreign labour, although not of seasonal workers. The immediate instigating factor was the threat of recession induced by OPEC's tripling of oil prices following the Arab–Israeli war in September 1973, but this simply accentuated more fundamental forces for change (White, 1986). First, de-industrialization was already affecting many of the manufacturing sectors which recruited unskilled immigrants; second, there was mounting public hostility towards immigrants, partly because of their growing visible concentration in cities and partly because of competitive pressure for jobs as the postwar baby boom cohorts in the host populations entered the workforce; and, third, there was significant political and socio-economic improvement, particularly in welfare state policies, in some of the traditional labour-supplying countries of southern Europe (Penninx, 1984).

Most governments in north-western Europe have responded with immigration policies which embrace varying mixes of recruitment restriction, repatriation and integration. *Recruitment restrictions* on labour have been rigorously applied, although there has been some relaxing of controls with the economic upturn from 1983. *Repatriation* has been a populist rallying call from the likes of Enoch Powell and Le Pen to resentful indigenous populations as soon as the bulk of immigrant workers stopped regarding themselves as temporary migrants; a mere 15 per cent of previously employed foreigners returned home from Germany during the 1974 recession (Kuhn, 1978). Clearly, many guests had outstayed their welcome, but refused to pack their bags. Financial assistance towards repatriation expenses has been offered in France and Germany, while France has also negotiated agreements with several countries of origin to facilitate the economic reintegration of those returning. It is widely thought, however, that such incentives do little more than support those who are returning anyway (Ogden, 1985; Salt, 1985).

Integration policies reflect a belated recognition of the hardships and discrimination suffered by immigrants, as well as the recognition that the immigrants are now 'here for good' (Castles, 1984). Accordingly, policies increasingly address issues of ethnic minority communities rather than of migrant workers. Considerations of social justice have permitted more liberal policies on family reunification at destination as well as some improvements in the provision of housing, education and social security. The acquisition of political rights has been more retarded, so that, during the 1980s, the proportion of foreign residents acquiring citizenship each year has invariably been under 1 per cent in Germany, Luxembourg and Belgium (van de Kaa, 1987; OECD, 1989).

The combined impact of these policies has meant that the size of foreign population in north-western Europe has grown somewhat rather than fallen from the early 1970s to the late 1980s, although the switch from workers to dependants has appreciably altered the composition of the population. There has also been a remarkable switch during the 1980s in the status of Italy and Spain from net emigration to net immigration. This reflects new economic growth, adjacency to the labour reservoirs of North Africa, and easier evasion of immigration restrictions than further north. It is widely thought that there were almost a million illegal immigrants in Italy in 1990, most of them entering initially on tourist visas.

Costs and benefits

Any evaluation of the European migrant-labour system must consider its impact on the societies at origin and destination and on the migrants themselves.

Since the system was initiated and organized by the 'core' developed countries, it can be expected to have conveyed benefits upon them. This was outstandingly the case in the days of strictly temporary and rotational immigration when the supply of immigrant labour could be turned on and off to match the cyclical pattern of labour demand in the core countries. Entrepreneurs also benefited from the lower cost of migrant workers, since lower wage grades could be created for them and they were less resistant to methods which reorganized the labour process more profitably. Guest workers were young, unaccompanied by any dependants and of good health, since they were often subject to medical examination before acceptance. In social security terms such workers had few liabilities, and since they produced considerably more goods and services than they consumed, they also served to check inflationary forces.

During the 1970s, however, many of the distinctive features of the migrant labour system were being eroded as immigrants and their dependants increasingly sought settlement and rights. In the words of a Swiss economist quoted by Power (1976): 'We called for workers, and there came human beings.' Employers found it more difficult to maintain job and wage differentials between immigrant and domestic labour, and much higher social infrastructural costs have had to be incurred by the state in the provision of family housing, education, health servies, unemployment, sickness and compensation benefits, and pensions. This has contributed to global economic restructuring in that European entrepreneurs have attempted to keep down production costs by locating new plant in peripheral countries where local workers could be paid even less than under the migrant-labour system and where the rising social costs of that system could be avoided (Hiemenz and Schatz, 1977; Paine, 1979).

In terms of migrant welfare, a major and, some would say, intolerable cost of the migrant-labour system has been family separation, invariably for periods of several months and often for several years. Family reunion in host countries has been hindered not only by entrance restrictions, but also by the shortage and high cost of housing, exemplified in the peripheral *bidonville* shanty towns of major French cities. Exploitation of immigrant labour has also been rife, as one UN report on the French building industry demonstrates. The industry

> has evolved complete networks of organization, from the recruitment of the workers in Africa to their work and their housing on building sites, sites which have all the aspects of real camps and where the laws are openly flouted: ridiculously low wages (sometimes agreed on in Africa when the employee does not have any benchmark to assess the pay offered); food and transport provided by the firm which charges excessively for its poor quality services; housing in huts; limitation of visits to certain hours, and the prohibition of women; suppression of all rights of trade unions and political expression.
>
> (Quoted by Power, 1976, p. 17)

It should not be assumed that such conditions are a thing of the past. Indeed, a particular form of exploitation in the 1980s has been among the growing numbers of newly settled, vulnerable female dependants – as, for example, among Bangladeshis working at home or in sweatshops in the clothing trades of the Spitalfields area of London. A particular subproletariat, notoriously susceptible to employer and landlord exploitation, are the illegal immigrants who gain entry clandestinely, or by forged papers, or by overstaying. They are vulnerable to abuse because they feel unable to complain to any legal authority.

It can be argued that first-generation immigrants migrated voluntarily, in response to even harsher conditions at origin. The dissatisfaction and resentment about working and living conditions in host countries are often greater among the rootless second generation who cannot identify with a host society which discriminates against them or with a homeland they may never have seen. This frustration has often expressed itself in so-called race riots, as in several British cities in 1981.

For the countries of origin, the costs and benefits of the migrant labour system are finely balanced. The fact that governments of such countries often initially encouraged, organized and financially subsidized the exodus of migrant workers implies that benefits are likely to accrue to the society at origin as well as to the migrants themselves. Pressures on domestic labour markets are reduced, and valuable foreign exchange is provided by emigrant worker remittances – to the extent that such receipts in Turkey in 1973 were comparable to 90 per cent of the value of the country's exports of merchandise. In addition, foreign workers have been thought to return home with modern technological skills, enabling them to act as agents of innovation and forces for development; an investment in human capital has taken place.

The reality of the situation is now known to be very different. It is not coincidental that as the inflow of remittances has increased, so too have the deficits on visible trade balances. Growing familiarization with foreign consumption styles leads to disdain for domestic products and a growing dependence on expensive foreign imports. Studies of return migration (King, 1986; J. Lewis, 1986) show time and time again that returning migrants are rarely bearers of initiative and generators of employment. Only a small number acquire appropriate vocational training – most are trapped in dead-end jobs – and their prime interest on return is to enhance their social status. This they attempt to achieve by disdaining manual employment, by early retirement, by the construction of a new house, by the purchase of land, a car and other consumer durables, or by taking over a small service establishment like a bar or taxi business; there is also a tendency for formerly rural dwellers to settle in urban centres. There is thus a reinforcement of the very conditions that promoted emigration in the first place. It is ironic that those migrants who are potentially most valuable for stimulating development in their home area – the minority who have secured valuable skills abroad – are the very ones who, because of successful adaptation abroad, are least likely to return.

There are also problems of demographic imbalance stemming from the selective nature of emigration. Many villages in southern Europe have been denuded of young men, with consequences not only for family formation and maintenance but also for agricultural production. In southern Italy, Wade (1979) has described how the agricultural workforce has become increasingly made up of women, old men and children. At its most acute, selective emigration can engender a social and economic malaise that George Bernard Shaw recognized in rural Ireland – 'a place of futility, failure and endless pointless talk'.

The common view, therefore, in the early days of the European migrant labour system, that migration, although often painful for individual migrants and their families, is a transitionary phase which aids the development of the sending country, is now discredited. 'Stripped to its bare essentials, the engagement of migrants creates value added which is largely, if not exclusively, internalized in the country of employment and thus further increases its wealth and power relative to the country of origin' (Böhning, 1979, p. 409). This is not to say that conditions at origin would be better without emigration; there is, after all, a reduction in unemployment and underemployment and a gain in income. What it does mean is that the gains from the migrant labour system have been demonstrably unequal.

North America

The United States has been a magnet to migrants from its less developed periphery in Mexico, Central America and the Caribbean. Nowhere in the world are gradients in material standards of living so great, but only Puerto Ricans have unrestricted access, profiting from their country's Commonwealth status in the United States. One consequence has been an escalation in illegal immigration, achieved by overstaying short-duration visas, by clandestine boat trips from Caribbean islands, or simply by walking over the border from Mexico. The 'educated guess' of the Immigration and Naturalization Service in 1986 was that the stock of settled illegal immigrants was rising by some half-million a year – very close to the number of legal immigrants (Bouvier and Gardner, 1986).

Decisive political action against illegal entry is impeded by the magnitude of the task in such a large country, but more fundamentally by pressure from employers of cheap labour and from co-ethnic voters, as well as by some need to sustain the ideology of the United States as a haven for the oppressed (Cohen, 1986). Consequently, penalties imposed on rogue employers of illegal immigrants have been far too slight to act as deterrents. American agribusiness has long been known as a major employer of legal and illegal immigrants (Figure 10.6), but a more recent appreciation is that some basic traits of advanced capitalism in association with the availability of immigrant labour are promoting job downgrading and informalization in some manufacturing sectors. The high incidence of low-wage jobs in the production and assembly areas of electronics and the growth of sweatshops and industrial homework in older industries have permitted southern California and New York to re-emerge as competitive in some industries with eastern Asia (Sassen-Koob, 1987).

Elsewhere in the region there is a complex series of movements between the many small states and islands, illustrated by the movement of sugar-cane harvesters from Haiti to the Dominican Republic and of construction workers from Colombia into the oil-boosted economy of Venezuela. The development of tourism and international tax havens in the Bahamas, Cayman Islands, Virgin Islands and elsewhere has also attracted construction and catering workers from adjacent islands as well as American and British expatriate professionals, entrepreneurs and retirees (McElroy and Albuquerque, 1988).

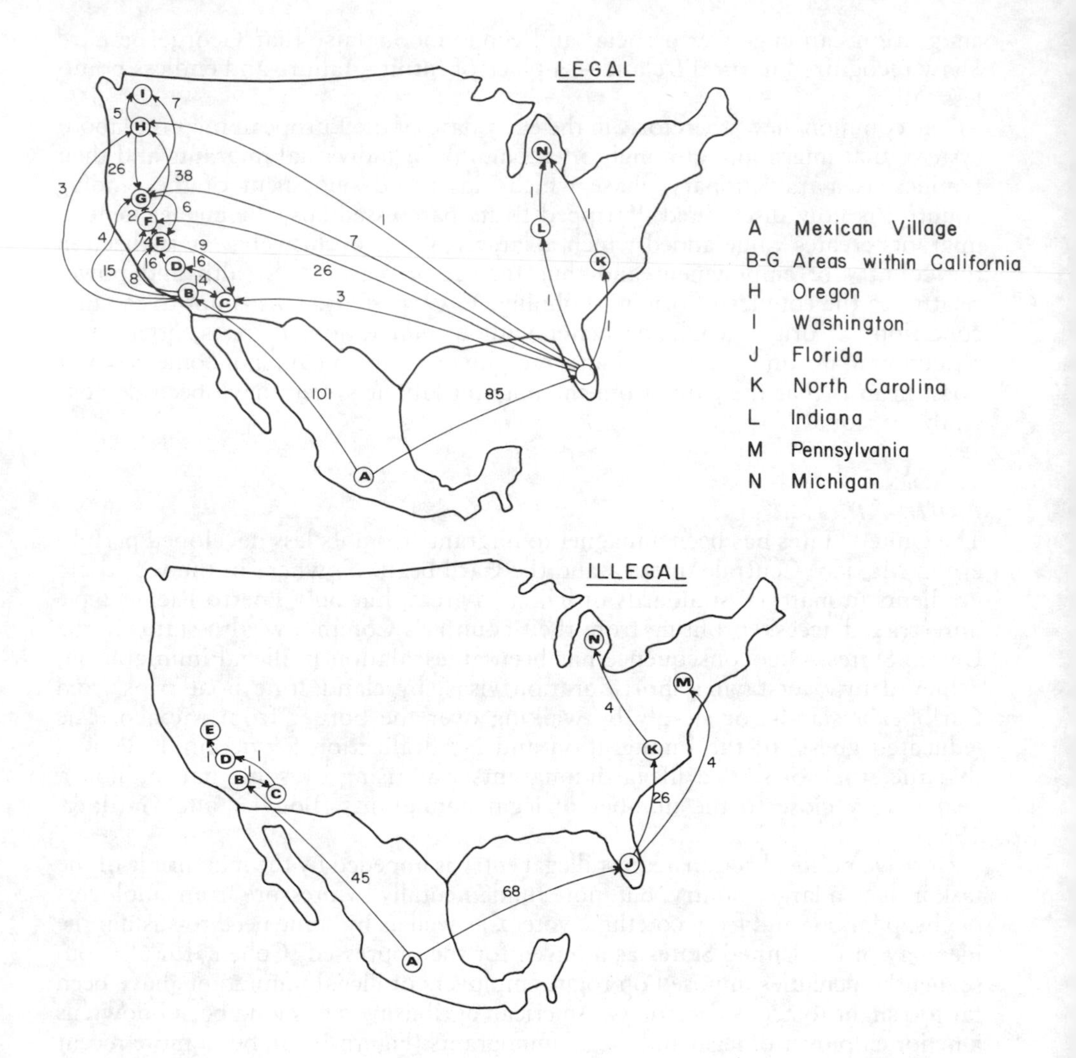

Figure 10.6
Migration routes by number of migrating units (worker and dependants) from a Mexican village to the United States in 1978
Source: Reichert and Massey (1979), Figure 1. By permission of Population Reference Bureau, Inc.

Southern Africa

As in other parts of the world, spatial inequalities and structural dependency are the basis of the migrant-labour system, but the southern African system is unique in the way in which it has been totally dominated by host country interests for a period extending over several decades.

Until the 1970s the policy of the South African Chamber of Mines was to recruit cheap labour throughout the southern half of the continent for arduous and dangerous labouring in the gold and diamond mines, thereby avoiding competition for local labour and inevitable wage increases (F. Wilson, 1972). Rigid control by government and employers ensures that migration is strictly temporary and that there is no accompaniment of workers by their families. There has never been any prospect that the migrant worker can become a permanent settler, as he serves his time in the degrading environment of mine, compound and barracks.

The number of foreign mineworkers in South Africa has fallen in recent years – from 392,000 in 1974 to 200,000 in 1983 (Taylor, 1986; Griffiths, 1987). Partly this reflects hostility from governments in adjacent states to exploitation of their citizens, partly the use of more capital-intensive production methods and partly South Africa's new policy of relying more on its most easily controlled sources of labour – its black 'homelands' and the vulnerable dependent states of Botswana, Lesotho and Swaziland. Between 1974 and 1983 these three states increased their share of all foreign mineworkers from 35 to 69 per cent, as former sources in Mozambique, Zambia, Malawi, Angola and Zimbabwe became unavailable or were excluded.

Middle East

Just as the 1973 rise in oil prices suppressed labour demand and migrant labour recruitment in the European labour market, so it stimulated demand and recruitment in the capital-rich but labour-deficient oil states of the Middle East. Until the mid-1980s there was a development boom in the oil states, all of them anxious to use their new-found wealth to press ahead quickly with infrastructure, social services and manufacturing. The major constraint has been labour shortage, the result of several factors: the indigenous populations are small – the combined population of the seven oil-rich countries is less than half that of Egypt; the role of women in Islamic societies is such that female participation rates in the labour force are very low; there is a widespread scorn for manual work; and the traditional paternal role of rulers often expresses itself in the creation of non-productive government posts for the indigenous population.

Importation of labour was encouraged by the *laissez-faire* attitudes, initially, of most countries in the region to international labour movement. Spontaneous movements have been facilitated by ethnic, linguistic and cultural affinities throughout the Arab world ('the circulation of Arab brethren'), and by typically Third World conditions of population pressure and labour surplus in many non-oil countries of the region. Consequently, by 1980 over half of the labour force was foreign in each of the oil-rich countries: Saudi Arabia, Libya, Bahrain, Oman, Qatar, Kuwait and the United Arab Emirates. The major suppliers of labour were Egypt, Jordan and North Yemen (with 10, 35 and 40 per cent respectively of their labour forces working abroad), while India and Pakistan were also significant contributors. There has been some occupational specialization by national origin. Thus, while the vast bulk of immigrants have been required for unskilled labour, particularly in construction, a significant proportion of Egyptians, Jordanians and Palestinians are employed throughout the Arab world as teachers, clerks and technicians – a reflection of higher education standards at origin. In addition, highly qualified manpower is provided by expatriate communities of West Europeans and North Americans.

During the 1980s there have been some significant changes in this macro-regional labour market, many of them paralleling the European experience of a decade earlier. As oil prices fell, construction work and therefore labour recruitment have fallen away (Birks, Seccombe and Sinclair, 1986). More migrant workers are overstaying and bringing in their families, prompting host governments to impose much stricter residential conditions on new labour contracts and to turn to non-Arab sources for construction work. Contracts have been given increasingly to construction firms in South Korea, the Philippines, Thailand and Turkey to bring in their own national workers, house them in isolated camps and ensure their removal on contract completion (Arnold and Shah, 1986). Another significant trend has been migrant-labour replacement in traditional labour-supplying countries. So many male workers have left that niches are often created for immigrants, most of them clandestine, from even poorer countries like Sudan and Somalia. Such labour replacement has been observed in Egypt (Sell, 1988), Jordan (Seccombe, 1986) and North Yemen (Findlay, 1987).

Other areas

Elsewhere within the Third World there are sometimes appreciable gradients in wages and living standards that stimulate labour migration. The Ivory Coast has long been an importer of labour from its poor inland neighbour, Burkina Faso. The stimulus of oil wealth in the Nigerian economy of the late 1970s attracted almost a million migrants from Ghana, Benin and Togo, but the subsequent slump in oil revenues and the need of the Nigerian government to find a scapegoat for growing economic problems led to mass expulsions of aliens in 1983 and 1985 (Afolayan, 1988). A third example has been the growing number of Indonesians, estimated by Hugo (1988) as at least half a million, working in the buoyant Malaysian economy. There, a significant facilitating factor has been the co-Malay bonds between the Indonesians and the politically dominant group in Malaysia.

Circulation of the highly skilled

As international labour migration by unskilled workers has subsided in recent decades – certainly within the European, Middle Eastern and southern African networks – so has the international circulation of highly trained professional, managerial and scientific personnel increased. This is related to global economic restructuring in which multi-locational transnational corporations are increasingly developing global internal labour markets for their skilled personnel (Salt, 1988; Findlay, 1988; Findlay and Gould, 1989). While the major flows remain those between highly developed countries (particularly their major metropolitan regions), the industrializing countries of the Third World are becoming increasingly involved in the networks.

Theoretical overviews

Classical theories of migration and of immigrant adaptation and assimilation are based on a functionalist perspective which views migration as an equilibriating response to spatial inequalities, as essentially voluntary in nature, as a rational attempt by migrants to maximize utility, and as a vehicle of upward social mobility. Although many international migrants enter a new society at the lowest economic and social strata, they are thought to have opportunities for improvement, certainly in subsequent generations, so that there is a progressive convergence of immigrant group characteristics towards those of the host community. But even when cultural pluralism persists, 'the functionalist approach tended to minimize the elements of conflict that pluralism created, emphasizing the importance of the dominant value system in promoting consensual forms of pluralistic integration' (Richmond and Verma, 1978, p. 5).

A quite different interpretation of international labour migration is provided by the Marxist conflict model, in which essentially coercive labour migration is a manifestation of continuing dependency which promotes underdevelopment in the periphery and overdevelopment at the core. Such analysis rejects the voluntary, rational and self-improving functionalist view of migration. Instead, the movement of labour from less developed peripheries to cores like north-western Europe is seen as a further expression of exploitation of labour under capitalism. Such labour is simply part of Marx's 'reserve army' at the beck and call of capitalist entrepreneurs, and suffering deprivations to sustain the living standards of the more privileged citizens of advanced countries (Castles and Kosack, 1973; Nikolinakos, 1975). Also integral to the conflict model is the view that migrant workers are prevented by discrimination from realizing the full range of labour market and housing market opportunities open to the indigenous population; they are trapped in occupational and residential ghettos and become increasingly alienated from, and not assimilated into, host societies.

This rather coarse model needs to be refined to take account of the role of the state and of complexities in the modern restructuring of capital. Thus, whereas the international mobility of capital remains comparatively unfettered, the international mobility of workers has become increasingly state-controlled. At the same time, the dynamics of international migration figure prominently in the increasingly worldwide division of labour (Portes and Walton, 1981). This is beginning to decompose the monolithic pattern of unskilled male migrants moving within macro-regional labour markets; highly skilled workers and female workers (e.g. Filipino and Sri Lankan domestics) are increasingly involved in worldwide movements. It has proved impossible for Third World states to sever colonial legacies and pursue economic and social development independent of the world capitalist system; and the energy of that system is provided by spatial flows of wealth, ideas and people.

11

POPULATION POLICIES

Recognizing that social and economic problems may be caused or aggravated by particular levels of population growth and by particular distributions of population, governments at various stages in the twentieth century have intervened in an attempt to regulate population dynamics. Strictly defined, population policies are those 'policies explicitly adopted by governments for their presumed demographic consequences' (Berelson, 1974, p. 6). Such policies attempt to alter the 'natural' course of population change, although the ultimate aim of demographic manipulation is the achievement of primary goals like the enhancement of national security and of economic and social welfare. There is little scope for any manipulation of the death rate to achieve demographic ends, so that it is political intervention in fertility and migration that forms the core of most population policies.

A straightforward initial distinction can be made between the pro-natalist policies adopted by several, but by no means the majority of, developed countries, and the anti-natalist policies now followed by the majority of Third World countries.

Pro-natalist policies

These have been adopted in some European countries in response to the fertility troughs reached in the 1930s and the 1970s. A pro-natalist policy can embrace two major strategies: first, restrictions on access to methods of fertility control; and second, a positive attempt to influence attitudes to child-bearing through the provision of financial incentives, child-care facilities and the creation of a moral climate approving of large families.

Perhaps the most explicit pro-natalist policies have been those pursued by Fascist regimes in the 1930s. The major elements in the Nazi programme in Germany (David, Fleischasker and Hahn, 1988) were: the criminal prosecution of induced abortions; the suppression of contraceptive information; a tax on unmarried adults; a marriage loan to young couples which could be partly written off by having children; family allowances, tax concessions and housing preferences for large families; and intensive propaganda on the building of a master race. The number of marriages and births did increase dramatically. The crude birth rate rose from 14.7 per 1,000 in 1933, when the Nazis came to power, to 18.0 in 1934 and 19.7 in 1938. But, in addition to policy influences, there was a stimulus to fertility from falling unemployment and growing national self-confidence. Moreover, most of the increase in births came from first births, suggesting that parents were encouraged to have their children sooner, but not necessarily in larger numbers. This is a common feature of pro-natalist policies – an immediate elevation of period fertility measures, but little effect on completed family size.

In Fascist Italy similar pro-natalist measures were introduced from 1926, but with little impact even on period measures. Other European countries where pro-natalist policies were widely discussed and variably implemented in response to the 1930s fertility trough were Belgium, Sweden, the United Kingdom and France. The interactive coincidence of stagnant economies and stagnant populations was a source of concern to many analysts (Spengler, 1938; Reddaway, 1939), and Royal Commissions to advise on population problems were established in Sweden and the United Kingdom. Little was done in Britain other than the institution of family allowances (and this for humanitarian or ethical, rather than demographic considerations), but in Sweden a conscious attempt was made to adopt an essentially democratic and egalitarian population policy (G. Myrdal, 1940; A. Myrdal, 1945). The guiding principles have been that the burden of national demographic replacement should be spread over the whole population, that individuals should be free to decide on the number of their children, and that quality of children should not be sacrificed to quantity. Therefore, liberal access to abortion and other birth control means has been provided, as well as a wide range of financial and social benefits for families with children.

It was France that exhibited the greatest government concern about low fertility, the outcome of an exceptional demographic history in which fertility had declined continuously from as early as the late eighteenth century and in which there was a mortality toll of over a million men during the First World War. Between 1800 and 1940 the populations of Britain and Germany had grown between two and three times, but that of France by only 50 per cent. Its military humiliations by Germany were widely attributed to differential growth rates of population. Accordingly in 1920 a law was enacted prohibiting not only induced abortion, but also the sale of contraceptives and propaganda for birth control; only after 1967 were the various provisions of this law repealed, although they had long been disregarded. Commissions on births were established in all *départements*, and a series of financial incentives to families with children was introduced, culminating in the *Code de Famille* of 1939. Family allowances and a *salaire unique* paid to mothers remaining at home have been appreciably higher than comparable allowances in other countries, to the extent that in the 1950s they comprised, for a three-child family, more than the average wage. Again it is difficult to assess the demographic effect of these measures and the pro-natalist atmosphere that they generated. The major problem is that fertility increased in all developed countries, regardless of the presence of pro-natalist policies, but it must be said that the increases in France were among the most marked (Figure 5.9). Calot and Hecht (1978) speculate that legislative measures may have raised French fertility by some 10 per cent.

These, then, have been the major government responses to the demographic stagnation of the 1930s. They may not comprise formal population policies ('there is in France no population policy if we mean by that a set of co-ordinated laws aimed at reaching some demographic goals', Bourgeois-Pichat, 1974, p. 546). Rather they should be regarded as 'mild ideologies and piecemeal programs leaning towards pro-natalism' (Stycos, 1977, p. 103). Indeed, governments in most developed countries found the arena of population regulation, at least for its own population, an altogether too sensitive area for government intervention.

Little was heard of pro-natalism in the 1950s and 1960s as fertility rates climbed to well above generation-replacement levels throughout the developed world. Instead, there was growing concern in the West about the pressure of growing populations on limited resources. The Commission on Population Growth and the American Future (1972, p. 7) reported: 'the time has come to challenge the tradition that population growth is desirable: What was unintended may turn out to be unwanted, in the society as in the family.' Similarly, a population panel established by the British government concluded (Ross, 1973, p. 6) that 'Britain would do better in future with a stationary rather than an increasing population'.

The widespread and appreciable fertility falls of the 1970s and 1980s (Figure 5.9) have, however, rekindled fears of population decline in several developed countries. By the late 1980s nearly every developed country had a total fertility rate below the generation-replacement level of 2.1 births per woman, and in several the rate had fallen to a remarkably low 1.4–1.5 (Austria, Belgium, Italy, Luxembourg, Switzerland and West Germany). Although the population of a few countries had consequently started to decline through natural decrease, in most developed countries the age structure resulting from past patterns of fertility and immigration can be expected to produce further population increases until at least the end of the century.

Policies in eastern Europe

It is in the countries of eastern Europe (although not Albania and Yugoslavia) that modern pro-natalism has been most evident and, indeed, most successful. In the mid-1960s these countries were exhibiting the world's lowest fertility rates, giving rise to considerable government concern about the future adequacy of labour force numbers. Most governments responded with packages of measures to restrict moderately what had hitherto been very liberal access to abortion and to provide appreciable financial incentives to child-bearing.

All eastern European countries already had in place social welfare programmes to ease the burden on parents of large families. They included maternity leave, family allowances, birth payments, income tax reductions, and subsidies for child-care and family housing. But they were not thought powerful enough to overcome women's double burden as mothers and workers in societies where female labour force participation outside the home is the highest in the world. Consequently, these programmes have been greatly strengthened from the mid-1960s. In East Germany, for example, 87 per cent of women aged 18–60 were in employment or apprenticeship training in 1980, and there were nursery places for 66 per cent of children under three years of age and for 96 per cent of the 3–6 year olds (David, 1982). In addition, some new measures have been aimed specifically at the promotion of child-bearing. One of the most attractive has been preferential access to subsidized housing granted to families with children in societies where urban housing shortages have been acute; for example, in Sofia, the capital city of Bulgaria, 40 per cent of newly married couples in 1975 had to live with relatives (David, 1982). Fertility does seem to have responded modestly to government policies, not in any sustained increases, but rather in the prevention of fertility reduction to the low levels of central and western Europe.

The *cause célèbre* of eastern European population policies has been Romania, where a remarkable degree of government coercion was used in pursuit of fertility goals. The crude birth rate had fallen to 14 per 1,000 in 1966, and state-provided abortion had become the dominant method of fertility control; the number of abortions per 100 live births rose from 30 in 1958 to a staggering 408 in 1965, accounting for about 80 per cent of all conceptions (Berelson, 1979). The government response was a 1966 decree making abortion legally available only under the most restrictive conditions and stopping the manufacture and import of birth control pills and intra-uterine devices. Thus the dominant means of fertility control was suddenly withdrawn, without any provision for – indeed the discouragement of – viable substitutes. Fertility levels responded dramatically, with the crude birth rate almost doubling to 27 per 1,000 in 1967. But as fertility fell subsequently, some commentators have tended to view the 1967 surge as a minor blip on the fertility curve, indicative of the inability of governments to alter fertility trends significantly.

But Berelson (1979) argues convincingly that the fertility effects of the 1967 Romanian measures were still being felt a decade later, that there was no sign of fertility falling to its 1966 level, and that there would be a positive echo effect on fertility in the 1990s. Some of his evidence is extended in Figure 11.1, where the average birth rate of five eastern European countries is used as a plausible estimate of Romania's rate in the absence of 1967 legislation. The Draconian measures against fertility restriction in Romania were actually strengthened in the 1980s (*Population Today*, February 1987). Thus, employed women up to age 40 were required to undergo a monthly gynaecological examination, officially as a health check, but in reality to ensure that any pregnancy was continued. Unmarried persons over 25 and voluntarily childless married couples were taxed punitively, and the minimum age at marriage for women was lowered to 15 in 1984.

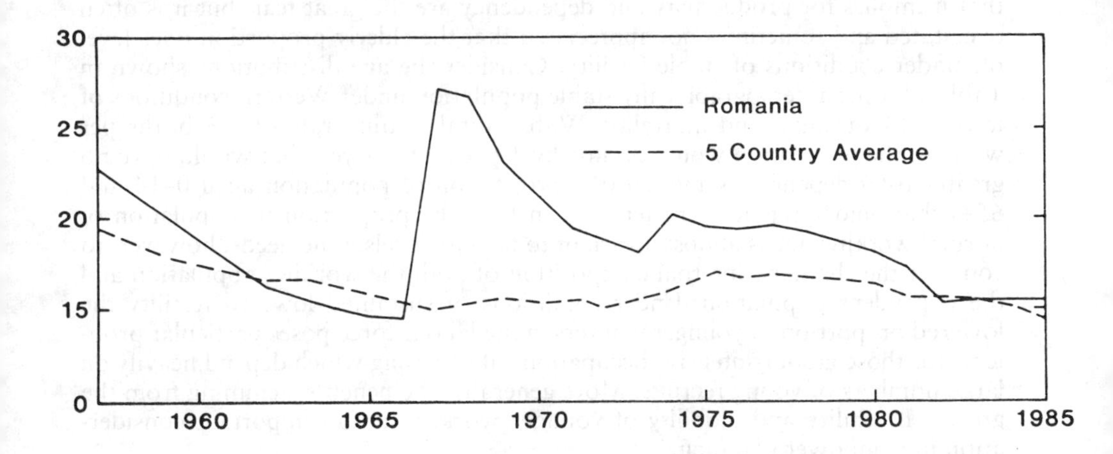

Figure 11.1
Crude birth rates (per 1,000), 1957–85: Romania and average of Bulgaria, Czechoslovakia, East Germany, Hungary and Poland
Source: Extended from Berelson (1979), Figure 1.

All this accorded with the megalomaniac views of the late President Ceausescu in a 1986 speech quoted by *Der Spiegel*:

> The foetus is the socialist property of the whole society. Giving birth is a patriotic duty, determining the fate of our country. Those who refuse to have children are deserters, escaping the laws of national continuity.

Figure 11.1 suggests that Romania's policy was moderately successful in that fertility falls were halted in the medium term, but there have been serious costs. These include the frustration of parental choice, an increased maternal death toll from illegal abortion, and the dislocation in education, housing and employment as the 1967–9 bulge moves through the age groups – an extreme and temporally concentrated form of the Western baby-boom cohort.

Policies in the West

Outside eastern Europe, government response to sub-replacement levels of fertility in the 1970s and 1980s has been somewhat muted. The implications of stationary or slowly declining populations have been widely considered (Council of Europe, 1978; Teitelbaum and Winter, 1985; Davis, Bernstam and Ricardo-Campbell, 1987; van de Kaa, 1987), in an atmosphere generally free from 1930s-type alarmist fears for national security and virility. Technological advances in armaments, manufacturing production and services have undermined the crude size role of military and labour manpower, so that the traditional view of a country's security, economic power and international prestige being a function of its population size has faded. The one exception is France (Huss, 1980), where many of the most powerful politicians from Left and Right (de Gaulle, Giscard d'Estaing, Debré, Chirac, Delors, Mitterrand) have continued to espouse the most traditional forms of nationalistic pro-natalism.

Elsewhere, greater attention and concern are being shown about the evolving age structure than the actual size of populations. An ageing population and all that it implies for productivity and dependency are the great fear, but it is often overstated and sometimes not appreciated that the elderly proportion does level off under conditions of stable fertility. Consider the age distributions shown in Table 11.1 for a demographically stable population under Western conditions of fertility (3 variants) and mortality. With a total fertility rate of 1.6 births per woman, the population would decline by 1 per cent per year but would have no greater total dependency burden (the proportion of population aged 0–14 and 65+) than under replacement fertility. In fact, the proportion of population of normal working age is almost invariant to fertility levels. One needs, however, to consider the changing internal composition of both the working population and the dependent population. Under conditions of sustained lowered fertility the lowered proportion of younger workers in the labour force poses particular problems for those labour-intensive occupations like nursing which depend heavily on large numbers of young recruits. More generally, the benefits stemming from the greater flexibility and mobility of younger workers are an important consideration in manpower planning.

As for the dependent population, there is a greater cost per capita of supporting the elderly than of children, mainly because of the rapidly escalating costs of health-care for the super-elderly. There is also a major problem concerning the funding of state pension systems (Weaver, 1987). These systems are generally

Table 11.1 Percentage distribution by age group in three demographically stable populations, all exhibiting a Western schedule of mortality (life expectancy at birth, 75 years)

	0–14	15–44	45–64	65+	Total
TFR = 2.1	20	38	25	17	100
TFR = 1.9	18	37	26	19	100
TFR = 1.6	14	34	28	24	100

Source: McNicoll (1987), p. 225.

based on an inter-generational transfer of resources from workers to the retired, rather than on workers paying into a fully funded pension scheme from which they will eventually be entitled to draw an annuity. In other words, current benefits to those retiring are financed by taxes on current workers. This has posed problems in the 1980s because of the association of growing numbers of the retired with the eroding tax base of sagging economies. A short-term response in several countries including Britain has been to increase taxes and trim benefits.

Despite these concerns about an ageing population and despite the likely extension of low fertility (given continuing trends in female employment, deferred marriage and divorce), there has been little explicit population policy response in Western countries. There are several reasons. Western governments are reluctant to infringe basic human rights in procreation and family-size choice, being sensitive to the association of pro-natalist 'social engineering' with pre-war Fascism and postwar communism. The discredited arguments of eugenics, concerned with the quality of population through heredity, also linger on to persuade some to oppose child-bearing incentives on the grounds, following Malthus on the Poor Law, that fertility would be enhanced predominantly among the poor and the inferior. A very different, politically orientated opposition comes from some feminists who see pro-natalism as a means of restoring male authority by forcing women back into the kitchen. Finally, there is particular reluctance and opposition in the Netherlands, with its crowded population and dependence on external resources, and in Britain, with its experience of appreciable unemployment in the 1970s and 1980s. It is not surprising that attempts by France in the 1980s to promote an EC pro-natalist initiative have been rejected.

It does seem, therefore, that the major response of Western governments to continuing sub-replacement fertility will be institutional mechanisms within the broad field of economic and social policies. Recent fertility trends have clearly been the outcome of changing norms within society, and governments are likely to concentrate more on adjustments to the consequences of such changes than on probably vain attempts to influence the norms which dictate them. Three examples of likely adjustments can be cited. Immigration could come back in favour, although the social and political costs could outweigh the advantages of a younger labour force and more substantial tax base; more retraining and redeployment of older workers could substitute for the flexibility that a younger labour force usually provides; and the funding problem of pension systems could be eased by reversing the trend towards younger retirement, which would also accord with health improvements at these ages.

However, there are signs of vaguely pro-natalist 'family policies' emerging in some Western countries (McIntosh, 1983, 1987). In reality, these represent a strengthening of the welfare state's partial socialization of the costs of children, which can be justified on the grounds of social justice alone. Increases in family benefits have been particularly prominent in one of the lowest fertility countries, West Germany (Köllmann and Rudenhausen, 1982). In some countries, particularly the United States, conservative 'pro-life' groups have campaigned vigorously against liberal legal abortion, but this has been on ethical rather than macro-demographic grounds.

In the West it does seem, then, that social policy and population policy are inextricably confounded (Demeny, 1987), confirming the prediction of Myrdal (1940, p. 205) that 'population policy will turn out to be simply an intensification of the important part of social policy which bears upon the family and children'.

Anti-natalist policies

Origin and growth

The first government formally to support a family planning programme with the aim of reducing national fertility was India in 1952, but such was the lack of government-perceived urgency throughout the Third World that by 1964 similar national programmes had been established only in Pakistan, South Korea, Fiji and China. It was the mid-1960s that witnessed a surge of fertility regulation policies, as a response to several factors:

1. The 1960 round of censuses revealed higher rates of population growth than had been expected, particularly in Latin America.
2. In 1965 and 1966 the monsoon rains failed in large parts of the Indian subcontinent, causing severe food shortages and widespread concern about the fragility of the population–resources balance. Major famines were averted only by heavy grain shipments on concessional terms from the United States.
3. Development economists became strongly influenced by Coale and Hoover's (1958) work on modelling the Indian and Mexican economies under different fertility assumptions. Projections over a thirty-year period suggested that per capita incomes would be considerably lower under the high, compared with the low, assumption. This was also the time when Rostow's (1960) stages model of economic growth held sway in development thinking orthodoxy. His critical 'take-off' of a national economy into self-sustained growth required a level of savings and productive investment that high population growth rates seemed to inhibit, essentially because a high dependency ratio diverts savings into welfare expenditure.
4. The development in the early 1960s of oral contraceptives and intra-uterine devices and their acceptance by lower status groups in pilot studies encouraged development planners for the first time to believe that Third World fertility rates could be regulated by intervention policies. A parallel was naïvely drawn with the role of imported medical and public health technology in reducing death rates.
5. There emerged, particularly through the work of Enke (1967), the economic concept of value of births averted by family planning programmes; this was the monetary value of what an additional child would consume over a lifetime (food, education, health care, etc.) minus the value of his or her lifetime production and of the family planning cost per participant. Accordingly, in 1965 President Lyndon Johnson urged a United Nations audience: 'Let us act on the fact that less than five dollars invested in population control is worth a

hundred dollars invested in economic growth', and in the same year a United States senator advocated family planning programmes 'to prevent American aid from being poured down a rat-hole' (both quoted by Stycos, 1971a, p. 115). The financial estimates of births averted may well have been wide of the mark, but there can be little doubt of the essentially modest cost of family planning programmes. In 1980 programme expenditure averaged only $0.5–1.0 per capita of total population in thirty-six less developed countries for which data are available (World Bank, 1984, table 7.4). Among countries with strong programmes, family planning spending (from domestic and foreign sources) accounted for only 0.5 per cent of total government expenditure in 1981 in India and Mauritius and only 0.2 per cent in South Korea.

A significant impetus was thus given in the mid-1960s to the formation and external funding of national family planning programmes in the Third World. In 1962 Sweden was the first developed country to earmark a major part of its foreign aid programme to birth control; in 1964 a population office was established within the US Agency for International Development, and AID missions in Latin America were advised to consider population programmes as a priority area; and in 1965 the United Nations began to provide advisory services to family planning programmes. By 1976 as many as 63 countries in the less developed world, embracing 92 per cent of its population, had launched their own programmes or endorsed those of private groups like the International Planned Parenthood Federation. In about half of the programmes, the explicit aim is to reduce fertility in the interest of national development planning, while in the others family planning is supported essentially on grounds of health, human rights and family welfare, regardless of any national demographic impact. It is remarkable how in the space of one generation Third World governments have shifted from almost universal indifference or condemnation of family planning to almost universal approval or acceptance.

Global pattern

Population geographers should be particularly interested in the overall global pattern of government stances on family planning. Figure 11.2 shows that in Asia, under conditions of very large populations and high densities, the great majority of governments are committed to a reduction in population growth rate. With the major exception of Egypt, African nations have generally not developed official anti-natalist positions. Their high death rates, international and tribal tensions, and low modernization levels clearly deter the adoption, let alone the implementation, of anti-natalist policies. Several former French territories in Africa still have in place anti-contraception laws from colonial times. Such countries (e.g. Chad, Niger, Ivory Coast, Guinea, Burkina Faso) have no tradition of private family planning bodies, which generally pave the way for government programmes.

In the Middle East the combination of Islam, labour immigration and military conflicts like the Iran–Iraq war has prevented the emergence of anti-natalist policies in the core of the region. The recent growth of Islamic fundamentalism has ensured that modern contraception is seen as a threat to the traditional power structures of the family, especially the male domination of wives and daughters. Thus, following the 1979 revolution in Iran, the country's embryonic family planning movement was dismantled, and even the much more firmly established programme in adjacent Pakistan has been severely compromised.

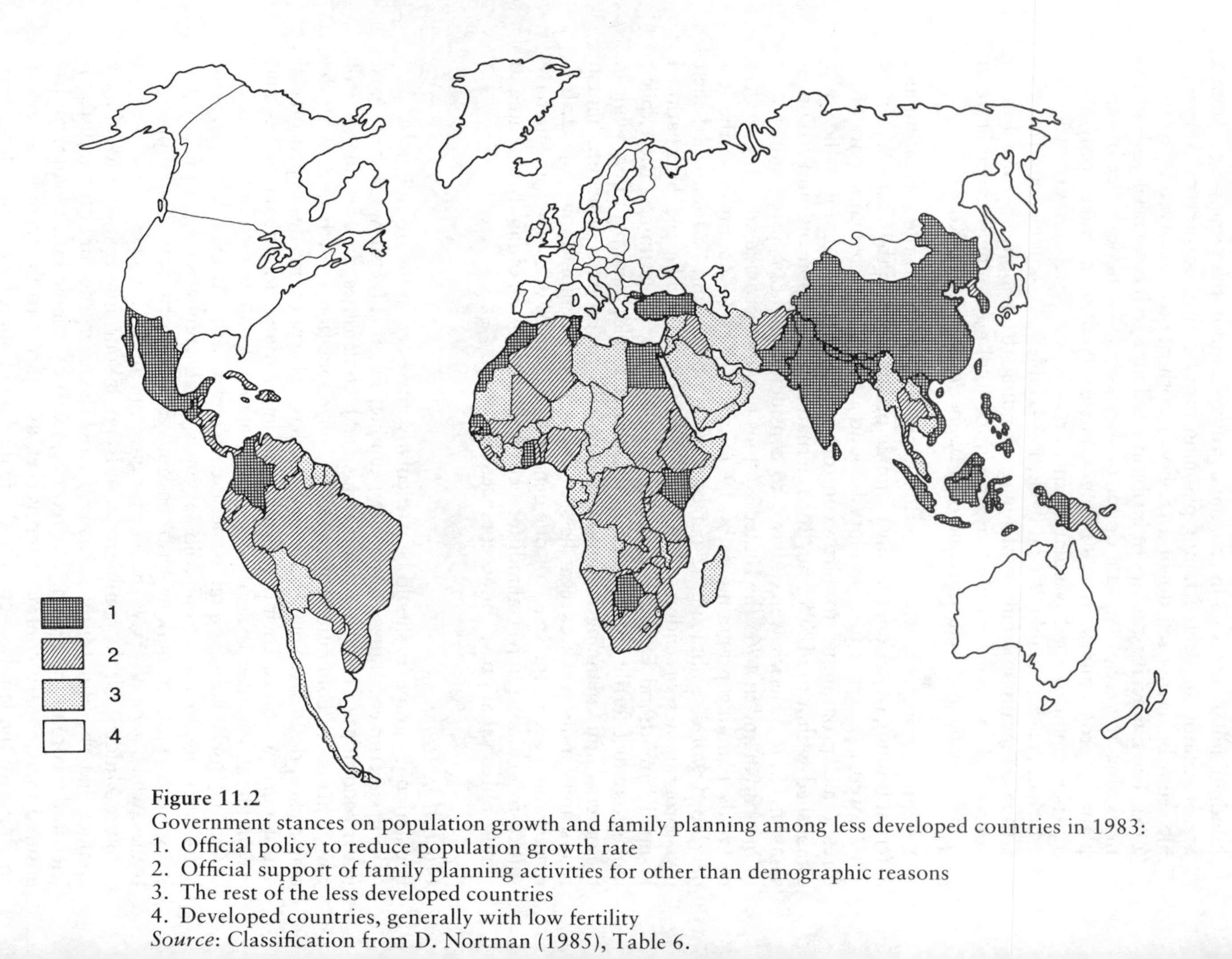

Figure 11.2
Government stances on population growth and family planning among less developed countries in 1983:
1. Official policy to reduce population growth rate
2. Official support of family planning activities for other than demographic reasons
3. The rest of the less developed countries
4. Developed countries, generally with low fertility

Source: Classification from D. Nortman (1985), Table 6.

Latin America is an enigma in governmental response to population growth. The majority of its governments have come, fairly late in the day, to support family planning activities on essentially humanitarian grounds, largely to lessen the excesses of back-street abortions and child abandonment. But despite possessing for some decades the world's highest rates of population growth, the continent has very few governments prepared to back anti-natalist policies.One potent contributory factor is the widespread cult of *machismo*. Another is the pattern of settlement, which provides vast areas of sparsely peopled 'outback' of popularly perceived development potential, even though most of the population lives under conditions of high density and even congestion. Then there was the suspicion that United States imperialism was imposing birth control as a means of limiting the emerging power of Latin America. This could be regarded as a further expression of exploitative dependency theory (Frank, 1969), with the international family planning movement viewed as a CIA wolf in sheep's clothing. Finally, there is the powerful role of the Roman Catholic Church in these former Spanish and Portuguese territories. There is a good deal of survey evidence that at the level of individual behaviour the Church's pronouncements on birth control are widely disregarded, but in the political arena the Church's influence on government decision-makers is often crucial (Stycos, 1971b).

In the particular case of Brazil, government reluctance to endorse an anti-natalist policy has been further explained by Leff (1982) and Daly (1985) in terms of a long-established cheap labour policy, which has involved, successively, slavery, immigration of poor southern Europeans and tacit approval of high fertility among an exploited working class.

Scepticism and opposition

Although family planning programmes have been an integral part of the development package orthodoxy preached by Western advisers for almost three decades, the appropriateness and effectiveness of their role continue to be questioned. The basic assumption of many programme planners that members of traditional societies desire smaller families but, in the absence of modern contraceptive knowledge and supplies, are unable to realize these desires, has come under increasing fire. The point has been put simply but powerfully by a president of the World Bank:

> It is the poor, as a generality, who have the most children. And it is the poorest countries, as a generality, that have the highest birth rates. But it is a mistake to think that the poor have children mindlessly, or without purpose, or – in the light of their own personal value systems – irresponsibly.
>
> Quite the contrary.
>
> The poor, by the very fact of their poverty, have little margin for error. The very precariousness of their existence habituates them to be cautious. They may be illiterate. They are seldom foolhardy. To survive at all they are forced to be shrewd . . .
>
> . . . Poor people have large families for many reasons. But the point is they do have reasons. Reasons of security for their old age. Reasons about additional help on the land. Reasons concerning the cultural preference for sons. Reasons related to the laws of inheritance. Reasons dictated by traditional religious behaviour. And reasons of personal pride.
>
> (McNamara, 1977, p. 170)

Some of these points have been demonstrated vividly by Mamdani (1972) in the context of a Punjab village which had failed to respond to an intensive family planning campaign essentially because the villagers 'want *larger* families . . . and, more important, they want them because they *need* them' (p. 43). A water carrier, mistaking Mamdani for a family planner who had contacted him some years previously, remarked:

> 'You were trying to convince me in 1960 that I shouldn't have any more sons. Now, you see, I have six sons and two daughters and I sit at home in leisure. They are grown up and they bring me money . . . You told me I was a poor man and couldn't support a large family. Now you see, because of my large family, I am a rich man.'
>
> (Mamdani, 1972, p. 109)

There is thus an increasing appreciation of the distinction between macro- and micro-consequences of high fertility. At a society level high fertility invariably entails high net costs, but at a family level net costs are likely to be low, and there may even be net benefits. This has become

> a source of some ambivalence in donor attitudes and local policymaker attitudes towards family planning programmes, which some interpret as vaguely coercive; if poor families benefit from large numbers of children, why persuade them to limit their fertility?
>
> (Birdsall, 1977, p. 84)

In other words, why blame and coerce the victim?

Opposition to the prominence of family planning programmes in development strategies has also developed at a government level. Socialist states are influenced by Marxist doctrine which declares that it is capitalism, not population growth *per se*, which creates a relative surplus population; overpopulation, as a maladjustment of the social system, is found only in the capitalist world; it can be solved only by the reorganization of society into a collective mode of production where the productive forces of the people would increase more rapidly than their numbers. This has prevented some Third World socialist states with problems of population pressure (Cuba, Burma, North Korea) from endorsing policies specifically to reduce population growth rates (Figure 11.2). On the other hand, China has shown an essentially pragmatic response to its massive demographic problems, enabling its government to pursue one of the world's most vigorous policies of fertility reduction.

More than anything, it was the 1974 World Population Conference at Bucharest that brought population issues generally, and Third World family planning programmes specifically, into a worldwide political arena. Western representatives were taken aback by the often savage criticism of their intervention policies in the Third World. The West was accused of placing too much emphasis in development strategies on the population problem generally and on birth control specifically, and far too little on the promotion of social and economic progress; if population policies had a role to play, they would have to be integrated into broader development programmes. Representatives of Third World governments called repeatedly, not for family planning programmes as the basic means of getting out of their poverty trap, but for 'a new international economic order' based on greater justice and equality in international systems of trade, banking, migration and information flow. They argued that high fertility was not the cause of their poverty and development problems, but the result. The point was driven home with catch-phrases like: 'The best contraceptive is development';

'Take care of the people, and the population will take care of itself'; 'To practise family planning, you first of all have to have something to lose.' Emphasis in the West on 'the world population problem' persuaded some Third World governments that they were being asked to deal with someone else's problems, and they were quick to retort that the greatest threat to the world's ecological balance was the profligacy and over-consumption of the West ('one American baby will consume in its lifetime 50 times more of the world's resources than an Indian baby').

The plan of action that finally emerged from the conference was far from the strong, unequivocal commitment to fertility reduction programmes that Western delegates had canvassed. The watered-down plan went little further than emphasizing the sovereignty of nations to determine their own population policies and recommending that governments should make available the information and facilities to enable couples to achieve their desired family size.

Yet, ten years later, there was a remarkable turnaround in many official attitudes at the second UN International Conference on Population, at Mexico City in 1984. In its much heralded 'neutralist' stance on population (see Chapter 7) the United States government now saw development, rather than population policies, as the Third World priority – or at least that form of development based on free markets, entrepreneurial initiative and reduced government interference; and, in response to its powerful pro-life lobby, it announced withdrawal of funding from any programmes which promoted abortion.

At Mexico City it was now the leading Third World nations that called for enhanced programme activity and international funding regardless of any fundamental restructuring of the world's economic relations. The global political climate had changed considerably from Bucharest (Finkle and Crane, 1985). OPEC no longer dominated, more conservative and intransigent governments had emerged in leading Western countries, there were growing disparities in wealth and development strategies within the Third World, and high levels of indebtedness led several of its formerly outspoken governments to more cautious political postures. It was now widely acknowledged that population growth certainly exacerbates, even if it does not fundamentally cause, the basic Third World problems of poverty and restricted opportunities.

Modern policy emphasis

Largely as a response to the criticism of, and the difficulties faced by, the earlier family planning programmes, there has been a growing awareness of the need to restructure programmes. More attention is now given to political and sociological factors, thereby broadening the perspective of programmes from largely supply systems organized and executed technocratically by medical workers and demographers. Four major changes in emphasis may be identified. Only the first embraces the supply side; the other changes relate to demand or motivation, recognizing rather belatedly that family planning works only for those who want it and that fertility behaviour can never be divorced from its socio-economic context.

Wider access

The early programmes were highly centralized, clinic-based, and catered almost exclusively for better-educated, more literate and higher-income urban groups; such policies were defended on the grounds of cost-effectiveness and the need to establish programmes where they had the best chance of success. But the principal policy challenge today is to design and implement strategies that will reach the less privileged, rural and illiterate people who form the majority in most less developed countries.

A useful demonstration of urban dominance within the early programmes is provided by a distribution map (Figure 11.3) of family planning field studies, meeting certain criteria of importance, undertaken in India between 1966 and 1972. Of a total of 144 studies, 79 (55 per cent) were located in large towns or cities, which as a group comprised less than 20 per cent of the country's population. Moreover, Figure 11.3 indicates that there were very few studies, urban or rural, in the poorer regions. The very real difficulties facing rural extension programmes have been demonstrated unequivocally in Blaikie's (1975) study of northern Bihar:

> Many FP workers have little knowledge of the village life they have firmly turned their backs on. They are faced with a suspicious target population, who can (and do) put forward cogent reasons for not taking up FP. Strategies of avoidance, derision, and sometimes open confrontation, make the work extremely difficult.
>
> FP workers simply do not bother to go far from their homes. Male FP workers are supposed to make twenty 'night-halts' a month in villages in their allocated zone to ensure that the more remote villages are not neglected. Although their monthly 'tour diaries' dutifully show that this regulation has been fulfilled, it is common knowledge that most invent their diaries in the same place in which they have spent most of the month – at home.
>
> Muzaffarpur District . . . is supposed to have a special FP doctor for each Block . . . Only 27 out of 40 sanctioned posts are filled. Reputedly, Patna [the state capital of Bihar] has about 300 underemployed or unemployed doctors who are unwilling to take up rural appointments.
>
> (Blaikie, 1975, pp. 82, 75 and 133 respectively)

What then have been some of the ways in which the more successful programmes have assessed and recruited clients extensively among the rural poor? The most effective strategy has been a labour-intensive motivational programme using low-level, local community workers to gain the confidence of eligible women and act as the essential link betwen community and clinic. In Indonesia, South Korea and the Philippines one fieldworker serves about 2,000 eligible couples (World Bank, 1984), and in Mauritius the figure is less than 1,000, enabling an average of just over one home visit per woman per year in the rural districts (Jones, 1989). In China the family planning role of the 'barefoot doctors' has been particularly notable; one of the seven chapters in their handbook is devoted to family planning, emphasizing motivation and late marriage as well as contraceptive methods.

The spatial allocation of scarce resources inevitably influences differential access to, and therefore take-up of, services like family planning. Geographers naturally focus on this issue, but their findings are contradictory (Fuller, 1984). In a working-class community on the fringes of Santiago in Chile, Fuller (1974) found that distance from home to clinic was more important than any socio-economic indicator in discriminating between users and non-users of contraception. However, at a very different scale and cultural setting, in northern Bihar, Blaikie (1975) found that the correlation coefficient between family planning knowledge and distance from family planning facility was a mere –0.01 for

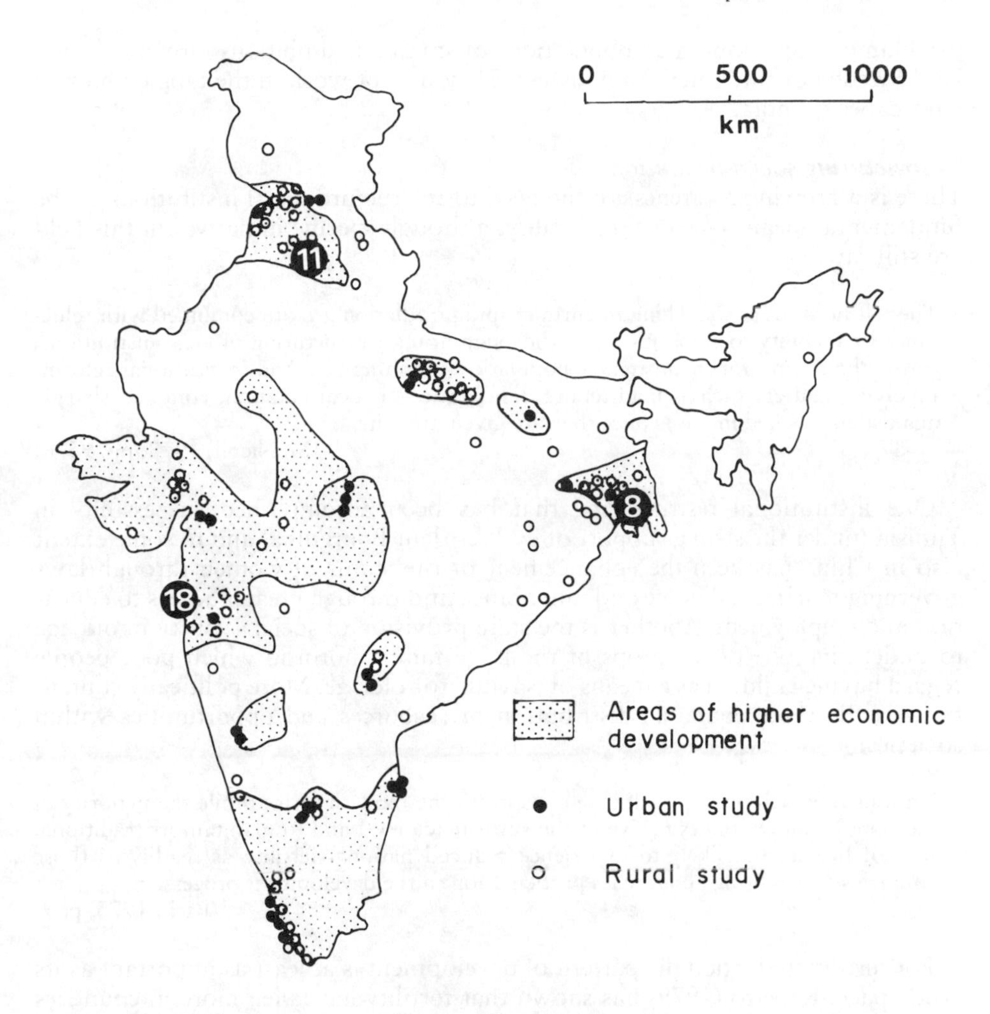

Figure 11.3
India: location of family planning field studies, 1966–72, and areas of higher economic development
Source: Blaikie (1975), Figure 7.1.

females. The female information network, in particular, is highly restricted both spatially and socially and is inaccessible to central-node extension methods.

It is caste, literacy and existing family size, rather than distance, that seem to determine the take-up of family planning in northern Bihar. This leads Blaikie to observe (1975, p. 138) that 'the functional reasons which make a particular innovation attractive or feasible at all are not constant from one decision-maker to another and often vary over space in a non-random manner'. Geographers should beware, therefore, of their intuitive bias towards the importance of contagious effects and the distance-decay function. The basic spatial diffusion models tend to be little more than information diffusion models which cater inadequately for differential resistance to innovation adoption imposed by cultural factors. Hawthorn (1976–7, p. 442) observes that such models 'may work among uncomplicated populations of small and ambitious farmers in lowland Sweden or the American midwest. They do not work in the tangled human landscapes of India.'

Restructuring social institutions

There is a growing awareness of the need to restructure social institutions as the fundamental means of reducing fertility, although specific initiatives in this field are still rare.

> The felt need to do something to curtail rapid population growth combined with reluctance or inability to embark on any thorough-going restructuring of local institutions, forms the setting for 'mainstream' population planning . . . And institutional reforms involving matters such as land tenure, inheritance or local taxation, come up sharply against entrenched interests once they are given any bite.
>
> (McNicoll, 1978, pp. 95–6)

One institutional restructuring that has been accomplished successfully in Tunisia (under the strong support of ex-President Bourguiba) and to some extent also in China has been the enhancement of the status of women through laws governing marriage, divorce and inheritance and through greater access to education and employment. Another is the state provision of social security in old age to undermine one of the props of the large family norm in which poor people regard having children as a means of 'saving' for old age. More politically contentious is the more equitable distribution of resources and opportunities within societies:

> Nations in which only a small élite constitutes the modern sector while the majority of the population continues to live at the subsistence level and to maintain its traditional way of life are not likely to experience reduced national fertility as readily as those countries which bring about mass participation in the development process.
>
> (Rich, 1973, p. 9)

For fertility reduction the pattern of development is at least as important as its crude pace. Repetto (1979) has shown that fertility has fallen more in countries where income distribution has improved (Costa Rica, Sri Lanka, Taiwan) than in those where it has not (Brazil, India, Puerto Rico). Bhattacharyya (1975) and Flegg (1979) provide similar findings. The implication is put graphically by S. George (1976, p. 63): 'the best way to go about it [reducing fertility] is not to distribute condoms and IUDs and hope for the best, but to give people effective land reform and more income'. It has to be said, however, that more recent, substantial fertility declines in countries like Brazil and Mexico have led even some observers with radical sympathies (Eberstadt, 1980; Corbridge, 1986) to conclude that reducing income inequalities is neither a necessary nor a sufficient condition for reducing fertility.

Incentives and disincentives

Governments can intervene, through their tax, welfare and other policies, to change the balance of incentives and disincentives bearing on fertility behaviour. The most explicit case of intervention to transfer the cost of fertility to those generating it is in Singapore, where Prime Minister Lee Kuan Yew pointed out in 1970: 'Beyond three children, the costs of subsidized housing, socialized medicine, and free education should be transferred to the parent' (quoted in McNamara, 1977, p. 171). Accordingly, no priority in the allocation of scarce housing was given to larger families, income tax relief was restricted to the first three children, paid maternity leave was granted for only two confinements, and charges in maternity hospitals rose for each successive birth (Fawcett and Koo, 1980).

Some of these measures may be regarded as punitive, since they unfairly burden the poor who gain most from children, so that they have rarely been adopted elsewhere. But a similar restriction on tax relief was introduced in South Korea in 1975, while in China a complex system of rewards and concessions, which varies from province to province, was introduced in 1979 for couples who limit themselves to a single child. Such couples are favoured in the allocation of urban housing, the granting of plots for private farming, the size of grain rations and the ancillary charges made for schooling. Additional benefits have sometimes been granted to Chinese parents whose only child is a daughter, in order to discourage the pro-natalist effect of son preference.

There are also some policies which seem to go beyond incentives and disincentives into real coercion, where the power of the state is used to overcome the resistance of individuals. Alleged coercion in India's vasectomy programme (accusations of beggars and youths being rounded up to meet sterilization quotas) led to the 1977 electoral defeat of Mrs Gandhi's Congress Party after thirty years in office. Rarely, if ever, can a population policy have had such dramatic, non-demographic impact.

It is in the Chinese programme that coercion has been particularly prominent. In the 1970s pressure was put on couples not to marry before the ages of 23 for females and 25 for males, although this was relaxed in 1980 when the legal minimum age for marriage was formalized at 20 and 22 respectively. Much more controversial has been the pressure imposed by a formidable party and state apparatus on couples, particularly in easily monitored urban settings, not to exceed one child. This has involved gynaecological monitoring of unsterilized women, considerable pressure exerted on those found pregnant to accept an abortion, and fines and continuing taxes for carrying an unauthorized pregnancy to term. However, it is important to realize that by far the major reduction in China's total fertility rate (from 5.8 births per woman in 1970 to 2.8 in 1979) had taken place *before* the one-child policy was initiated in 1979, and also that the most recent indications (Hardee-Cleaveland and Banister, 1988; Tien, 1989) are that the one-child policy is no longer being actively promoted, let alone enforced, especially in remoter rural areas.

Mobilizing community interests

The need to use existing social networks to mobilize community interests in the support of family planning programmes is now more widely appreciated. The problem is to convince the population in general, often initially through community leaders, that the fertility behaviour of individuals does have a very real public impact, not least upon the local community. There is nothing revolutionary in this, since historical demographers have shown that in pre-industrial Europe and Japan social pressures at village level often regulated the fertility behaviour of individuals to ensure that the demands of growing populations did not outstrip local resources.

Indonesia provides a good example of how an anti-natalist social consensus can be successfully developed even in a very large and very poor country. Its programme was initiated in 1970, and its success can be measured by the achievement in the late 1980s of a total fertility rate of 3.3 births per woman and a contraceptive prevalence rate of some 50–55 per cent among eligible couples. From the outset the programme benefited from the active endorsement of President Suharto, and it very quickly appreciated the need to consult, persuade and mobilize opinion leaders at local level – the village heads, Muslim leaders and even traditional birth attendants (Warwick, 1986, 1988).

The wife of the village head was often put in charge of the village contraceptive distribution centre. The initial hostility of religious leaders was removed when the programme, in deference to their views, dropped sterilization and abortion services and also, so far as possible, gynaecological examination of women by male doctors. A major training programme for poorly educated midwives, while obviously focusing on hygienic birth procedures, also included discussion of family planning methods. These confidantes of rural women became more favourably disposed to the family planning programme, in marked contrast to the situation in Egypt, where the ignoring of untrained traditional birth attendants by the Ministry of Health in the 1970s led to them effectively sabotaging the family planning programme by spreading damaging rumours (Gadalla, Mehanna and Tennant, 1977).

The island of Bali is probably the greatest success story of the Indonesian programme. Certainly its Hindu rather than Muslim religion, its tradition of women working outside the home in farming, construction and roadwork, and the sheer density of its crowded rural population have made its people receptive to the programme, but the key factor seems to have been the long-established role of the *banjar* (hamlet) as a mutual-aid community organization (Streatfield, 1986). Key members of Bali's 3,500 *banjar*s were lobbied by family planning personnel, so that promotion of family planning came to figure prominently in the monthly meetings of household heads on community affairs. By endorsing the programme as beneficial to its members, the *banjar* creates a climate in which people are prepared to accept family planning as a social obligation. Hull, Hull and Singarimbun (1977) describe how maps are displayed in the *banjar* halls, with houses of IUD-users outlined in blue, pill-users in red and condom-users in green. They also indicate that a good deal of the agricultural land is worked collectively, undermining the basic need of a peasant family to increase its labour force by child-bearing.

It seems clear, then, that anti-natalist policies are most likely to succeed when the interests of individuals and societies can be brought into congruence, so that the fertility behaviour of individuals contributes to, rather than detracts from, the improvement of common welfare.

Policy options after the demographic transition

A few Third World countries are beginning to restructure their family planning programmes now that their major demographic goals have been attained. Subreplacement fertility has been achieved in Hong Kong, Singapore, Taiwan, South Korea, Mauritius, Cuba, a few small Caribbean territories and possibly China. Emphasis should now be placed on the social welfare element of their programmes, including child spacing, family counselling, the reduction of abortion and premature sterilization, and a modification of the contraceptive 'method-mix' in the interests of clients' health and convenience rather than of aggregate demographic attainment.

There is also concern being shown in a few countries about the future ageing of populations (Warnes, 1986; L. Martin, 1988; Tout, 1989). This is because a rapid transition from high rates of fertility and mortality to low rates will result in a particularly rapid ageing of the population, as is now occurring in Japan (L. Martin, 1989). Thus in Singapore the 'Stop at Two' official slogan of the 1960s and 1970s has been replaced by 'At least two. Better three. Four if you can afford it' (Dwyer, 1987). However, in the other low-fertility countries with more vulnerable economies, social policy adjustments rather than pro-natalism are the likely response, as indeed they have been in the West. In Taiwan Freedman (1986) shows how the extended family has survived the demographic transition. Therefore, as a means of supporting the aged in the future, he argues that existing family support systems should be reinforced as a matter of policy, rather than being replaced indiscriminately by new institutions on the Western model. He is advocating a hybrid system of support in which much less responsibility than in the West is given to impersonal institutions.

Population redistribution in developed countries

In discussions of population policy, no competent geographer would ignore, as some demographers have done, policies of population redistribution. Indeed, this has been the principal theme in the activities of the International Geographical Union Commission on Population Geography in the 1970s and 1980s (e.g. Clarke and Kosiński, 1982; Kosiński and Elahi, 1985). It has to be said, however, that the great majority of population redistribution policies have had only limited effectiveness (Fuchs, 1984).

Planned deconcentration

Very few developed countries have formulated and adopted explicit, coherent policies on population redistribution, but between the 1950s and 1970s fairly amorphous policies evolved in many developed countries to counteract, or at least restrain, what was thought to be cumulative economic and population growth in favoured metropolitan areas like Greater London and Greater Paris at the expense of peripheral regions (Klaassen and Drewe, 1973; Sundquist, 1975; Vanhove and Klaassen, 1980). A case for government intervention could be made on several grounds: economically, there was concern for the under-use of social capital stock and infrastructure in areas of net out-migration and for the inflationary effects on housing and other sectors in areas of appreciable net in-migration; socially, there was concern for the debilitation and possible disintegration of some communities, particularly when culturally distinctive societies like rural Wales and rural Brittany were at risk; and, strategically, there was obvious vulnerability in large concentrations of industry and population.

Government attempts at that time to promote deconcentration operated not so much through specific migration measures as through more general economic and regional planning policies. These attempted to restrain the physical and industrial growth of metropolitan core regions and to encourage growth in the peripheral regions by the provision of financial and infrastructural incentives for job creation. In Britain, for example, some elements of this 'carrot-and-stick' policy included: government restrictions on manufacturing and office growth in the South East through the withholding of industrial development certificates and office development permits; a package of incentives (grants, loans, tax concessions, industrial premises) for job creation in development areas; and, at an intra-regional scale, the decanting of population from congested inner cities to a range of new towns and expanded towns where public housing and new jobs were provided.

Within the more depressed regions in Britain and elsewhere, governments often promoted a concentration of employment and population at the more favoured locations. This accorded with the 'growth pole' orthodoxy of the time which held that injections of regional aid could achieve most impact through spatial concentration and that 'ripples of growth' would emanate from the favoured centres. In this, and in wider deconcentration strategies, the regional planning and population redistribution objectives of socialist countries in eastern Europe broadly paralleled those of the West (Fuchs and Demko, 1979), although much greater direct control of their economies offered more policy opportunities, although not necessarily successes.

During the 1970s it became apparent that inter-regional population dispersal or counter-urbanization was occurring *spontaneously* in a host of developed countries, even including those like the United States, Australia and Japan with very weak or non-existent deconcentration policies. There was growing appreciation that the vital factor promoting deconcentration was not planning policy but the diseconomies of metropolitan scale that seem to set in at an advanced stage of economic development. It became clear that deconcentration would occur regardless of government intervention, and that such intervention could exacerbate some of the problems which are now known to accompany deconcentration. The major problem is the selective nature of out-migration, so that metropolitan cores are left with more than their fair share of the old, the sick, the unskilled, the unemployed and the alienated (Berry, 1980; Frey, 1980). At the same time, the revenue of city authorities falls, making it ever more difficult to maintain community services (Stone, 1978).

Consequently, during the 1980s many Western countries dampened their deconcentration policies and transferred more attention and resources to problems of the inner cities. In Britain deconcentration policies were effectively abandoned by a newly elected Conservative government wedded to the unrestrained operation of market forces and antipathetic to the spatial social engineering, equity and welfare basis of traditional regional policy.

Migration: an equilibriating or cumulative process?

In neo-classical equilibrium theory, migration is regarded as an adjustment mechanism which directs the spatial economy to a state of equilibrium. It is seen as enabling people to respond to changing spatial patterns of economic opportunities, so as to achieve a more optimal pattern of labour allocation. The theory is that migration, stimulated by inter-regional disparities in unemployment, labour vacancies and wage rates, eventually destroys such disparities by siphoning off surplus labour in some areas and eliminating labour shortages in others. As an apparent corrector of imbalances, migration would therefore seem to merit government approval, particularly since, at the level of individual welfare, migration has traditionally been an important vehicle of social mobility; consider, for example, how cityward migration from rural areas has given people access to the training needed for high-income employment.

More sophisticated monitoring of regional economies has, however, shown that migration is far from the ideal homeostatic process that equilibrium theory suggests. Segmentation in the labour market and constraints like housing prevent anything like a free flow of labour. More fundamentally, migration may very well maintain, and even accentuate, the inter-regional disparities of which it is initially a function. Drawing on Keynesian theory, Myrdal (1957) and Hirschman (1958) were among the first to demonstrate this systematically, particularly with the cumulative-causation or cumulative-disequilibrium model in which areas consolidate initial economic advantage through inter-regional flows of goods, finance, information and people. The model demonstrates the essential interdependence between supply of and demand for labour; thus an influx of migrants will increase demand for local goods and services, create more jobs and draw in more migrants, while the opposite downward spiral operates in areas of net out-migration. Then there is the selective nature of migration, with migrants drawn disproportionately from the younger, more skilled and more ambitious elements in the labour force, which further increases the productivity and employment-generation potential of areas of in-migration at the expense of areas of out-migration through the transfer of human capital.

The destabilizing role of migration has been very evident in the poorer and more remote upland areas of Europe. Initially, rural exodus reduces population pressure by creating a more tolerable relationship between population numbers and resources, but because of its self-perpetuating nature, migration may proceed eventually to undermine the demographic and economic health of the community (White, 1980). So many young men may leave that the agricultural land cannot be worked effectively, and the population may fall below the level necessary to support vital community services, like shops, churches, schools and doctors. Beaujeu-Garnier (1966) talks of a vitality threshold, a level of population below which decline becomes irreversible. An extreme example is the remote Scottish island of St Kilda, where demographic seepage over generations finally toppled it over the threshold of survival, leading to the state evacuation of its debilitated and demoralized population in 1930.

Assistance to migrants

In those Western developed countries like the United States and 1980s Britain where regional development policies have either been poorly developed in the first place or effectively dismantled, and where government ideology favours the unconstrained operation of market forces, there has been much official interest (e.g. Roberts, 1987) in stimulating labour migration to improve the efficiency and flexibility of the labour market and therefore the competitiveness of firms. Favoured strategies include:

1. national job information systems;
2. more retraining facilities;
3. subsidized job-search and relocation assistance for the unemployed;
4. initiatives like the National Mobility Scheme in Britain in which local authorities and housing associations make a small proportion of their public-sector new lettings available to incomers who have found a job locally;
5. an encouragement to investors to provide more private rented accommodation by the introduction of tenancy conditions that are more favourable to landlords;
6. most controversially, the pegging or even reduction in welfare support for the unemployed to force them out of what, under Thatcherism in Britain, has been termed the ‘dependency culture’.

Therefore, in countries adopting *laissez-faire* economic policies, the jobs-to-people strategies of traditional regional development policy are now being replaced by people-to-jobs initiatives. But there is no realistic prospect that such initiatives could ever rival the relocation packages increasingly offered by private employers to their higher-status personnel. In migration, as in many things, individual need is rarely matched by available resources.

Population redistribution in less developed countries

A 1983 UN survey of 129 governments of less developed countries revealed that only six considered the distribution of their populations as 'appropriate'; three-quarters stated that they were pursuing policies to moderate or even reverse existing streams of internal migration (Gugler, 1986). There are three major types of policy being pursued, with varying degrees of commitment and success: the inter-regional redistribution of rural population in relation to natural resources; the local reorganization of rural settlement forms; and the slowing of capital city growth.

Inter-regional redistribution of rural population

The standard explicit aim of such policies is to relieve population pressure in congested areas and to develop remote, sparsely peopled but resource-rich areas. Invariably, however, there are further, more controversial and often unacknowledged aims. One is the geopolitical need to secure vulnerable borders from military and/or settlement incursion. Another is the use of land-colonization schemes as an external 'spatial fix' to internal structural problems in much the same way as overseas colonialism was used by European nations a century or so ago. In particular, land colonization can be regarded as an alternative to land reform and therefore as a means of perpetuating rural social inequality.

The major examples of such population redistribution by land settlement schemes are to be found in Indonesia and the Latin American interior. In Indonesia, where the island of Java has only 7 per cent of the country's land area but a remarkable two-thirds of its population, there is a long history of government-sponsored migration to the outer islands, dating back to the Dutch colonial regime in the early years of this century. Until the 1970s the demographic scale of this so-called 'transmigration' was very limited, but there has been much greater commitment, supported by oil revenues, from the beginning of Indonesia's Five Year Plans in 1969. Between 1969 and 1984 a quarter of Java's natural increase was siphoned off by transmigration: 2.4 million people were resettled in the outer islands, especially southern Sumatra, together with a million who moved spontaneously – many of these doubtless stimulated by relatives and friends among official transmigrants (Hugo, 1988). The opening of new reception areas in the 1980s in Irian Jaya has been a particularly controversial development because of the impact on Melanesian indigenous communities and the provocative planned siting of many new settlements along the frontier with Papua New Guinea (Monbiot, 1989).

In Latin American resettlement schemes political considerations have been even more prominent. Ostensibly the aims have been to connect 'men without land to land without men' (the Brazilian president in 1970 describing the planned movement of poor people from the north-east to Amazonia subsequent to the opening up of the region by the Transamazon Highway; quoted by Wood and Carvalho,

1988, p. 244). In reality, the prime purpose of such schemes has been to divert the political pressure by groups acutely disadvantaged by oppressive socio-economic systems. More often than not, any state-supported colonization by small farmers has been swamped sooner or later by large-scale capitalist ventures of ranching, forestry or commercial agriculture attracted by the 'open' character of frontier zones. The associated land speculation, land grabbing, violent social conflict and ecological devastation have been widely observed in eastern Paraguay (Kleinpenning and Zoomers, 1988) and in the Amazonian frontier zones of Brazil (Wood and Carvalho, 1988) and of several other states (Schmink and Wood, 1984; Collins, 1988).

Reorganization of local settlement

Some governments have adopted so-called 'villagization' policies explicitly to stimulate collectivization and the provision of basic amenities, but implicitly to facilitate political control. Sutton and Lawless (1978), Thomas (1982) and Luling (1989) have discussed the often traumatic results of such policies in Algeria, Tanzania and Ethiopia respectively. Such is the scale of the movement in Ethiopia that over 12 million people had been moved to the new regimented settlements by the end of 1988.

Slowing capital city growth

Capital cities doubling in population every dozen years would impose burdens on even the most efficient and affluent developed countries, but for some time such growth has been the norm in Third World countries with few and severely stretched resources. Inevitable results have been growing unemployment and underemployment, mushrooming shanty towns of squatters on the urban periphery or in gullies, swamps and hillsides within the city, and grossly overextended systems of transport, water and sanitation (Dwyer, 1975; Beier, 1976; Lloyd, 1979). Many of these social costs of cityward migration are simply not recognized in the balance of costs and benefits considered by individual migrants, but they are of central concern to government decision-makers – not least because of the way in which they foment highly visible social and political unrest through an 'explosion of despair' when great deprivation and conspicuous consumption are juxtaposed within the confines of the city (Cohen, 1974; Gutkind, 1975).

Some governments have attempted to stem the cityward tide through the direction and sometimes coercion of migrants. In some cities, including Djakarta, Manila, Tunis, Dar-es-Salaam, Brazzaville and Kinshasa, an array of bureaucratic controls has been used sporadically in the 1970s and 1980s to return poorly integrated incomers and the unemployed to their regions of origin. Such policies have little success since administrative requirements can be circumvented by petty corruption, and even those who are expelled often return immediately. In some authoritarian socialist states a policy of enforced de-urbanization has sometimes been pursued, partly to develop rural resources and partly to curb the influence of the disaffected middle classes. The most sinister example was the Killing Fields of Cambodia under the Khmer Rouge regime, but much personal hardship also characterized China's 'rustication' programme during 1969–73 and Vietnam's decanting of population from war-swollen Ho Chi Minh City after reunification of the country in 1976 (Desbarats, 1987). But the essential weakness of all these attempts to redirect migration is that they address the symptoms of problems rather than their basic causes.

More fundamental policies are those which try to redress the spatial imbalance of opportunities within countries. The relocation of national capitals to more central but less developed regions is a dramatic but often isolated policy measure (Turkey, Pakistan, Brazil, Malawi, Nigeria, Tanzania, Ivory Coast). Attempts in some countries to divert public and private investment down the urban hierarchy away from primate cities have generally been unsuccessful, not least because multinational companies insist on investing in metropolitan nodes with infrastructural facilities and networks. One exception has been South Korea, where much of the industrial and population growth of the 1970s was directed to settlements outside Seoul but near enough to it to draw upon its specialized services. But this decentralization was entirely dependent on a stable authoritarian government with a competent administration and well-developed system of central economic planning (Oberai, 1982; Nelson, 1983) – in other words, a rarely replicated set of facilitating conditions.

Rural development as a policy does not necessarily restrain cityward migration, since the provision of modern transportation and particularly of education often accelerates rural exodus because of broadening horizons, rising material aspirations and a growing contempt for rural life. In any case, Third World governments rarely have the commitment to, or resources for, the sort of rural development strategies of 'growth with equity' that have been widely canvassed as the new development-planning orthodoxy of the 1970s and 1980s. In the great majority of Third World countries the metropolitan dominance and spatial disparities created under colonialism have been perpetuated and even reinforced by post-independence governments furthering the interests of the élite groups that they represent.

1992 WORLD POPULATION DATA SHEET

Reproduced from a sheet published by the Population Reference Bureau, Inc.

Prepared by Carl Haub and Machiko Yanagishita, Demographers, Population Reference Bureau.

Region or Country	Population Estimate mid-1992 (millions)	Birth Rate (per 1,000 pop.)	Death Rate (per 1,000 pop.)	Natural Increase (annual, %)	Population "Doubling Time" in Years (at current rate)	Population Projected to 2010 (millions)	Population Projected to 2025 (millions)	Infant Mortality Rate[a]	Total Fertility Rate[b]	%Population Under Age 15/65+	Life Expectancy at Birth Male/Female (years)	Urban Population (%)	Data Availability Code[c]	%Married Women Using Contraception (Total/Modern)	Government View of Fertility Level (H = too high, S = satisfactory, L = too low)	Per Capita GNP, 1990 (US$)
WORLD	5,420	26	9	1.7	41	7,114	8,545	68	3.3	33/ 6	63/67	43		55/47		$ 3,790
MORE DEVELOPED	1,224	14	9	0.5	148	1,333	1,392	18	1.9	21/12	71/78	73		72/47		17,900
LESS DEVELOPED	4,196	30	9	2.0	34	5,781	7,153	75	3.8	36/ 4	61/64	34		51/47		810
LESS DEVEL. (Excl. China)	3,031	33	10	2.3	30	4,361	5,562	84	4.4	39/ 4	58/61	37		43/36		1,000
AFRICA	654	43	14	3.0	23	1,085	1,540	99	6.1	45/ 3	52/55	30		–/–		630
NORTHERN AFRICA	**147**	**35**	**8**	**2.6**	**27**	**216**	**274**	**72**	**4.8**	**42/ 4**	**59/62**	**43**		**32/28**		**1,070**
Algeria	26.0	35	7	2.4	28	37.9	47.1	61	4.9	44/ 4	65/67	50	B	36/31	H	2,060
Egypt	55.7	32	7	2.4	28	81.3	103.1	73	4.4	41/ 4	58/61	45	B	38/35	H	600
Libya	4.5	37	7	3.0	23	7.1	9.3	64	5.2	50/ 2	65/70	76	B	–/–	S	–
Morocco	26.2	33	8	2.4	29	36.0	43.9	73	4.2	41/ 4	62/65	46	B	36/29	H	950
Sudan	26.5	45	14	3.1	22	42.2	57.3	87	6.5	46/ 2	52/53	20	B	9/ 6	S	–
Tunisia	8.4	27	6	2.1	33	11.3	13.4	44	3.4	38/ 5	65/66	53	B	50/40	H	1,420
Western Sahara	0.2	49	21	2.7	25	0.3	0.4	–	–	–/–	–/–	–	D	–/–	–	–
WESTERN AFRICA	**182**	**47**	**17**	**3.0**	**23**	**312**	**449**	**111**	**6.7**	**46/ 3**	**48/50**	**23**		**7/ 3**		**410**
Benin	5.0	49	19	3.1	23	8.9	12.8	88	7.1	46/ 3	45/49	39	C	9/ 1	S	360
Burkina Faso	9.6	50	17	3.3	21	17.0	26.0	121	7.2	48/ 4	51/52	18	C	–/–	H	330
Cape Verde	0.4	41	8	3.3	21	0.7	0.9	41	5.4	45/ 5	59/63	33	B	–/–	H	890
Côte d'Ivoire	13.0	50	14	3.6	19	25.5	39.3	92	7.4	48/ 3	52/55	43	C	3/ 1	S	730
Gambia	0.9	46	21	2.6	27	1.6	2.4	138	6.3	44/ 3	42/46	22	C	–/–	H	260
Ghana	16.0	44	13	3.2	22	26.9	35.4	86	6.4	45/ 3	52/56	32	B	13/ 5	H	390
Guinea	7.8	47	22	2.5	28	11.6	16.1	148	6.1	44/ 3	40/44	22	C	–/–	H	480
Guinea – Bissau	1.0	43	23	2.0	35	1.5	1.9	151	5.8	41/ 4	40/43	27	C	–/–	H	180
Liberia	2.8	47	15	3.2	22	5.5	8.3	144	6.8	46/ 4	53/56	44	B	6/ 6	H	–
Mali	8.5	52	22	3.0	23	14.2	21.7	113	7.3	47/ 4	43/46	22	B	5/ 1	H	270
Mauritania	2.1	46	18	2.8	25	3.5	5.0	122	6.5	44/ 3	46/49	41	B	1/ 0	S	500
Niger	8.3	52	20	3.2	22	15.1	24.3	124	7.1	49/ 3	43/46	15	B	–/–	H	310
Nigeria	90.1	46	16	3.0	23	152.2	216.2	114	6.5	45/ 2	48/49	16	B	6/ 4	H	370
Senegal	7.9	45	17	2.8	25	13.1	17.4	84	6.3	46/ 3	47/49	37	B	11/ 2	H	710
Sierra Leone	4.4	48	23	2.6	27	7.3	10.2	147	6.5	44/ 3	41/44	30	C	–/–	H	240
Togo	3.8	50	13	3.7	19	7.1	11.3	99	7.2	49/ 2	53/57	24	B	34/ 3	H	410
EASTERN AFRICA	**206**	**47**	**15**	**3.2**	**22**	**359**	**528**	**110**	**7.0**	**47/ 3**	**50/53**	**19**		**–/–**		**230**
Burundi	5.8	47	15	3.2	21	10.1	14.9	111	7.0	46/ 3	50/54	5	C	9/ 1	H	210
Comoros	0.5	48	12	3.5	20	0.9	1.4	89	7.1	48/ 3	54/58	26	C	–/–	H	480
Djibouti	0.4	46	17	2.9	24	0.7	1.1	117	6.6	45/ 3	46/49	79	D	–/–	S	–
Ethiopia	54.3	47	20	2.8	25	94.0	140.2	139	7.5	46/ 3	46/48	12	C	4/ 3	H	120
Kenya	26.2	45	9	3.7	19	44.8	62.3	62	6.7	49/ 2	59/63	22	B	27/18	H	370
Madagascar	11.9	45	13	3.2	22	21.3	31.7	115	6.6	47/ 3	53/56	23	C	–/–	H	230
Malawi	8.7	53	18	3.5	20	14.9	23.1	137	7.7	48/ 3	48/50	15	B	7/ 1	H	200
Mauritius	1.1	21	7	1.5	48	1.3	1.4	20.4	2.2	30/ 5	65/72	41	A	75/46	S	2,250
Mozambique	16.6	45	18	2.7	26	26.6	35.6	136	6.3	44/ 3	46/49	23	C	–/–	S	80
Reunion	0.6	24	6	1.8	38	0.8	0.9	13	2.3	33/ 5	67/75	62	B	–/–	–	–
Rwanda	7.7	51	16	3.4	20	14.4	23.2	117	8.0	48/ 3	48/51	7	C	10/ 1	H	310
Seychelles	0.1	24	8	1.6	44	0.1	0.1	*13.0*	2.6	35/ 6	65/74	52	A	–/–	H	4,670
Somalia	8.3	49	19	2.9	24	13.9	17.8	127	6.6	46/ 3	44/48	24	C	–/–	S	150
Tanzania	27.4	50	15	3.5	20	50.2	77.9	105	7.1	48/ 3	49/54	21	C	–/–	H	120
Uganda	17.5	52	15	3.7	19	32.5	49.6	96	7.4	49/ 2	50/52	10	B	5/ 3	H	220
Zambia	8.4	51	13	3.8	18	15.5	24.2	76	7.2	49/ 2	51/54	49	B	–/–	H	420
Zimbabwe	10.3	41	10	3.1	22	17.0	22.6	61	5.6	45/ 3	58/61	26	B	43/36	H	640
MIDDLE AFRICA	**72**	**45**	**15**	**3.0**	**23**	**122**	**182**	**97**	**6.1**	**44/ 3**	**49/53**	**38**		**–/–**		**460**
Angola	8.9	47	19	2.8	25	14.9	21.6	132	6.4	45/ 3	42/46	26	D	–/–	H	–
Cameroon	12.7	44	12	3.2	22	23.1	36.3	85	6.4	46/ 3	54/59	42	C	16/ 4	H	940
Central African Republic	3.2	44	18	2.6	27	4.9	6.9	141	5.6	42/ 3	45/48	43	D	–/–	H	390
Chad	5.2	44	19	2.5	28	7.7	10.3	127	5.8	43/ 4	45/47	30	D	–/–	S	190
Congo	2.4	43	14	2.9	24	3.9	5.5	114	5.8	45/ 3	52/55	41	C	–/–	L	1,010
Equatorial Guinea	0.4	43	16	2.6	26	0.6	0.8	112	5.5	43/ 4	48/52	28	C	–/–	L	330
Gabon	1.1	41	16	2.5	28	1.4	1.8	99	5.2	33/ 6	51/54	43	D	–/–	L	3,220
Sao Tome and Principe	0.1	35	10	2.5	28	0.2	0.3	71.9	5.4	42/ 5	64/67	38	A	–/–	S	380
Zaire	37.9	46	14	3.1	22	65.6	98.2	83	6.1	43/ 4	50/54	40	C	–/–	H	230
SOUTHERN AFRICA	**47**	**35**	**8**	**2.7**	**26**	**76**	**106**	**57**	**4.6**	**40/ 4**	**60/66**	**52**		**45/43**		**2,390**
Botswana	1.4	40	9	3.1	23	2.4	3.3	45	4.8	45/ 3	55/62	24	B	33/32	H	2,040
Lesotho	1.9	41	12	2.9	24	3.1	4.4	95	5.8	43/ 4	53/62	19	B	5/ 2	H	470
Namibia	1.5	43	11	3.1	22	2.9	4.1	102	5.9	46/ 3	59/61	27	D	26/26	–	–
South Africa	41.7	34	8	2.6	26	66.0	92.0	52	4.5	40/ 4	61/67	56	B	48/46	H	2,520
Swaziland	0.8	44	12	3.2	22	1.5	2.2	101	6.2	46/ 3	51/59	23	B	20/17	H	820
	Pop. 1992	Birth Rate	Death Rate	Nat. Incr.	Doub. Time	Pop. 2010	Pop. 2025	Inf. Mort.	TFR	<15/65+	Life Expect.	% Urban	Data Avail.	% Fam. Plan.	Pop. Policy	GNP p.c.

Region or Country	Population Estimate mid-1992 (millions)	Birth Rate (per 1,000 pop.)	Death Rate (per 1,000 pop.)	Natural Increase (annual, %)	Population "Doubling Time" in Years (at current rate)	Population Projected to 2010 (millions)	Population Projected to 2025 (millions)	Infant Mortality Rate[a]	Total Fertility Rate[b]	%Population Under Age 15/65+	Life Expectancy at Birth Male/Female (years)	Urban Population (%)	Data Availability Code[c]	%Married Women Using Contraception (Total/Modern)	Government View of Fertility Level (H=too high, S=satisfactory, L=too low)	Per Capita GNP, 1990 (US$)
ASIA	3,207	26	9	1.8	39	4,207	4,998	68	3.2	33/ 5	63/66	31		56/52		1,680
ASIA (Excl. China)	2,042	30	10	2.0	34	2,787	3,407	81	3.9	36/ 4	60/63	34		47/41		2,520
WESTERN ASIA	**139**	**36**	**8**	**2.8**	**24**	**226**	**313**	**63**	**4.7**	**41/ 4**	**64/68**	**62**		**–/–**		**–**
Bahrain	0.5	27	3	2.4	29	0.8	1.0	20	3.9	35/ 2	70/74	81	C	–/–	S	–
Cyprus	0.7	19	9	1.1	66	0.8	0.9	11	2.4	26/10	74/78	62	B	–/–	L	8,040
Gaza	0.7	55	9	4.6	15	1.3	1.9	45	–	–/–	–/–	–	C	–/–	–	–
Iraq	18.2	45	8	3.7	19	34.1	51.9	67	7.0	45/ 3	66/68	73	C	–/–	L	–
Israel	5.2	21	6	1.5	45	6.9	8.0	8.7	2.9	31/ 9	75/78	91	A	–/–	L	10,970
Jordan	3.6	39	5	3.4	20	6.4	9.2	39	5.6	48/ 3	69/73	70	B	35/27	H	1,240
Kuwait	1.4	32	2	3.0	23	3.2	4.6	16	4.4	45/ 1	72/76	–	–	–/–	S	–
Lebanon	3.4	28	7	2.1	33	4.9	6.1	46	3.7	40/ 5	66/70	84	D	–/–	S	–
Oman	1.6	42	7	3.5	20	3.0	4.9	44	6.7	47/ 3	64/68	11	D	–/–	S	–
Qatar	0.5	27	2	2.5	28	0.7	0.9	26	4.5	28/ 1	69/74	90	C	–/–	S	15,860
Saudi Arabia	16.1	42	7	3.5	20	31.1	47.1	65	7.1	45/ 3	63/66	77	D	–/–	S	–
Syria	13.7	45	7	3.8	18	25.6	38.7	48	7.1	49/ 4	64/66	50	C	–/–	S	990
Turkey	59.2	29	7	2.2	32	81.2	98.1	59	3.6	35/ 4	64/69	59	B	63/31	H	1,630
United Arab Emirates	2.5	31	3	2.8	25	4.9	6.6	25	4.9	35/ 1	69/73	78	C	–/–	S	19,860
West Bank	1.6	44	8	3.6	19	2.4	3.0	40	–	–/–	–/–	–	C	–/–	–	–
Yemen	10.4	51	17	3.5	20	19.0	29.9	124	7.5	49/ 3	48/51	25	C	–/–	H	–
SOUTHERN ASIA	**1,231**	**33**	**11**	**2.2**	**31**	**1,725**	**2,151**	**95**	**4.3**	**38/ 4**	**58/58**	**26**		**43/37**		**440**
Afghanistan	16.9	48	22	2.6	27	34.5	48.5	172	6.9	46/ 4	41/42	18	D	–/–	H	–
Bangladesh	111.4	37	13	2.4	29	165.1	211.6	120	4.9	44/ 3	54/53	14	B	31/24	H	200
Bhutan	0.7	39	19	2.0	35	1.0	1.4	142	5.9	39/ 4	46/49	13	D	–/–	S	190
India	882.6	30	10	2.0	34	1,172.1	1,383.1	91	3.9	36/ 4	58/59	26	B	49/43	H	350
Iran	59.7	41	8	3.3	21	105.0	159.2	43	6.1	46/ 3	63/66	54	C	–/22	H	2,450
Maldives	0.2	41	6	3.4	20	0.4	0.6	34	5.8	47/ 3	62/59	28	B	–/–	S	440
Nepal	19.9	42	17	2.5	28	[illegible]	40.8	112	6.1	42/ 3	50/50	8	B	14/14	H	170
Pakistan	121.7	44	13	3.1	23	195.1	281.4	109	6.1	44/ 4	56/57	28	B	12/ 9	H	380
Sri Lanka	17.6	21	6	1.5	46	21.4	24.0	19.4	2.4	35/ 4	68/73	22	A	62/40	H	470
SOUTHEAST ASIA	**451**	**28**	**8**	**1.9**	**36**	**592**	**696**	**61**	**3.4**	**37/ 4**	**60/64**	**29**		**46/43**		**–**
Brunei	0.3	28	3	2.5	28	0.4	0.5	9	3.5	36/ 3	69/72	59	B	–/–	S	–
Cambodia	9.1	38	16	2.2	32	10.5	13.4	127	4.5	36/ 3	47/50	13	D	–/–	L	–
Indonesia	184.5	26	8	1.7	40	238.8	278.2	70	3.0	37/ 4	58/63	31	B	50/47	H	560
Laos	4.4	46	17	2.9	24	7.2	9.8	112	6.8	44/ 4	48/51	16	C	–/–	S	200
Malaysia	18.7	30	5	2.5	27	27.1	34.9	29	3.6	37/ 4	69/73	35	B	51/30	S	2,340
Myanmar (Burma)	42.5	30	11	1.9	36	57.7	69.9	72	3.9	37/ 4	56/60	24	C	5/ 0	S	–
Philippines	63.7	32	7	2.4	28	85.5	100.8	54	4.1	39/ 4	63/66	43	B	36/22	H	730
Singapore	2.8	19	5	1.4	51	3.2	3.3	6.7	1.8	23/ 6	72/77	100	A	74/73	L	12,310
Thailand	56.3	20	6	1.4	48	69.2	76.4	39	2.4	34/ 4	64/69	18	B	66/64	H	1,420
Viet Nam	69.2	30	8	2.2	31	92.4	108.2	45	4.0	39/ 5	62/66	20	B	53/38	H	–
EAST ASIA	**1,386**	**19**	**7**	**1.2**	**57**	**1,664**	**1,839**	**32**	**2.1**	**27/ 6**	**69/73**	**34**		**71/69**		**2,910**
China	1,165.8	20	7	1.3	53	1,420.3	1,590.8	34	2.2	28/ 6	68/71	26	B	71/70	H	370
Hong Kong	5.7	12	5	0.7	99	6.3	6.2	6.7	1.2	21/ 9	75/80	–	A	81/75	–	11,540
Japan	124.4	10	7	0.3	217	129.4	124.1	4.6	1.5	18/13	76/82	77	A	64/60	L	25,430
Korea, North	22.2	24	6	1.9	37	28.5	32.1	31	2.5	29/ 4	66/72	64	D	–/–	S	–
Korea, South	44.3	16	6	1.1	65	51.7	54.8	15	1.6	26/ 5	67/75	74	B	77/70	S	5,400
Macao	0.5	17	3	1.3	52	0.6	0.6	10	1.5	25/ 6	77/81	97	B	–/–	–	–
Mongolia	2.3	36	8	2.8	25	3.5	4.6	64	4.6	44/ 4	62/67	42	C	–/–	H	–
Taiwan	20.8	16	5	1.1	62	24.0	25.4	6.2	1.7	27/ 6	71/76	71	A	78/62	S	–
	Pop. 1992	Birth Rate	Death Rate	Nat. Incr.	Doub. Time	Pop. 2010	Pop. 2025	Inf. Mort.	TFR	<15/65+	Life Expect.	% Urban	Data Avail.	% Fam. Plan.	Pop. Policy	GNP p.c.

(–)indicates data unavailable or inapplicable
[a] Infant deaths per 1,000 live births
[b] Average number of children born to a woman during her lifetime
[c] A=complete data . . . D=little or no data
[d] Estonia, Latvia, and Lithuania are shown under Northern Europe
[e] Former republics of Yugoslavia
[f] On April 27, 1992, Serbia and Montenegro formed a new state, the Federal Republic of Yugoslavia

Region or Country	Population Estimate mid-1992 (millions)	Birth Rate (per 1,000 pop.)	Death Rate (per 1,000 pop.)	Natural Increase (annual, %)	Population "Doubling Time" in Years (at current rate)	Population Projected to 2010 (millions)	Population Projected to 2025 (millions)	Infant Mortality Rate[a]	Total Fertility Rate[b]	%Population Under Age 15/65+	Life Expectancy at Birth Male/Female (years)	Urban Population (%)	Data Availability Code[c]	%Married Women Using Contraception (Total/Modern)	Government View of Fertility Level (H = too high, S = satisfactory, L = too low)	Per Capita GNP, 1990 (US$)
EUROPE	511	12	10	0.2	338	523	516	11	1.6	20/14	71/78	75		72/47		12,990
NORTHERN EUROPE	**93**	**14**	**11**	**0.3**	**242**	**96**	**97**	**9**	**1.9**	**19/15**	**72/79**	**83**		**72/66**		**17,930**
Denmark	5.2	13	12	0.1	753	5.1	4.8	7.5	1.7	17/16	72/78	85	A	63/60	S	22,090
Estonia	1.6	14	12	0.2	365	1.7	1.8	25	2.0	22/11	66/75	71	B	–/26	–	–
Finland	5.0	13	10	0.3	224	5.0	4.8	5.8	1.8	19/13	71/79	62	A	80/77	S	26,070
Iceland	0.3	19	7	1.2	58	0.3	0.3	*5.9*	2.3	26/11	75/80	90	A	–/–	S	21,150
Ireland	3.5	15	9	0.6	122	3.4	3.3	8.0	2.2	27/11	71/77	56	A	60/–	S	9,550
Latvia	2.7	14	13	0.1	630	2.9	3.0	19	2.0	21/12	65/75	71	B	–/19	–	–
Lithuania	3.7	15	11	0.4	158	4.1	4.4	18	2.0	23/11	67/76	69	B	–/12	–	–
Norway	4.3	14	11	0.4	193	4.5	4.7	6.9	1.9	19/16	73/80	71	A	71/65	S	23,120
Sweden	8.7	14	11	0.3	210	8.9	9.0	6.0	2.1	18/18	75/80	83	A	78/71	S	23,860
United Kingdom	57.8	14	11	0.3	257	59.9	61.0	7.9	1.8	19/16	73/79	90	A	72/71	S	16,070
WESTERN EUROPE	**178**	**12**	**10**	**0.2**	**398**	**179**	**174**	**7**	**1.6**	**18/14**	**73/79**	**82**		**–/–**		**–**
Austria	7.9	12	11	0.1	495	8.2	8.2	7.4	1.5	17/15	73/79	55	A	71/56	S	19,240
Belgium	10.0	13	11	0.2	347	9.7	9.3	7.9	1.6	18/15	72/79	95	A	81/63	S	15,440
France	56.9	13	9	0.4	169	58.8	58.6	7.3	1.8	20/14	73/81	73	A	81/67	L	19,480
Germany	80.6	11	11	−0.1	(–)	78.2	73.7	7.5	1.4	16/15	72/78	90	A	–/–	L	–
Liechtenstein	0.03	13	7	0.6	110	0.03	0.03	*2.7*	1.4	20/10	66/73	–	A	–/–	L	–
Luxembourg	0.4	13	10	0.3	239	0.4	0.4	*7.4*	1.6	17/13	71/78	78	A	–/–	L	28,770
Netherlands	15.2	13	9	0.5	147	16.6	16.7	6.8	1.6	18/13	74/80	89	A	76/72	S	17,330
Switzerland	6.9	13	10	0.3	231	6.9	6.9	6.8	1.6	16/15	74/81	60	A	71/65	L	32,790
EASTERN EUROPE	**96**	**13**	**11**	**0.2**	**369**	**101**	**103**	**17**	**1.9**	**23/11**	**67/75**	**63**		**69/23**		**2,080**
Bulgaria	8.9	12	12	−0.0	(–)	8.8	8.6	14.8	1.7	20/13	68/75	68	A	76/ 8	L	2,210
Czechoslovakia	15.7	14	12	0.2	347	16.8	17.2	11.3	2.0	23/12	68/75	76	A	66/25	S	3,140
Hungary	10.3	12	14	−0.2	(–)	10.5	10.4	15.4	1.8	20/14	65/74	63	A	73/62	L	2,780
Poland	38.4	14	11	0.4	187	41.3	42.7	15.9	2.0	25/10	67/76	61	A	75/26	S	1,700
Romania	23.2	12	11	0.1	578	24.0	24.4	25.7	1.6	23/11	67/73	54	B	58/ 5	L	1,640
SOUTHERN EUROPE	**144**	**11**	**9**	**0.2**	**344**	**146**	**141**	**12**	**1.5**	**20/13**	**72/79**	**68**		**70/34**		**12,860**
Albania	3.3	25	6	1.9	36	3.9	4.5	30.8	3.0	33/ 5	70/76	36	A	–/–	S	–
Bosnia-Hercegovina[e]	4.2	14	6	0.8	90	4.4	4.3	15.2	1.7	28/ 6	69/75	36	A	–/–	–	–
Croatia[e]	4.6	12	11	0.1	1,386	4.8	4.8	10.0	1.7	21/12	68/76	51	A	–/–	–	–
Greece	10.3	10	9	0.1	990	10.4	10.0	10.0	1.5	19/14	73/78	58	A	–/–	L	6,000
Italy	58.0	10	9	0.1	1,386	56.4	51.9	8.6	1.3	17/14	73/80	72	A	78/32	L	16,850
Macedonia[e]	1.9	17	7	1.0	70	2.2	2.3	35.3	2.1	29/ 7	70/74	54	A	–/–	–	–
Malta	0.4	15	8	0.7	92	0.4	0.4	11.3	2.0	23/11	74/78	85	A	–/–	S	6,630
Portugal	10.5	11	10	0.1	533	10.8	10.5	11.0	1.4	21/13	71/78	30	A	66/33	S	4,890
San Marino	0.02	12	7	0.5	144	0.03	0.03	*3.8*	1.3	17/13	73/79	91	A	–/–	–	–
Slovenia[e]	1.9	13	10	0.3	267	2.1	2.2	8.9	1.6	23/11	69/77	49	A	–/–	–	–
Spain	38.6	10	9	0.2	433	40.1	39.3	7.6	1.3	20/13	73/80	91	A	59/38	S	10,920
Yugoslavia[f]	10.0	15	9	0.5	131	10.8	11.0	24.4	2.1	24/ 9	69/74	47	A	–/–	–	–
FORMER USSR[d]	284	17	10	0.7	104	328	362	39	2.2	26/ 9	65/74	66		–/19		
Armenia	3.5	24	7	1.8	40	4.5	5.0	35	2.9	30/ 5	69/75	68	B	–/12	–	–
Azerbaijan	7.1	26	6	2.0	36	9.5	11.4	45	2.7	33/ 5	67/74	53	B	–/ 7	–	–
Belarus	10.3	14	11	0.3	217	11.1	11.5	20	1.9	23/10	67/76	67	B	–/13	S	–
Georgia	5.5	17	9	0.9	80	6.1	6.5	33	2.2	25/ 9	68/76	56	B	–/ 8	–	–
Kazakhstan	16.9	22	8	1.4	50	21.9	26.8	44	2.7	32/ 6	64/73	58	B	–/22	–	–
Kyrgyzstan	4.5	29	7	2.2	31	6.6	8.7	35	3.7	37/ 5	64/72	38	B	–/25	–	–
Moldova	4.4	18	10	0.8	88	5.2	5.8	35	2.3	28/ 8	66/72	48	B	–/15	–	–
Russia	149.3	14	11	0.2	301	162.3	170.7	30	1.9	23/10	64/75	74	B	–/22	–	–
Tajikistan	5.5	38	6	3.2	22	9.1	12.2	73	5.0	43/ 4	67/72	31	B	–/15	–	–
Turkmenistan	3.9	34	7	2.7	26	5.5	6.8	93	4.2	41/ 4	62/68	45	B	–/12	–	–
Ukraine	52.1	13	12	0.1	1,155	53.3	52.9	22	1.9	22/12	66/75	68	B	–/15	S	–
Uzbekistan	21.3	33	6	2.7	25	32.8	43.1	64	4.0	41/ 4	66/72	40	B	–/19	–	–
OCEANIA	28	20	8	1.2	57	35	39	33	2.6	26/ 9	69/75	71		63/68		13 190
Australia	17.8	15	7	0.8	83	21.5	23.9	8.0	1.9	22/11	73/80	85	A	76/72	S	17,080
Federated States of Micronesia	0.1	29	7	2.3	31	0.1	0.1	41	4.2	–/–	–/–	–	C	–/–	–	–
Fiji	0.8	27	7	2.0	35	0.9	1.1	20	3.1	38/ 3	62/60	39	B	41/35	H	1,770
French Polynesia	0.2	28	5	2.3	31	0.3	0.4	16	3.4	37/ 3	66/71	58	B	–/–	–	–
Marshall Islands	0.1	48	9	3.9	18	0.1	0.2	54	7.1	51/ 3	61/64	–	C	–/–	–	–
New Caledonia	0.2	24	6	1.8	39	0.2	0.3	18	2.8	33/ 5	69/76	59	B	–/–	–	–
New Zealand	3.4	18	8	1.0	71	3.8	4.0	7.6	2.1	23/11	72/78	84	A	70/62	S	12,680
Papua-New Guinea	3.9	34	11	2.3	31	5.7	7.3	99	5.4	40/ 3	53/55	13	B	5/–	H	860
Solomon Islands	0.4	41	5	3.6	20	0.6	0.8	32	6.3	47/ 3	60/61	9	B	–/23	H	580
Vanuatu	0.2	36	5	3.1	22	0.3	0.4	32	5.4	45/ 3	67/72	18	B	13/–	S	1,060
Western Samoa	0.2	34	7	2.8	25	0.3	0.4	43	4.7	40/ 4	64/69	21	B	–/19	H	730
	Pop. 1992	Birth Rate	Death Rate	Nat. Incr.	Doub. Time	Pop. 2010	Pop. 2025	Inf. Mort.	TFR	<15/65+	Life Expect.	% Urban	Data Avail.	% Fam. Plan.	Pop. Policy	GNP p.c.

Region or Country	Population Estimate mid-1992 (millions)	Birth Rate (per 1,000 pop.)	Death Rate (per 1,000 pop.)	Natural Increase (annual, %)	Population "Doubling Time" in Years (at current rate)	Population Projected to 2010 (millions)	Population Projected to 2025 (millions)	Infant Mortality Rate[a]	Total Fertility Rate[b]	%Population Under Age 15/65+	Life Expectancy at Birth Male/Female (years)	Urban Population (%)	Data Availability Code[c]	%Married Women Using Contraception (Total/Modern)	Government View of Fertility Level (H = too high, S = satisfactory, L = too low)	Per Capita GNP, 1990 (US$)
NORTH AMERICA	283	16	8	0.8	89	328	363	9	2.0	21/12	72/79	75		74/69		21,580
Canada	27.4	15	7	0.8	89	32.1	35.0	7.1	1.8	21/11	73/80	78	A	73/69	S	20,450
United States	255.6	16	9	0.8	89	295.5	327.5	9.0	2.0	22/13	72/79	75	A	74/69	S	21,700
LATIN AMERICA	453	28	7	2.1	34	609	729	54	3.4	36/ 5	64/70	70		57/48		2,170
CENTRAL AMERICA	**118**	**31**	**6**	**2.5**	**28**	**166**	**204**	**50**	**4.1**	**40/ 4**	**65/71**	**64**		**49/42**		**2,170**
Belize	0.2	37	5	3.1	22	0.3	0.4	32	4.5	45/ 6	67/72	52	B	47/42	S	1,970
Costa Rica	3.2	27	4	2.4	29	4.5	5.6	15.3	3.3	36/ 5	75/79	45	A	70/58	H	1,910
El Salvador	5.6	36	8	2.9	24	7.8	9.6	55	4.6	44/ 4	61/68	48	C	47/44	H	1,100
Guatemala	9.7	39	7	3.1	22	15.8	21.6	61	5.2	45/ 3	60/65	39	B	23/19	H	900
Honduras	5.5	40	8	3.2	22	8.7	11.5	69	5.6	46/ 3	62/66	44	B	41/33	H	590
Mexico	87.7	29	6	2.3	30	119.5	143.3	47	3.8	38/ 4	66/72	71	B	53/45	H	2,490
Nicaragua	4.1	38	8	3.1	23	6.4	8.2	61	5.0	47/ 4	59/65	57	C	27/23	H	–
Panama	2.4	24	5	1.9	37	3.2	3.7	21	2.9	35/ 5	71/75	53	B	58/54	S	1,830
CARIBBEAN	**35**	**26**	**8**	**1.8**	**38**	**43**	**49**	**54**	**3.1**	**33/ 7**	**67/71**	**58**		**51/48**		**–**
Antigua and Barbuda	0.1	14	6	0.8	84	0.1	0.1	*24.4*	1.7	27/ 6	70/74	58	A	53/51	S	4,600
Bahamas	0.3	19	5	1.5	47	0.3	0.4	26.3	2.2	30/ 5	69/76	75	A	–/–	S	11,510
Barbados	0.3	16	9	0.7	102	0.3	0.3	*9.0*	1.8	25/11	70/76	32	A	55/53	S	6,540
Cuba	10.8	18	6	1.1	62	12.3	12.9	11.1	1.9	23/ 9	74/78	73	A	70/67	S	–
Dominica	0.1	20	7	1.2	55	0.1	0.1	*18.4*	2.5	33/ 7	73/79	–	A	50/48	H	1,940
Dominican Republic	7.5	30	7	2.3	30	9.9	11.4	61	3.6	39/ 3	66/69	58	B	50/47	H	820
Grenada	0.1	33	8	2.5	28	0.1	0.1	*15.9*	4.9	43/ 6	69/74	–	A	31/27	H	2,120
Guadeloupe	0.4	20	6	1.4	50	0.4	0.4	9.9	2.4	27/ 8	71/78	48	A	44/31	–	–
Haiti	6.4	45	16	2.9	24	9.4	12.3	106	6.0	45/ 4	53/56	29	B	10/10	H	370
Jamaica	2.5	25	5	2.0	35	3.1	3.6	17	2.6	34/ 8	71/75	51	B	55/51	H	1,510
Martinique	0.4	18	6	1.2	59	0.4	0.5	9	2.0	22/ 9	74/81	82	B	51/38		–
Netherlands Antilles	0.2	19	6	1.2	55	0.2	0.2	*6.3*	2.1	26/ 7	72/76	53	A	–/–	–	–
Puerto Rico	3.7	19	7	1.2	59	3.9	4.2	14.3	2.2	29/10	70/78	72	A	70/62	–	6,470
St. Kitts-Nevis	0.04	23	11	1.2	60	0.05	0.1	*22.2*	2.5	32/ 9	63/69	45	A	41/37	H	3,330
Saint Lucia	0.2	23	6	1.7	40	0.2	0.3	20.8	3.3	44/ 6	69/74	46	A	47/46	H	1,900
St. Vincent and the Grenadines	0.1	23	6	1.6	43	0.1	0.2	21.7	2.6	38/ 5	70/73	21	A	58/55	H	1,610
Trinidad and Tobago	1.3	21	7	1.4	50	1.5	1.7	10.2	2.5	34/ 5	67/73	64	A	53/44	H	3,470
SOUTH AMERICA	**300**	**26**	**7**	**1.9**	**36**	**399**	**476**	**56**	**3.2**	**35/ 5**	**64/70**	**74**		**62/50**		**2,180**
Argentina	33.1	21	8	1.2	56	40.2	45.5	25.7	2.7	30/ 9	66/73	86	A	–/–	S	2,370
Bolivia	7.8	36	10	2.7	26	11.3	14.2	89	4.9	41/ 4	58/64	51	C	30/12	H	620
Brazil	150.8	26	7	1.9	37	200.2	237.2	69	3.1	35/ 5	62/68	74	B	66/57	S	2,680
Chile	13.6	23	6	1.8	39	17.2	19.8	17.1	2.7	31/ 6	71/76	85	A	–/–	S	1,940
Colombia	34.3	26	6	2.0	35	45.6	54.2	37	2.9	36/ 4	68/73	68	B	66/55	S	1,240
Ecuador	10.0	31	7	2.4	29	14.5	17.9	57	3.8	41/ 4	65/69	55	B	53/41	H	960
Guyana	0.8	25	7	1.8	39	1.0	1.2	52	2.6	33/ 4	61/67	35	B	31/28	S	370
Paraguay	4.5	34	7	2.7	25	6.9	9.2	34	4.7	40/ 4	65/69	43	B	48/35	S	1,110
Peru	22.5	31	9	2.2	32	31.0	37.4	76	4.0	39/ 4	60/63	70	B	46/23	H	1,160
Suriname	0.4	26	6	2.0	34	0.6	0.7	31	2.8	34/ 4	67/72	48	B	–/–	S	3,050
Uruguay	3.1	18	10	0.8	83	3.5	3.7	20.4	2.4	26/12	68/75	89	A	–/–	L	2,560
Venezuela	18.9	30	5	2.5	27	27.3	34.6	24.2	3.6	38/ 4	67/73	84	A	–/–	S	2,560
	Pop. 1992	Birth Rate	Death Rate	Nat. Incr.	Doub. Time	Pop. 2010	Pop. 2025	Inf. Mort.	TFR	<15/65+	Life Expect.	% Urban	Data Avail.	% Fam. Plan.	Pop. Policy	GNP p.c.

- (–) indicates data unavailable or inapplicable
- [a] Infant deaths per 1,000 live births
- [b] Average number of children born to a woman during her lifetime
- [c] A=complete data . . . D=little or no data
- [d] Estonia, Latvia, and Lithuania are shown under Northern Europe
- [e] Former republics of Yugoslavia
- [f] On April 27, 1992, Serbia and Montenegro formed a new state, the Federal Republic of Yugoslavia

ACKNOWLEDGEMENTS

The authors gratefully acknowledge the assistance and cooperation of staff members of the Center for International Research of the U.S. Bureau of the Census, the Population Division and the Statistical Office of the United Nations (UN), and the World Bank in the preparation of this year's *Data Sheet*. At PRB, Zuali Malsawma made valuable contributions in the data collection process and Jackie Guenther and Aichin Jones assisted in production.

NOTES

The *Data Sheet* lists all geopolitical entities with populations of 150,000 or more and all members of the UN. These include sovereign states, dependencies, overseas departments, and some territories whose status or boundaries may be undetermined or in dispute. **More developed countries**, following the UN classification, comprise all of Europe and North America, plus Australia, Japan, New Zealand, and the former USSR. All other regions and countries are classified as **less developed**. As a result of recent political developments, the republics of the former USSR and of Yugoslavia are shown separately. The names of the former Soviet republics considered to lie wholly or partly in Europe are shown in italics. Within the former USSR, Russia lies on both the Asian and European continents, as does Turkey.

World and Regional Totals: Regional population totals are independently rounded and include small countries or areas not shown. Regional and world rates and percentages are weighted averages of countries for which data are available; regional averages are shown when data or estimates are available for at least two-thirds of the region's population.

World Population Data Sheets from earlier years should **not be used as a time series**. Fluctuations in values from year to year often reflect revisions based on new data or estimates rather than actual changes in levels. Since data often refer to different years and are of greatly varying reliability, caution should be exercised both in the ranking of countries and in the relationship among variables. Additional information on likely trends and time series data can be obtained from PRB, and are available in UN, World Bank, and U.S. Census Bureau publications.

SOURCES

The rates and figures are primarily compiled from the following sources: official yearbooks and statistical bulletins of countries; UN *Demographic Yearbook 1990* (forthcoming) and *Population and Vital Statistics Report, Data Available as of 1 January 1992* of the UN Statistical Office; *World Population Prospects as Assessed in 1990* of the UN Population Division; data files of the Center for International Research, U.S. Bureau of the Census; data from publications of the Council of Europe and the European Communities; and long-term population projections of the World Bank. Other sources include recent demographic surveys, special studies and direct communication with demographers and statistical bureaus in the U.S. and abroad. Specific data sources may be obtained by contacting the authors of the *Data Sheet*.

For countries with complete registration of births and deaths, rates are those most recently reported. For developed countries, nearly all vital rates refer to 1990 or for a 12 month period ending at some point in 1991. For less developed countries, the reference date is usually for some point in the late 1980s or early 1990s. The complete registration of vital statistics (births and deaths) is indicated by an infant mortality rate shown to one decimal place.

DEFINITIONS

Mid-1992 Population: Estimates are based on a recent census or on official national data or on UN, U.S. Census Bureau, or World Bank projections. The effects of refugee movements, large numbers of foreign workers, and population shifts due to contemporary political events are taken into account to the extent possible.

Birth and Death Rate: These rates are often referred to as 'crude rates' since they do not take a population's age structure into account. Thus, crude death rates in more developed countries, with a relatively large proportion of older persons, are often higher than those in less developed countries.

Rate of Natural Increase (RNI): Birth rate minus the death rate, implying the annual rate of population growth without regard for migration. Expressed as a percentage.

Population 'Doubling Time': The number of years until the population will double assuming a *constant* rate of natural increase (RNI). Based upon the *unrounded* RNI, this column provides an indication of potential growth associated with a given RNI. It is not intended to forecast the actual doubling of any population.

Population in 2010 and 2025: Population projections are based on reasonable assumptions on the future course of fertility, mortality, and migration. Projections are based on official country projections, or on series issued by the UN, the U.S. Bureau of the Census, World Bank, or PRB projections.

Infant Mortality Rate: The annual number of deaths of infants under age one year per 1,000 live births. Rates shown with decimals are completely registered national statistics while those without are estimates from sources cited above. Rates shown in italics are based upon less than 50 annual infant deaths and, as a result, are subject to considerable yearly variability.

Total Fertility Rate (TFR): The average number of children a woman will have assuming that current age-specific birth rates will remain constant throughout her childbearing years (usually considered to be ages 15–49).

Population Under Age 15/Age 65+: The percentage of the total population in those age groups, often considered the 'dependent ages.'

Life Expectancy at Birth: The average number of years a newborn infant can be expected to live under *current* mortality levels.

Urban Population: Percentage of the total population living in areas termed urban by that country or estimated in a source listed above.

Data Availability: Availability of data is graded from 'A' to 'D'. An 'A' indicates a country with both complete vital statistics (birth and death data) and a published national-level census within 15 years or a continuous population register. Countries rated 'B' have one of those two sources plus either a usable census more than 15 years old or a national survey or sample registration system within 10 years, or both. 'C' indicates that at least one census, survey, or sample registration system is available. 'D' indicates that little or no reliable demographic information is available and that estimates are based on fragmentary data or demographic models. There is considerable variation in the quality of data even within the same category.

Contraceptive Use: The percentage of currently married or 'in-union' women of reproductive age (15–49) who use any form of contraception. 'Modern' methods include clinic and supply methods such as the pill, condom, IUD, and sterilization. Data are the most recent available from sources such as the Demographic and Health Survey program or official country estimates.

Government View of Current Fertility Level: This population policy indicator presents the officially stated position of country governments on the level of the national birth rate. Most indicators are from the UN Population Division, *Global Population Policy Database, 1989*, supplemented by recent reports from individual countries.

Per Capita GNP: Gross National Product includes the value of all domestic and foreign output. Estimates are from the *World Bank Atlas, 1991*. Per capita GNP for Nigeria was adjusted by PRB to reflect the new, lower 1991 Census count in that country.

REFERENCES

Abel-Smith, B., with Leiserson, A. (1978) *Poverty, Development and Health Policy*, Public Health Papers No. 69, WHO, Geneva.

Abler, R., Adams, J. and Gould, P. (1971) *Spatial Organisation: The Geographer's View of the World*, Prentice-Hall, Englewood Cliffs, N. J.

Abu-Lughod, J., Foley, M. and Winnick, L. (1960) *Housing Choice and Housing Constraints*, McGraw-Hill, New York.

Adegbola, O. (1977) New estimates of fertility and child mortality in Africa south of the Sahara, *Population Studies*, Vol. 31, pp. 467–86.

Adelman, I. (1963) An econometric analysis of population growth, *American Economic Review*, Vol. 53, pp. 314–39.

Afolayan, A. (1988) Immigration and expulsion of ECOWAS aliens in Nigeria, *International Migration Review*, Vol. 22, pp. 4–27.

Akhtar, R. and Learmonth, A. (1977) The resurgence of malaria in India 1965–76, *GeoJournal*, Vol. 1, no. 5, pp. 69–79.

Alderson, M. (1987) The use of area mortality, *Population Trends*, Vol. 47, pp. 24–33.

Allen-Price, E. D. (1960) Uneven distribution of cancer in West Devon, *Lancet*, Vol. 1, pp. 1235–8.

Alonso, W. (1973) *National Interregional Demographic Accounts*, University of California, Berkeley.

Amin, S. (ed.) (1974) *Modern Migrations in Western Africa*, Oxford University Press, London.

Anderson, T. (1955) Intermetropolitan migration: a comparison of the hypotheses of Zipf and Stouffer, *American Sociological Review*, Vol. 20, pp. 287–91.

Andorka, R. (1978) *Determinants of Fertility in Advanced Societies*, Methuen, London.

Andorka, R. (1982) Lessons from studies on differential fertility in advanced societies, in C. Hohn and R. Mackensen (eds.) *Determinants of Fertility Transition*, International Union for the Scientific Study of Population, Liège, pp. 21–33.

Appleby, A. (1978) *Famine in Tudor and Stuart England*, Liverpool University Press.

Armitage, R. (1987) English regional fertility and mortality patterns, 1975–1985, *Population Trends*, Vol. 47, pp. 16–23.

Arnold, F. and Shah, N. (eds.) (1986) *Asian Labour Migration to the Middle East*, Westview, Boulder.

Arriaga, E. and Davis, K. (1969) The pattern of mortality change in Latin America, *Demography*, Vol. 6, pp. 223–42.

Arriaga, E. (1970) *Mortality Decline and its Demographic Effects in Latin America*, University of California Press, Berkeley.

Arriaga, E. (1981) The deceleration of the decline of mortality in LDCs: the case of Latin America, in International Union for the Scientific Study of Population *International Population Conference, Manila 1981*, Vol. 2, Liège, pp. 21–50.

Askham, J. (1975) *Fertility and Deprivation: A Study of Differential Fertility amongst Working-class Families in Aberdeen*, Cambridge University Press.

Ashton, B., Hill, K., Piazza, A. and Zeitz, R. (1984) Famine in China, 1958–61, *Population and Development Review*, Vol. 10, pp. 613–45.

Aydolot, P. and Keeble, D. (eds.) (1988) *High Technology Industry and Innovative Environments: The European Experience*, Routledge, London.

Baines, D. (1972) The use of published census data in migration studies, in Wrigley (ed.) (1972), pp. 311–35.

Bairagi, R. (1986) Food crisis, nutrition and female children in rural Bangladesh, *Population and Development Review*, Vol. 12, pp. 307–15.

Balan, J. (1983) Agrarian structures and internal migration in historical perspective: Latin American case studies, in Morrison (ed.), pp. 151–85.

Banks, J.A. (1954) *Prosperity and Parenthood*, Routledge, London.

Barbara, D. (1981) *The Endangered Sex: Neglect of Female Children in Rural North India*, Cornell University Press, Ithaca.

Barber, M. (1984) Refugees and super-power rivalry, in J. Crisp and C. Nettleton (eds.) *Refugee Report 1984*, British Refugee Council, London.

Barclay, G. (1958) *Techniques of Population Analysis*, Wiley, New York.

Barney, G. (1980) *The Global 2000 Report to the President of the US*, Vol. 1, The Summary Report, Pergamon, New York.

Barrett, F. (1974) *Residential Search Behaviour: A Study of Intra-Urban Relocation in Toronto*, York University, Toronto.

Basu, A. (1989) Is discrimination in food really necessary for explaining sex differentials in childhood mortality?, *Population Studies*, Vol. 43, pp. 193–210.

Baumol, W. and Blinder, A. (1979) *Economics: Principles and Policy*, Harcourt Brace, New York.

Beaujeu-Garnier, J. (1966) *Geography of Population*, Longman, London.

Beaver, M. W. (1973) Population, infant mortality and milk, *Population Studies*, Vol. 27, pp. 243–54.

Beaver, S. E. (1975) *Demographic Transition Theory Reinterpreted: An Application to Recent Natality Trends in Latin America*, Lexington Books, Lexington.

Becker, G. (1960) An economic analysis of fertility, in National Bureau Commitee for Economic Research, *Demographic and Economic Change in Developed countries*, Princeton University Press, pp. 209–31.

Becker, G. (1981) *A Treatise on the Family*, Harvard University Press, Cambridge, Mass.

Behm, H. (1979) Socioeconomic determinants of mortality in Latin America, *Proceedings of the Meeting of Socioeconomic Determinants and Consequences of Mortality, Mexico City*, WHO, Geneva, pp. 139–65.

Beier, G. J. (1976) Can Third World cities cope?, *Population Bulletin*, Vol. 31, no. 4.

Bell, D. (1976) *The Cultural Contradictions of Capital*, Heinemann, London.

Bentham, C. (1984) Mortality rates in the more rural areas of England & Wales, *Area*, Vol. 16, pp. 219–26.

Bentham, C. and Haynes, R. (1985) Health, personal mobility and use of health services in rural Norfolk, *Journal of Rural Studies*, Vol. 1, pp. 231–9.

Berelson, B. (ed.) (1974) *Population Policy in Developed Countries*, McGraw-Hill, New York.

Berelson, B. (1979) Romania's 1966 anti-abortion decree: the demographic experience of the first decade, *Population Studies*, Vol. 33, pp. 209–2.

Berkner, L. and Mendels, F. (1978) Inheritance systems, family structure and demographic patterns in Western Europe 1700–1900, in Tilly (ed.), pp. 209–4.

Berry, B. J. (1980) Inner city futures: an American dilemma revisited, *Transactions of Institute of British Geographers*, Vol. 5, pp. 1–28.

Berry, B. M. and Schofield, R. (1971) Age at baptism in pre-industrial England, *Population Studies*, Vol. 25, pp. 453–63.
Bhattacharyya, A. (1975) Income inequality and fertility, *Population Studies*, Vol. 29, pp. 5–19.
Biggar, J. (1979) The sunning of America: migration to the Sunbelt, *Population Bulletin*, Vol. 34, no. 1.
Birdsall, J. (1978) Spacing mechanisms and adaptive behaviour of Australian aborigines, in F. Ebling and D. Stoddart (eds.) *Population Control by Social Behaviour*, Institute of Biology, London, pp. 213–44.
Birdsall, N. (1977) Analytical approaches to the relationship of population growth and development, *Population and Development Review*, Vol. 3, pp. 63–102.
Birks, S., Seccombe, I. and Sinclair, C. (1986) Migrant workers in the Arab Gulf: the impact of declining oil revenues, *International Migration Review*, Vol. 20, pp. 799–814.
Blaikie, P. (1975) *Family Planning in India: Diffusion and Policy*, Arnold, London.
Blake, J. (1961) *Family Structure in Jamaica*, Free Press, Glencoe, Ill.
Blake, J. (1968) Are babies consumer durables?, *Population Studies*, Vol. 22, pp. 5–25.
Blake, J. (1984) Catholicism and fertility: on attitudes of young Americans, *Population and Development Review*, Vol. 10, pp. 329–40.
Blaxter, M. and Patterson, E. (1982) *Mothers and Daughters: A Three-Generational Study of Health Attitudes and Behaviour*, Heinemann, London.
Bloom, D. and Freeman, R. (1986) The effects of rapid population growth on labour and employment, *Population and Development Review*, Vol. 12, pp. 381–414.
Blum, A. and Monnier, A. (1989) Recent mortality trends in the USSR, *Population Studies*, Vol. 43, pp. 211–41.
Bogue, D. (1959) Internal migration, in Hauser and Duncan (eds.), pp. 486–509.
Bogue, D. (1969) *Principles of Demography*, Wiley, New York.
Böhning, W. R. (1979) International migration in Western Europe: reflections on the past five years, *International Labour Review*, Vol. 118, pp. 401–14.
Bongaarts, J. (1988) *Modeling the spread of HIV and the demographic impact of AIDS in Africa*, Working Paper 140, Centre for Policy Studies, The Population Council, New York.
Bongaarts, J., Frank, O. and Lesthaeghe, R. (1984) The proximate determinants of fertility in sub-Saharan Africa, *Population and Development Review*, Vol. 10, pp. 511–37.
Bongaarts, J. and Potter, R. (1983) *Fertility, Biology, and Behaviour: An Analysis of the Proximate Determinants*, Academic Press, New York.
Boserup, E. (1965) *The Conditions of Agricultural Growth: The Economics of Agrarian Change under Population Pressure*, Allen & Unwin, London.
Boserup, E. (1981) *Population and Technology*, Blackwell, Oxford.
Boserup, E. (1987) Population and technology in preindustrial Europe, *Population and Development Review*, Vol. 13, pp. 691–701.
Boughey, A. (1968) *Ecology of Populations*, Macmillan, New York.
Bourgeois-Pichat, J. (1974) France, in Berelson (ed.), pp. 545–91.
Bouvier, L. (1977) International migration: yesterday, today and tomorrow, *Population Bulletin*, Vol. 32, no. 4.
Bouvier, L. (1980) America's baby boom generation: the fateful bulge, *Population Bulletin*, Vol. 35, no. 1.
Bouvier, L. and Gardner, R. (1986) Immigration to the US, *Population Bulletin*, Vol. 41, no. 4.
Bradbury, J. (1985) Regional and industrial restructuring processes in the new international division of labour, *Progress in Human Geography*, Vol. 9, pp. 38–63.
Brass, W., Coale, A., Demeny, P., Heisal, D., Lotimer, F., Romaniuk, A. and Van de Walle, E. (1968) *The Demography of Tropical Africa*, Princeton University Press.
Braun, R. (1978) Early industrialization and demographic change in the Canton of Zurich, in Tilly (ed.), pp. 289–334.
Bright, M. and Thomas, D. (1941) Interstate migration and intervening opportunities, *American Sociological Review*, Vol. 6, pp. 773–83.

Brookfield, H. (1973) On one geography and a Third World, *Transactions of Institute of British Geographers*, Vol. 58, pp. 1–20.
Brookfield, H. (1978) Third world development, *Progress in Human Geography*, Vol. 2, pp. 121–32.
Brown, D. and Wardwell, J. (eds.) (1980) *New Directions in Urban–Rural Migration: The Population Turnaround in Rural America*, Academic Press, New York.
Brown, L., (1985) *State of The World 1985*, Worldwatch Institute, Norton, New York.
Brown, L. and Moore, E. (1970) The intra-urban migration process: a perspective, *Geografiska Annaler*, Vol. 52B, pp. 1–13.
Brownlea, A. (1972) Modelling the geographic epidemiology of infectious hepatitis, in McGlashan (ed.), pp. 279–300.
Buksmann, P. (1980) Migration and land as aspects of underdevelopment: an approach to agricultural migration in Bolivia, in White and Woods (eds.), pp. 108–28.
Burnley, I. (1982) *Population, Society and Environment in Australia*, Shillington House, Melbourne.
Byrne, D., Harrisson, P., Keithley, J. and McCarthy, P. (1986) *Housing and Health*, Gower, Aldershot.
Byrne, J. (1972) *Levels of Fertility in Commonwealth Caribbean 1921–65*, Institute for Social and Economic Research, University of West Indies, Jamaica.
Cadwallader, M. (1985) Structural-equation models of migration, *Environment and Planning*, Vol. A17, pp. 101–13.
Cadwallader, M. (1986) Migration and intra-urban mobility, in M. Pacione (ed.) pp. 257–83.
Caldwell, J. (1978) A theory of fertility: from high plateau to destabilisation, *Population and Development Review*, Vol. 4, pp. 553–77.
Caldwell, J. (1980) Mass education as a determinant of the timing of fertility decline, *Population and Development Review*, Vol. 6, pp. 225–55.
Caldwell, J. (1981) The mechanisms of demographic change in historical perspective, *Population Studies*, Vol. 35, pp. 5–27.
Caldwell, J. (1982) *Theory of Fertility Decline*, Academic Press, London.
Caldwell, J. (1986) Routes to low mortality in poor countries, *Population and Development Review*, Vol. 12, pp. 171–220.
Caldwell, J. and Caldwell, P. (1987) The cultural context of high fertility in sub-Saharan Africa, *Population and Development Review*, Vol. 13, pp. 409–37.
Caldwell, J. and Caldwell, P. (1988) Is the Asian family planning program model suited to Africa?, *Studies in Family Planning*, Vol. 19, pp. 19–28.
Caldwell, J., Caldwell, P. and Quiggin, P. (1989) The Social context of AIDS in sub-Saharan Africa, *Population and Development Review*, 15, 185–234.
Caldwell, J., Gaminiratne, K., Caldwell, P., de Silva, S., Caldwell, B., Weeraratne, N. and Silva, P. (1987) The role of traditional fertility regulation in Sri Lanka, *Studies in Family Planning*, Vol. 18, pp. 1–21.
Caldwell, J., Reddy, P. and Caldwell, P. (1982) The causes of demographic change in rural south India: a micro approach, *Population and Development Review*, Vol. 8, pp. 689–727.
Caldwell, J., Reddy, P. and Caldwell, P. (1983) The social component of mortality decline: an investigation in South India, *Population Studies*, Vol. 37, pp. 185–205.
Caldwell, J., Reddy, P. and Caldwell, P. (1985) Educational transition in rural South India, *Population and Development Review*, Vol. 11, pp. 29–51.
Caldwell, J. and Ruzicka, L. (1978) The Australian fertility transition, *Population and Development Review*, Vol. 4, pp. 81–103.
Calhoun, J. (1963) Population density and social pathology, in L. Duhl (ed.) *The Urban Condition*, Basic Books, New York.
Calot, G. and Hecht, J. (1978) The control of fertility trends, in Council of Europe (1978), pp. 178–96.
Campbell, A. (1974) Beyond the demographic transition, *Demography*, Vol. 11, pp. 549–61.

Cannon, H. and Hopps, H. (1972) *Geochemical Environment in Relation to Health and Disease*, Geological Society of America, Special Paper 140, Boulder, Colorado.
Caprio, R. et al. (1975) Residential location, ambient lead pollution and lead absorption in children, *Professional Geographer*, Vol. 27, pp. 37–42.
Carr-Hill, R. (1988) Time trends in inequalities in health, *Journal of Biosocial Science*, Vol. 20, pp. 265–73.
Carrothers, G. (1956) An historical review of the gravity model and potential concepts of human interaction, *Journal of American Institute of Planners*, Vol. 22, pp. 94–102.
Carstairs, V. (1981) Small area analysis and health service research, *Community Medicine*, Vol. 3, pp. 131–9.
Cartwright, F. (1972) *Disease and History*, Crowell, New York.
Castles, S. (1984) *Here for Good: Western Europe's New Ethnic Minorities*, Pluto, London.
Castles, S. and Kosack, G. (1973) *Immigrant Workers and Class Structure in Western Europe*, Oxford University Press, London.
Cavanaugh, J. (1979) Is fertility declining in less developed countries? *Population Studies*, Vol. 33, pp. 283–293.
Chambers, J. D. (1965) Three essays on the population and economy of the Midlands, in Glass and Eversley (eds.), pp. 308–53.
Chambers, J. D. (1972) *Population, Economy and Society in Pre-Industrial Britain*, Oxford University Press, London.
Chambers, R. (1986) Hidden losers: the impact of rural refugees and refugee programs on poorer hosts, *International Migration Review*, Vol. 20, pp. 245–63.
Champion, A. (1983) Population trends in the 1970s, in Goddard and Champion (eds.), pp. 187–214.
Champion, A. (ed.) (1989) *Counterurbanisation*, Arnold, London.
Champion, A. and Congdon, P. (1988) An analysis of the recovery of London's population change rate, *Built Environment*, Vol. 13, pp. 193–211.
Champion, A., Fielding, A. and Keeble, D. (1989) Counterurbanisation in Europe, *Geographical Journal*, Vol. 155, pp. 52–80.
Chang, M.-C., Freedman, R. and Sun, T.-H. (1987) Trends in fertility family size preferences, and family planning practice: Taiwan 1961–85, *Studies in Family Planning*, Vol. 18, pp. 320–37.
Chapman, M. (1975) Mobility in a non-literate society: method and analysis for two Guadalcanal communities, in Kosiński and Prothero (eds.), pp. 129–45.
Chapman, M. and Prothero, R. M. (1985) Themes on circulation in the Third World, in Prothero and Chapman (eds.), pp. 1–26.
Chen, L., Huq, E. and D'Souza, S. (1981) Sex bias in the family allocation of food and health care in rural Bangladesh, *Population and Development Review*, Vol. 7, pp. 55–70.
Chen, L., Rahman, M. and Sardar, A. (1980) Epidemiology and causes of death among children in a rural area of Bangladesh, *International Journal of Epidemiology*, Vol. 9, pp. 25–33.
Cherry, S. (1980) The hospitals and population growth, *Population Studies*, Vol. 34, pp. 59–75.
Childe, V. G. (1936) *Man Makes Himself*, Watts, London.
Christensen, H. (1983) *Sustaining Afghan Refugees in Pakistan*, Report 83.3, UN Research Institute for Social Development, Geneva.
Cipolla, C. (1965) Four centuries of Italian demographic development, in Glass and Eversley (eds.), pp. 570–87.
Clark, C. (1967), *Population Growth and Land Use*, Macmillan, London.
Clark, P. and Souden, D. (1987) *Migration and Society in Early Modern England*, Hutchinson, London.
Clark, W. (1982) Recent research on migration and mobility: a review and interpretation, *Progress in Planning*, Vol. 18, pp. 1–56.

Clark, W. and Huff, J. (1977) Some empirical tests of duration-of-stay effects in intraurban migration, *Environment and Planning A*, Vol. 9, pp. 1357–74.
Clark, W. and Moore, E. (eds.) (1978) *Population Mobility and Residential Change*, Studies in Geography 25, Northwestern University, Evanston.
Clark, W. and Onaka, J. (1983) Life cycle and housing adjustment as explanations of residential mobility, *Urban Studies*, Vol. 20, pp. 47–57.
Clarke, J. I. (1965) *Population Geography*, Pergamon, Oxford.
Clarke, J. I. (1976) Population and scale: some general considerations, in Kosiński and Webb (eds.), pp. 21–9.
Clarke, J. I. (ed.) (1984) *Geography and Population: Approaches and Applications*, Pergamon, Oxford.
Clarke, J. I. (1985) Islamic populations: limited demographic transition, *Geography*, Vol. 70, pp. 118–28.
Clarke, J. I. and Kosiński, L. (eds.) (1982) *Redistribution of Population in Africa*, Heinemann, London.
Cleland, J. and Hobcraft, J. (eds.) (1985) *Reproductive Change in Developing Countries: Insights from the World Fertility Survey*, Oxford University Press.
Cleland, J. and Rodriguez, G. (1988) The effect of parental education on marital fertility in developing countries, *Population Studies*, Vol. 44, pp. 419–42.
Cliff, A. and Haggett, P. (1989) Spatial aspects of epidemic control, *Progress in Human Geography*, Vol. 13, pp. 315–47.
Cliff, A., Haggett, P., Ord, J. and Versey, G. (1981) *Spatial Diffusion: An Historical Geography of Epidemics in an Island Community*, Cambridge University Press.
Cliffe, L. (1978) Labour migration and peasant differentiation: Zambian experience, *Journal of Peasant Studies*, Vol. 5, pp. 326–46.
Coale, A. and Demeny, P. (1983) *Regional Model Life Tables and Stable Populations* (2nd edn), Academic Press, New York.
Coale, A. and Watkins, S. (1986) *The Decline of Fertility in Europe*, Princeton University Press.
Coale, A. J. (1969) The decline of fertility in Europe from the French Revolution to World War II, in S. Behrman *et al.* (eds.) *Fertility and Family Planning: A World View*, University of Michigan Press, Ann Arbor, pp. 3–24.
Coale, A. J. (1974) The history of the human population, *Scientific American*, Vol. 231, no. 3, pp. 41–51.
Coale, A. J. and Hoover, E. (1958) *Population Growth and Economic Development in Low-Income Countries*, Princeton University Press.
Coale, A. J. and Zelnick, M. (1963) *New Estimates of Fertility and Population in the United States*, Princeton University Press.
Cohen, M. (1974) *Urban Policy and Political Conflict in Africa: A Study of the Ivory Coast*, University of Chicago Press.
Cohen, M. (1977) *The Food Crisis in Prehistory: Overpopulation and the Origins of Agriculture*, Yale University Press, New Haven.
Cohen, M. and Armelagos, G. (eds.) (1984) *Paleopathology at the Origins of Agriculture*, Academic Press, London.
Cohen, R. (1986) Policing the frontiers: the state and the migrant in the international division of labour, in J. Henderson and M. Castells (eds.) *Global Restructuring and Territorial Development*, Sage, London, pp. 86–111.
Cole, H., Freeman, C., Jahoda, M. and Pavitt, K. (1973) *Thinking about the Future: A Critique of The Limits to Growth*, Chatto and Windus, London.
Coleman, D. (1986) Population regulation: a long-range view, in Coleman and Schofield (eds.), pp. 14–41.
Coleman, D. and Schofield, R. (eds.) (1986) *The State of Population Theory*, Blackwell, Oxford.
Collins, J. (1983) Fertility determinants in a High Andes community, *Population and Development Review*, Vol. 9, pp. 61–75.

Collins, J. (1988) *Unseasonal Migrations*, Princeton University Press.

Collver, O. (1965) *Birth Rates in Latin America: New Estimates of Historical Trends and Fluctuations*, Institute of International Studies, Berkeley.

Commission on Population Growth and the American Future (1972) *Final Report on Population and the American Future*, Signet, New York.

Committee on National Urban Policy, National Research Council (1983) *Rethinking Urban Policy*, National Academy of Sciences Press, Washington, D.C.

Compton, P. (1978) Fertility differentials and their impact on population distribution and composition in Northern Ireland, *Environment and Planning*, Vol. 10A, pp. 1397–411.

Compton, P. (1982) The demographic dimension, in F. Boal and N. Douglas (eds.) *Integration and Division: Geographical Perspectives on the Northern Ireland Problem*, Academic Press, London, pp. 75–104.

Compton, P. (1985) Rising mortality in Hungary, *Population Studies*, Vol. 39, pp. 71–86.

Compton, P. and Coward, J. (1989) Fertility and Family Planning in Northern Ireland, Avebury, Aldershot.

Compton, P., Coward, J. and Power, J. (1986) Regional differences in attitude to abortion in Northern Ireland, *Irish Geography*, Vol. 19, pp. 58–68.

Condran, G. and Crimmins-Gardner, E. (1978) Public health measures and mortality in US cities in the late nineteenth century, *Human Ecology*, Vol. 6, pp. 27–54.

Congdon, P. and Batey, P. (eds.) (1989) *Advances in Regional Demography*, Pinter/Belhaven, London.

Connell, K. (1950) *The Population of Ireland 1750–1845*, Clarendon Press, Oxford.

Coombs, L. (1979) Underlying family-size preferences and reproductive behaviour, *Studies in Family Planning*, Vol. 10, pp. 25–36.

Corbridge, S. (1986) *Capitalist World Development: A Critique of Radical Development Geography*, Macmillan, London.

Corsini, C. (1977) Self-regulating mechanisms of traditional populations before the demographic revolution, in International Union for the Scientific Study of Population, *International Population Conference Mexico 1977*, Vol. 3, Liège.

Council of Europe (1978) *Population Decline in Europe: Implications of a Declining or Stationary Population*, Arnold, London.

Cousens, S. (1960) The regional pattern of emigration during the Great Irish Famine 1846–51, *Transactions of Institute of British Geographers*, Vol. 28, pp. 119–34.

Coward J. (1978) Changes in the pattern of fertility in the Republic of Ireland, *Tijdschrift voor Economische en Sociale Geografie*, Vol. 69, pp. 353–61.

Coward, J. (1986a) Fertility patterns in the modern world, in Pacione (ed.), pp. 58–94.

Coward, J. (1986b) The analysis of regional fertility patterns, in Woods and Rees (eds.), pp. 45–67.

Cowgill, D. (1949) The theory of population growth cycles, *American Journal of Sociology*, Vol. 55, pp. 163–70.

Cox, B. et al. (1987) *Health and Lifestyle Survey*, Health Promotion Research Trust, London.

Crafts, N. (1978) Average age at first marriage for women in mid-nineteenth century England and Wales: a cross-section study, *Population Studies*, Vol. 32, pp. 21–5.

Crawford, M. D., Gardner, M. and Morris, J. (1968) Mortality and hardness of local water supplies, *Lancet*, Vol. 1, pp. 827–33.

Crawford, M. D., Gardner, M. and Morris, J. (1971) Changes in water hardness and local death rates, *Lancet*, Vol. 2, p. 327.

Cronjé, G. (1984) Tuberculosis and mortality decline in England and Wales 1851–1910, in Woods and Woodward (eds.), pp. 79–101.

Cunliffe, B. (1978) *Rome and her Empire*, Bodley Head, London.

Curson, P. (1985) *Times of Crises*, Sydney University Press.

Curson, P. (1986) Mortality patterns in the modern world, in Pacione (ed.), pp. 95–131.

Curtin, P. (1969) *The Atlantic Slave Trade: A Census*, University of Wisconsin Press, Madison.

Cutright, P. and Kelly, W. (1981) The role of family planning programs in fertility declines in less developed countries, *International Family Planning Perspectives*, Vol. 7, pp. 145–51.

Daly, H. (1985) Marx and Malthus in north-east Brazil, *Population Studies*, Vol. 39, pp. 329–38.

Das Gupta, M. (1977) From a closed to an open system: fertility behaviour in a changing Indian village, in Epstein and Jackson (eds.), pp. 97–121.

Das Gupta, M. (1987) Selective discrimination against female children in India, *Population and Development Review*, Vol. 13, pp. 77–100.

DaVanzo, J. (1980) *Micro Economic Approaches to Studying Migration Decisions*, Rand Corporation, Santa Monica, California.

David, H. (1982) Eastern Europe: pronatalist policies and private behaviour, *Population Bulletin*, Vol. 36, no. 6.

David, H., Fleischhasker, J. and Hohn, C. (1988) Abortion and eugenics in Nazi Germany, *Population and Development Review*, Vol. 14, pp. 81–112.

Davies, J. and Chilvers, C. (1980) The study of mortality in small administrative areas of England & Wales, *Journal of Epidemiology and Community Health*, Vol. 34, pp. 86–95.

Davis, K. (1967) Will current programs succeed?, *Science*, Vol. 158, pp. 730–9.

Davis, K. (1974) The migrations of human populations, *Scientific American*, Vol. 231, no. 3, pp. 93–105.

Davis, K., Bernstam, M. and Ricardo-Campbell, R. (eds.) (1987) *Below-Replacement Fertility in Industrial Societies*, Cambridge University Press.

David, K. and Blake, J. (1956) Social structure and fertility: an analytical framework, *Economic Development and Cultural Change*, Vol. 4, pp. 211–35.

Davis, Karen and Marshall, R. (1979) New developments in the market for rural health care, *Research in Health Economics*, Vol. 1, pp. 57–110.

Day, L. (1968) Natality and ethnocentrism: some relationships suggested by an analysis of Catholic–Protestant differentials, *Population Studies*, Vol. 22, pp. 27–50.

Day, L. (1984) *Analysing Population Trends: Differential Fertility in a Pluralistic Society*, Croom Helm, London.

De Castro, J. (1952) *The Geography of Hunger*, Little Brown, Boston.

Deevey, E. (1950) The probability of death, *Scientific American*, Vol. 182, no. 4, April, pp. 58–60.

Deevey, E. (1960) The human population, *Scientific American*, Vol. 203, pp. 194–204.

de Janvry, A. (1981) *The Agrarian Question and Reformism in Latin America*, Johns Hopkins University Press, Baltimore.

Demeny, P. (1972) Early fertility decline in Austria-Hungary: a lesson in demographic transition, in Glass and Revelle (eds.), pp. 153–72.

Demeny, P. (1974) The populations of the underdeveloped countries, *Scientific American*, Vol. 231, no. 3, pp. 149–59.

Demeny, P. (1985) Bucharest, Mexico City and beyond, *Population and Development Review*, Vol. 11, pp. 99–106.

Demeny, P. (1987) Pronatalist policies in low-fertility countries, in Davis *et al.* (eds.), pp. 335–58.

Dennis, R. (1977) Intercensal mobility in a Victorian city, *Transactions of Institute of British Geographers*, Vol. 2, pp. 349–63.

Department of the Environment (1987) *Handling Geographic Information, Report of Committee Chaired by Lord Chorley*, HMSO, London.

Desbarats, J. (1987) Population redistribution in Vietnam, *Population and Development Review*, Vol. 13, pp. 43–76.

de Soto, H. (1989) *The Other Path*, Harper & Row, New York.

Devine, T. (1979) Temporary migration and the Scottish Highlands in the nineteenth century, *Economic History Review*, Vol. 32, pp. 344–59.

de Vries, J. (1984) *European Urbanisation 1500–1800*, Harvard University Press, Cambridge, Mass.

de Vries, J. (1986) The population and economy of preindustrial Netherlands, in R. Rotberg and T. Rabb (eds.) *Population and History*, Cambridge University Press, pp. 101–22.
Diaz-Briquets, S. and Perez, L. (1982) Fertility decline in Cuba, *Population and Development Review*, Vol. 8, pp. 513–37.
Dillman, D. (1979) Residential preferences, quality of life and the population turnaround, *Journal of Agricultural Economics*, Vol. 61, pp. 960–6.
Dinkel, R. (1985) The seeming paradox of increasing mortality in a highly industrialised nation: the example of the Soviet Union, *Population Studies*, Vol. 39, pp. 87–97.
Doenges, C. and Newman, J. (1989) Impaired fertility in tropical Africa, *Geographical Review*, Vol. 79, pp. 99–111.
Drake, M. (1969) *Population and Society in Norway 1735–1865*, Cambridge University Press.
D'Souza, S. and Chen, L. (1980) Sex differentials in mortality in rural Bangladesh, *Population and Development Review*, Vol. 6, pp. 257–70.
Dumond, D. (1975) The limitation of human population: a natural history, *Science*, Vol. 187, pp. 713–21.
Duncan, O. (1959) Human ecology and population studies, in Hauser and Duncan (eds.), pp. 678–716.
Duncan, O. (1964) Residential areas and differential fertility, *Eugenics Quarterly*, Vol. 11, pp. 82–9.
Duncan, O., Cuzzort, R. and Duncan, B. (1961) *Statistical Geography*, Free Press of Glencoe, Glencoe, Ill.
Dunn, R., Forrest, R. and Murie, A. (1987) The geography of council house sales in England 1979–85, *Urban Studies*, Vol. 24, pp. 47–59.
Durand, J. (1960) Mortality estimates from Roman tombstone inscriptions, *American Journal of Sociology*, Vol. 65, pp. 365–73.
Dwyer, D. (1975) *People and Housing in Third World Cities*, Longman, London.
Dwyer, D. (1987) New population policies in Malaysia and Singapore, *Geography*, Vol. 72, pp. 248–50.
Dyson, T. and Crook, N. (eds.) (1984) *India's Demography*, South Asian Publishers, New Delhi.
Dyson, T. and Moore, M. (1983) On kinship structure, female autonomy and demographic behaviour in India, *Population and Development Review*, Vol. 9, pp. 35–60.
Dyson, T. and Murphy, M. (1985) The onset of fertility transition, *Population and Development Review*, Vol. 11, pp. 399–440.
Dyson, T. and Murphy, M. (1986) Rising fertility in developing countries, in Woods and Rees (eds.), pp. 68–94.
Easterlin, R. (1976a) Population change and farm settlement in the northern United States, *Journal of Economic History*, Vol. 36, pp. 45–75.
Easterlin R. (1976b) The conflict between aspirations and resources, *Population and Development Review*, Vol. 2, pp. 417–25.
Easterlin, R. (1978) The economics and sociology of fertility: a synthesis, in Tilly (ed.), pp. 57–133.
Easterlin, R. (1980) *Birth and Fortune*, Basic Books, New York.
Eberstadt, N. (1980) Recent declines in fertility in less developed countries, *World Development*, Vol. 8, pp. 37–60.
Economic Commission for Europe (1976) *Fertility and Family Planning in Europe around 1970*, UN, New York.
Ehrlich, P. R. (1971) *The Population Bomb*, Pan, London.
Ehrlich, P. R. and Ehrlich, A. (1972) *Population, Resources, Environment*, Freeman, San Francisco.
Ekpenyong, S. (1984) The effect of mining activities in a peasant community, *Development and Change*, Vol. 15, pp. 251–73.
Elizaga, J. (1972) Internal migration: an overview, *International Migration Review*, Vol. 6, pp. 121–46.

Elliot, B. and McCrone, D. (1980) Urban development in Edinburgh: a contribution to the political economy of place, *Scottish Journal of Sociology*, Vol. 4, pp. 1–26.

Enke, S. (1967) *Raising Per Capita Income Through Fewer Births*, General Electric-TEMPO, Santa Barbara.

Epstein, T. S. and Jackson, D. (eds.) (1977) *The Feasibility of Fertility Planning: Micro Perspectives*, Pergamon, Oxford.

Ermisch, J. (1982) Investigations into the causes of the postwar fertility swings, in Eversley and Köllmann (eds.), pp. 141–55.

Ermisch, J. (1983) *The Political Economy of Demographic Change: Causes and Implications of Population Trends in Great Britain*, Heinemann, London.

Evans, M. (1986) American fertility patterns: white–nonwhite comparisons, *Population and Development Review*, Vol. 12, pp. 267–93.

Eversley, D. and Köllmann, W. (eds.) (1982) *Population Change and Social Planning*, Arnold, London.

Farmer, B. (1986) Perspectives on the Green Revolution in South Asia, *Modern Asian Studies*, Vol. 20, pp. 175–99.

Faulkingham, R. and Thorbahn, P. (1975) Population dynamics and drought: a village in Niger, *Population Studies*, Vol. 29, pp. 463–77.

Fawcett, J. and Khoo, S.–E. (1980) Singapore: rapid fertility transition in a compact society, *Population and Development Review*, Vol. 6, pp. 549–79.

Federici, N. (1968) The influence of women's employment on fertility, in E. Szabady (ed.) (1968) *World Views of Population Problems*, Akadémiai Kaidó, Budapest, pp. 77–82.

Federici, N. (1976) Urban/rural differences in mortality 1950–70, *WHO Statistics Report*, Vol. 29, pp. 249–378.

Feshbach, M. (1982) The Soviet Union: population trends and dilemmas, *Population Bulletin*, Vol. 37, no. 3.

Fielding, A. (1971) *Internal Migration in England and Wales*, Centre for Environmental Studies UWP 14, London.

Fielding, A. (1986) Counterurbanisation, in Pacione (ed.), pp. 224–56.

Fielding, A. (1989) Inter-regional migration and social change: a study of S.E. England based upon data from the Longitudinal Study, *Transactions of Institute of British Geographers*, Vol. 14, pp. 24–36.

Findlay, A. (1987) *The Role of International Labour Migration in the Transformation of an Economy: The Yemen Arab Republic*, MIG WP.35, International Labour Office, Geneva.

Findlay, A. (1988) From settlers to skilled transients: the changing structure of British international migration, *Geoforum*, Vol. 19, pp. 401–10.

Findlay, A. and Findlay, A. (1980) *Migration Studies in Tunisia and Morocco*, Geography Occasional Paper 3, University of Glasgow.

Findlay, A. and Gould, W. (1989) Skilled international migration: a research agenda, *Area*, Vol. 21, pp. 3–11.

Findlay, A. and White, P. (eds.) (1986) *West European Population Change*, Croom Helm, London.

Finkle, J. and Crane, B. (1985) The United States at the International Conference on Population, *Population and Development Review*, Vol. 11, pp. 1–28.

Firth, R. (1957) *We the Tikopia: A Sociological Study of Kinship in Primitive Polynesia* (2nd edn), Allen & Unwin, London.

Flegg, A. (1979) The role of inequality of income in the determination of birth rates, *Population Studies*, Vol. 33, pp. 457–77.

Flegg, A. (1982) Inequalities of income, illiteracy and medical care as determinants of infant mortality in underdeveloped countries, *Population Studies*, Vol. 36, pp. 441–58.

Flinn, M. (ed.) (1977) *Scottish Population History from the 17th Century to the 1930s*, Cambridge University Press.

Flinn, M. (1981) *The European Demographic System, 1520–1820*, Johns Hopkins University Press, Baltimore.

Flowerdew, R. and Salt, J. (1979) Migration between labour market areas in Great Britain 1970–71, *Regional Studies*, Vol. 13, pp. 211–31.

Foggin, P. (1983) A spatial analysis of factors relating to family planning in Thailand, *Singapore Journal of Tropical Geography*, Vol. 4, pp. 11–24.

Forbes, D. (1984) *The Geography of Underdevelopment*, Croom Helm, London.

Forbes, J., Lamont, D. and Robertson, I. (1979) *Intraurban Migration in Greater Glasgow*, Scottish Development Department, Edinburgh.

Foreit, K., Koh, K. and Suh, M. (1980) Impact of the national family planning program on fertility in rural Korea: a multivariate areal analysis, *Studies in Family Planning*, Vol. 11, pp. 79–90.

Forrest, R. (1987) Spatial mobility, tenure mobility, and emerging social divisions in the UK housing market, *Environment and Planning A*, Vol. 19, pp. 1611–30.

Forster, C. and Tucker, G. (1972) *Economic Opportunity and White American Fertility Ratios 1800–60*, Yale University Press, New Haven.

Fotheringham, A. (1981) Spatial structure and distance-decay parameters, *Annals of the Association of American Geographers*, Vol. 71, pp. 425–36.

Fox, A. J. (1977) Occupational mortality 1970–72, *Population Trends*, Vol. 9, pp. 8–15.

Fox, A. J. (ed.) (1988) *Inequalities in Health within Europe*, Gower, Aldershot.

Fox, A. J. and Goldblatt, P. (1982) *Longitudinal Study: Socio-demographic Mortality Differentials*, HMSO, London.

Fox, A. J., Goldblatt, P. and Jones, D. (1985) Social class mortality differentials, *Journal of Epidemiology and Community Health*, Vol. 39, pp. 1–8.

Fox, A. J., Jones, D. and Goldblatt, P. (1984) Approaches to studying the effect of socio-economic circumstances on geographic differences in mortality in England & Wales, *British Medical Bulletin*, Vol. 40, pp. 309–14.

Fox, A. J., Jones, D. and Moser, K. (1985) Socio-demographic differentials in mortality 1971–81, *Population Trends*, Vol. 40, pp. 10–16.

Frank, A. G. (1969) *Capitalism and Underdevelopment in Latin America*, Monthly Review Press, New York.

Frank, O. (1983) Infertility in sub-Saharan Africa, *Population and Development Review*, Vol. 9, pp. 137–44.

Frank, O. (1987) The demand for fertility control in sub-Saharan Africa, *Studies in Family Planning*, Vol. 18, pp. 181–201.

Frank, O. and McNicoll, G. (1987) Fertility and population policy in Kenya, *Population and Development Review*, Vol. 13, pp. 209–43.

Frankel, F. (1971) *India's Green Revolution: Economic Gains and Political Costs*, Princeton University Press.

Freedman, D. (1963) The relation of economic status to fertility, *American Economic Review*, Vol. 53, pp. 414–21.

Freedman, D. and Thornton, A. (1982) Income and fertility: the elusive relationship, *Demography*, Vol. 19. pp. 65–78.

Freedman, J. (1980) Human reactions to population density, in M. Cohen, R. Malpass and H. Klein (eds.) *Biosocial Mechanisms of Population Regulation*, Yale University Press, New Haven, pp. 189–208.

Freedman, R. (1986) Policy options in Taiwan after the demographic transition, *Population and Development Review*, Vol. 12, pp. 77–100.

Freedman, R. (1987) The contributions of social science research to population policy and family planning program effectiveness, *Studies in Family Planning*, Vol. 18, pp. 57–82.

Freedman, R., Khoo, S.–A., and Supraptilah, B. (1981) Use of modern contraceptives in Indonesia: a challenge to the conventional wisdom, *International Family Planning Perspectives*, Vol. 7, pp. 3–15.

Freedman, R., Whelpton, P. and Smit, J. (1961) Socio-economic factors in religious differentials in fertility, *American Sociological Review*, Vol. 26, pp. 608–14.

Frey, W. (1980) Status selective white flight and central city population change, *Journal of Regional Science*, Vol. 20, pp. 71–89.

Friedlander, D. (1973–4) Demographic patterns and socio-economic characteristics of the coal-mining population in England and Wales in the nineteenth century, *Economic Development and Cultural Change*, Vol. 22, pp. 39–51.

Friedlander, D. and Roshier, R. (1966) A study of internal migration in England and Wales, *Population Studies*, Vol. 19, pp. 239–79 (Part I) and Vol. 20, pp. 45–60 (Part II).

Frisch, R. (1978) Nutrition, fatness and fertility, in W. Moseley (ed.) *Nutrition and Human Reproduction*, Macmillan, London.

Fuchs, R. (1984) Government policy and population distribution, in Clarke (ed.), pp. 127–37.

Fuchs, R. and Demko, G. (1978) The postwar mobility transition in Eastern Europe, *Geographical Review*, Vol. 68, pp. 171–82.

Fuchs, R. and Demko, G. (1979) Population distribution policies in developed socialist and western nations, *Population and Development Review*, Vol. 5, pp. 439–67.

Fuller, G. (1974) On the spatial diffusion of fertility decline: the distance-to-clinic variable in a Chilean community, *Economic Geography*, Vol. 50, pp. 324–2.

Fuller, G. (1984) Population geography and family planning, in Clarke (ed.), pp. 103–9.

Gadalla, S., Mehanna, S. and Tennant, C. (1977) *Cultural Values and Population Policies in Egypt*, American University in Cairo.

Galbraith, V. and Thomas, D. S. (1941) Birth rates and the interwar business cycles, *Journal of the American Statistical Association*, Vol. 36, pp. 465–76.

Galle, O. and Taeuber, K. (1966) Metropolitan migration and intervening opportunities, *American Sociological Review*, Vol. 31, pp. 5–13.

Galloway, P. (1986) Long-term fluctuations in climate and population in the preindustrial era, *Population and Development Review*, Vol. 12, pp. 1–24.

Galloway, P. (1988) Basic patterns in annual variations in fertility, nuptiality, mortality and prices in pre-industrial Europe, *Population Studies*, Vol. 42, pp. 275–303.

Gardner, M., Winter, P. and Barker, D. (1984) *Atlas of Mortality from Selected Diseases in England & Wales 1968–1978*, Wiley, Chichester.

Gatrell, A. (1988) *Handling Geographic Information for Health Studies*, Northern Regional Research Laboratory Report 15, University of Lancaster.

George, P. (1951) *Introduction a l'Étude Géographique de la Population du Monde*, Institut National d'Études Démographiques, Paris.

George, P. (1959) *Questions de Géographie de la Population*, Presses Universitaires de France, Paris.

George, S. (1976) *How the Other Half Dies: The Real Reasons for World Hunger*, Penguin, Harmondsworth.

Germani, G. (1965) Migration and acculturation, in P. Hauser (ed.) *Handbook for Social Research in Urban Areas*, UNESCO, pp. 159–78.

Ghallab, M. (1984) Population theory and policy in the Islamic World, in Clarke (ed.), pp. 233–41.

Giddens, A. (1979) *Central Problems in Social Theory*, Macmillan, London.

Gilbert, A. and Gugler, J. (1982) *Cities, Poverty and Development: Urbanisation in the Third World*, Oxford University Press.

Gilbert, A. and Kleinpenning, J. (1986) Migration, regional inequality and development in the Third World, *Tijdschrift voor economische en sociale geografie*, Vol. 77, pp. 2–6.

Gille, H. (1949–50) The demographic history of the northern European countries in the eighteenth century, *Population Studies*, Vol. 3, pp. 3–65.

Ginsberg, R. (1978) Probability models of residence histories, in Clark and Moore (eds.), pp. 233–65.

Gittinger, J., Leslie, J. and Hoisington, C. (eds.) (1987) *Food Policy*, Johns Hopkins University Press, Baltimore.

Glaeser, B. (ed.) (1987) *The Green Revolution Revisited: Critique and Alternatives*, Allen & Unwin, London.

Glantz, M. (ed.) (1987) *Drought and Hunger in Africa*, Cambridge University Press.

Glaser, W. (1978) *The Brain Drain: Emigration and Return*, Pergamon, Oxford.

Glass, D. (1963–4) Some Indicators of differences between urban and rural mortality in England and Wales and Scotland, *Population Studies*, Vol. 17, pp. 1241–52.
Glass, D. and Eversley, D. (eds.) (1965) *Population History*, Arnold, London.
Glass, D. and Revelle, R. (eds.) (1972) *Population and Social Change*, Arnold, London.
Gober-Meyers, P. (1978) Interstate migration and economic growth: a simultaneous equations approach, *Environment and Planning A*, Vol. 10, pp. 1241–52.
Goddard, J. and Champion, A. (eds.) (1983) *The Urban and Regional Transformation of Britain*, Methuen, London.
Goeller, H. and Weinberg, A. (1976) The age of substitutability, *Science*, 20 February, pp. 683–9.
Goldscheider, C. (1971) *Population Modernization and Social Structure*, Little Brown, Boston.
Goldscheider, C. and Mosher, W. (1988) Religious affiliation and contraceptive usage: changing American patterns, 1955–82, *Studies in Family Planning*, Vol. 19, pp. 48–57.
Goliber, T. (1985) Sub-Saharan Africa: population pressures on development, *Population Bulletin*, Vol. 40, no. 1.
Goubert, P. (1960) *Beauvais et le Beauvaisis de 1600 à 1730*, Paris.
Gould, W. (1982) Emigrants from fear, *Geographical Magazine*, Vol. 54, pp. 494–8.
Gould, W. T. and Prothero, R. M. (1975) Space and time in African population mobility, in Kosiński and Prothero (eds.), pp. 39–49.
Gray, R. (1974) The decline of mortality in Ceylon and demographic effects of malaria control, *Population Studies*, Vol. 28, pp. 205–29.
Greenhalgh, S. (1988) Fertility as mobility: Sinic transitions, *Population and Development Review*, Vol. 14, pp. 629–74.
Greenwood, M. (1970) Lagged response in the decision to migrate, *Journal of Regional Science*, Vol.10, pp. 375–84.
Greenwood, M. (1981) *Migration and Economic Growth in the United States*, Academic Press, New York.
Griffin, K. (1974) *The Political Economy of Agrarian Change: An Essay on the Green Revolution*, Macmillan, London.
Griffith, G. T. (1926) *Population Problems of the Age of Malthus*, Cass, London.
Griffiths, I. (1987) The traffic in labour, *Geographical Magazine*, Vol. 59, pp. 13–18.
Griffiths, M. (1971) A geographical study of mortality in an urban area, *Urban Studies*, Vol. 8, pp. 111–20.
Grigg, D. (1976) Population pressure and agricultural change, *Progress in Geography*, Vol. 8, pp. 133–76.
Grigg, D. (1979) Ester Boserup's theory of agrarian change: a critical review, *Progress in Human Geography*, Vol. 3, no. 1, pp. 64–84.
Grigg, D. (1980) Migration and overpopulation, in White and Woods (eds), pp. 60–83.
Grigg, D. (1982) Modern population growth in historical perspective, *Geography*, Vol. 67, pp. 97–108.
Gugler, J. (1976) Migrating to urban centres of unemployment in tropical Africa, in Richmond and Kubat (eds.), pp. 184–204.
Gugler, J. (1986) Internal migration in the Third World, in Pacione (ed.), pp. 194–223.
Guillet, E. (1963) *The Great Migration*, University of Toronto Press.
Guralnick, L. (1963) *Mortality by Industry and Cause of Death*, Vital Statistics Special Reports 53, US Department of Health, Education and Welfare.
Gutkind, P. (1975) The view from below: political consciousness of the urban poor in Ibadan, *Cahiers d'études africaines*, Vol. 15, pp. 5–35.
Gwatkin, D. (1980) Indications of change in developing country mortality trends: the end of an era?, *Population and Development Review*, Vol. 6, pp. 615–44.
Gwynne, T. and Sill, M. (1976) Census enumeration books: a study of mid-nineteenth century immigration, *Local Historian*, Vol. 12, pp. 74–9.
Habbakuk, H. J. (1971) *Population Growth and Economic Development since 1750*, Leicester University Press.

Hagerstrand, T. (1957) Migration and area: survey of a sample of Swedish migration fields and hypothetical considerations on their genesis, *Lund Studies in Geography*, Vol. 13B, pp. 27–158.
Hagerstrand, T. (1967) *Innovation Diffusion as a Spatial Process*, University of Chicago Press.
Haggett, P. (1976) Hybridizing alternative models of an epidemic diffusion process, *Economic Geography*, Vol. 52, pp. 136–46.
Haines, M. (1977) Fertility, nuptiality and occupation: a study of coalmining populations and regions in England and Wales in the mid-nineteenth century, *Journal of Interdisciplinary History*, Vol. 8, pp. 245–80.
Hajnal, J. (1965) European marriage patterns in perspective, in Glass and Eversley (eds.), pp. 101–43.
Hall, R. (1986) Household trends within western Europe, in Findlay and White (eds.), pp. 18–34.
Halstead, S., Walsh, J. and Warren, K. (eds.) (1985) *Good Health at Low Cost*, Rockefeller Foundation, New York.
Hanley, S. (1977) The influence of economic and social variables on marriage and fertility in eighteenth and nineteenth century Japanese villages, in Lee (ed.), pp. 165–99.
Hanley, S. and Wolf, A. (eds.) (1985) *Family and Population in East Asian History*, Stanford University Press.
Hansen, J. C. and Kosiński, L. (1973) *Population Geography 1973*, IGU Commission on Population Geography, Bergen.
Hardee-Cleaveland, K. and Banister, J. (1988) Fertility policy and implementation in China, *Population and Development Review*, Vol. 14, pp. 245–86.
Harris, A. and Clausen, R. (1966) *Labour Mobility in Great Britain 1953–63*, Government Social Survey, London.
Harris, J. and Todaro, M. (1970) Migration, unemployment and development, *American Economic Review*, Vol. 60, pp. 126–42.
Harris, M. and Ross, E. (1987) *Death, Sex and Fertility: Population Regulation in Preindustrial and Developing Societies*, Columbia University Press, New York.
Harrison, B. and Kluver, J. (1989) Reassessing the 'Massachusetts miracle', *Environment and Planning A*, Vol. 21, pp. 771–802.
Hart, N. (1988) Sex differentials in mortality, in Fox (ed.),
Hart, T. (1971) The inverse care law, *Lancet*, Vol. 1, pp. 405–12.
Hartshorne, R. (1939) *The Nature of Geography*, Association of American Geographers, Lancaster, Pa.
Harvey, D. (1985) *Consciousness and the Urban Experience*, Blackwell, Oxford.
Hauser, P. and Duncan, O. (eds) (1959) *The Study of Population: An Inventory and Appraisal*, University of Chicago Press.
Hausfater, G. and Hrdy, S. (eds.) (1984) *Infanticide: Comparative and Evolutionary Perspectives*, Aldine, New York.
Hawthorn, G. (1976–7) A review of: Family Planning in India (P. Blaikie), *Journal of Development Studies*, Vol. 13, pp. 442–3.
Hawthorn, G. (ed.) (1978) *Population and Development, High and Low Fertility in Poorer Countries*, Cass, London.
Hawthorn, G. (1982) The paradox of the modern: determinants of fertility in N. and W. Europe since 1950, in C. Hohn and R. Mackensen (eds.) *Determinants of Fertility Transition*, International Union for the Scientific Study of Population, Liège, pp. 283–96.
Hayami, Y. (1988) Asian development: a view from the paddy fields, *Asian Development Review*, Vol. 6, pp. 50–63.
Haynes, R. (1987) *The Geography of Health Services in Britain*, Croom Helm, London.
Haynes, R. (1988) The urban distribution of lung cancer mortality in England & Wales 1980–83, *Urban Studies*, Vol. 25, pp. 497–506.
Haynes, R. M. (1974) Application of exponential distance decay to human and animal activities, *Geografiska Annaler*, Vol. 56B, pp. 90–104.

Heenan, B. (1967) Rural–urban distribution of fertility in South Island, New Zealand, *Annals of Association of American Geographers*, Vol. 57, pp. 713–35.
Heenan, B. (1975) Some spatial aspects of differential mortality in New Zealand, *New Zealand Geographer*, Vol. 31, pp. 29–53.
Heenan, B. (1983) Cigarette smoking among New Zealanders, in N. McGlashan and J. Blunden (eds.) *Geographical Aspects of Health*, Academic Press, London, pp. 241–56.
Heer, D. (1966) Economic development and fertility, *Demography*, Vol. 3, pp. 423–44.
Helleiner, K. (1957) The vital revolution reconsidered, *Canadian Journal of Economics and Political Science*, Vol. 23, pp. 1–9.
Henry, L. (1956) *Anciennes Familles Genevoises*, Institut National d'Études Démographiques, Paris.
Henry, L. (1957) La mortalité d'après les incriptions funéraires, *Population*, Vol. 12, pp. 149–52.
Henry, L. (1967) *Manuel de Démographie Historique*, Librarie Droz, Geneva.
Henry, L. and Lévy, C. (1960) Ducs et pairs sous l'ancien régime, *Population*, Vol. 15, pp. 807–30.
Herbert, D. (1973) The residential mobility process: some empirical observations, *Area*, Vol. 5, pp. 44–8.
Hermalin, A. (1975) Regression analysis of areal data, in C. Chandrasekharan and A. Hermalin (eds.) *Measuring the Effect of Family Planning Programs on Fertility*, International Union for the Scientific Study of Population, Dolhain, Belgium, pp. 245–99.
Hermalin, A. and Van de Walle, E. (1977) The Civil Code and nuptiality: empirical investigation of a hypothesis, in Lee (ed.), pp. 71–111.
Hiemenz, U. and Schatz, K. (1977) *Transfer of Employment Opportunities as an Alternative to the International Migration of Workers*, Migration for Employment Project Working Paper 9, International Labour Office, Geneva.
Hirschman, A. (1958) *The Strategy of Economic Development*, Yale University Press, New Haven.
Hobcraft, J., McDonald, J. and Rutstein, S. (1984) Socio-economic factors in infant and child mortality, *Population Studies*, Vol. 38, pp. 193–223.
Hoem, B. and Hoem, J. (1989) The impact of women's employment on second and third order births in modern Sweden, *Population Studies*, Vol. 43, pp. 46–67.
Hofsten, E. (1966) Population registers and computers, *Review of the International Statistical Institute*, Vol. 34, pp. 186–93.
Hogarth, T. and Daniel, W. (1987) The long-distance commuters, *New Society*, 29 May, pp. 11–13.
Holland, W. (1988) *European Community Atlas of Avoidable Death*, Oxford University Press.
Hollingsworth, T. (1970) *Migration: A Study Based on Scottish Experience Between 1939 and 1954*, Oliver and Boyd, Edinburgh.
Hollingsworth, T. (1972) The importance of the quality of the data in historical demography, in Glass and Revelle (eds.), pp. 71–86.
House, J. (1968) *Mobility of the Northern Business Manager*, Papers on Migration and Mobility, Department of Geography, University of Newcastle upon Tyne.
House, J. and Knight, E. (1965) *Migrants of North East England 1951–61*, Papers on Migration and Mobility, Geography Department, University of Newcastle upon Tyne.
Howe, G. M. (1970), *National Atlas of Disease Mortality in the United Kingdom*, Nelson, London.
Howe, G. M. (1972) *Man, Environment and Disease in Britain*, David and Charles, Newton Abbot.
Howe, G. M. (1986) Does it matter where I live?, *Transactions of Institute of British Geographers*, Vol. 11, pp. 387–414.
Howe, G. M. and Lorraine, J. (eds.) (1973) *Environmental Medicine*, Heinemann, London.
Howell, N. (1979) *Demography of the Dobe !Kung*, Academic Press, London.

Howell, N. (1986) Feedbacks and buffers in relation to scarcity and abundance: studies of hunter-gatherer populations, in Coleman and Schofield (eds.), pp. 157–87.
Hufton, O. (1975) Women and the family economy in eighteenth-century France, *French Historical Studies*, Vol. 9, pp. 1–22.
Hugo, G. (1986) *Australia's Changing Population*, Oxford University Press.
Hugo, G. (1988) Population movement in Indonesia since 1971, *Tijdschrift voor Economische en Sociale Geografie*, Vol. 79, pp. 242–56.
Hull, T., Hull, V. and Singarimbun, M. (1977) Indonesia's family planning story, *Population Bulletin*, Vol. 32, no. 6.
Hunter, J. M. (1976) *The Making of the Crofting Community*, Donald, Edinburgh.
Hunter, J., Rey, L. and Scott, D. (1983) Man-made lakes, man-made diseases, *World Health Forum*, Vol. 4, pp. 177–82.
Huss, M. (1980) *Demography, Public Opinion and Politics in France, 1974–80*, Occasional Paper 16, Department of Geography, Queen Mary College, London.
Illsley, R. (1986) Occupational class, selection and the production of inequalities in health, *Quarterly Journal of Social Affairs*, Vol. 2, pp. 151–65.
Isbell, E. (1944) Internal migration in Sweden and intervening opportunities, *American Sociological Review*, Vol. 9, pp. 627–39.
Isiugo-Abanihe (1985) Child fosterage in West Africa, *Population and Development Review*, Vol. 11, pp. 53–73.
Jackson, J. (ed.) (1969) *Migration*, Cambridge University Press.
Jain, A. (1985) The impact of development and population policies on fertility in India, *Studies in Family Planning*, Vol. 16, pp. 181–98.
James, P. (1954) The geographic study of population, in P. James and C. Jones (eds.) *American Geography: Inventory and Prospect*, Syracuse University Press, pp. 106–22.
Janowitz, B. (1971) An empirical study of the effects of socio-economic development on fertility rates, *Demography*, Vol. 8, pp. 319–30.
Janowitz, B. (1973) An econometric analysis of trends in fertility rates, *Journal of Development Studies*, Vol. 9, pp. 413–25.
Jansen, C. and King, R. (1968) Migrations et 'occasions intervenantes' en Belgique, *Recherches Économiques de Louvain*, Vol. 4, pp. 519–26.
Japan Health Promotion Foundation (1981) *National Atlas of Major Disease Mortalities for Cities, Towns and Villages in Japan 1969–78*, Tokyo.
Jerome, H. (1926) *Migration and Business Cycles*, National Bureau of Economic Research, New York.
Johnson, B. L. (1969) *South Asia*, Heinemann, London.
Johnson, J. H. and Salt, J. (1980) Labour migration within organisations, *Tijdschrift voor Economische en Sociale Geografie*, Vol. 71, pp. 277–84.
Johnson, J. H., Salt, J. and Wood, P. (1974) *Housing and the Migration of Labour in England and Wales*, Saxon House, Farnborough.
Johnson, N. and Lean, S. (1985) Relative income, race and fertility, *Population Studies*, Vol. 39, pp. 99–112.
Johnson, P., Conrad, C. and Thomson, D. (1989) *Workers Versus Pensioners*, Manchester University Press.
Johnston, R. (1986) *Philosophy and Human Geography* (2nd ed), Arnold, London.
Johnston, R. (1989) The southwards drift: preliminary analysis of career patterns of 1980 graduates in Great Britain, *Geography*, Vol. 74, pp. 239–44.
Jones, E. F. (1971) Fertility decline in Australia and New Zealand 1861–1936, *Population Index*, Vol. 37, pp. 301–38.
Jones, E. F. and Westoff, C. (1979) The end of 'Catholic' fertility, *Demography*, Vol. 16, pp. 209–17.
Jones, H. (1970) Migration to and from Scotland since 1960, *Transactions of Institute of British Geographers*, Vol. 49, pp. 145–59.
Jones, H. (1976) The structure of the migration process: findings from a growth point in Mid-Wales, *Transactions of Institute of British Geographers*, Vol. 1, pp. 421–32.

Jones, H. (1977) Metropolitan dominance and family planning in Barbados, *Social and Economic Studies*, Vol. 26, pp. 327–38.
Jones, H. (1979) Modern emigration from Scotland to Canada, *Scottish Geographical Magazine*, Vol. 95, pp. 4–12.
Jones, H. (1986) Evolution of Scottish migration patterns: a social-relations-of-production approach, *Scottish Geographical Magazine*, Vol. 102, pp. 151–64.
Jones, H. (1989) Fertility decline in Mauritius: the role of Malthusian population pressure, *Geoforum*, Vol. 20, pp. 315–27.
Jones, H., Caird, J., Berry, W. and Dewhurst, J. (1986) Peripheral counterurbanisation: an integration of census and survey data in northern Scotland, *Regional Studies*, Vol. 20, pp. 15–26.
Jones, R. E. (1976) Infant mortality in rural North Shropshire 1561–1810, *Population Studies*, Vol. 30, pp. 305–17.
Joseph, G. (1975) A Markov analysis of age/sex differences in inter-regional migration in Great Britain, *Regional Studies*, Vol. 9, pp. 69–78.
Jowett, J. (1989) Mao's man-made famine, *Geographical Magazine*, Vol. 61, no. 4, pp. 16–21.
Katz, M. (1976) *The People of Hamilton, Canada West: Family and Class in a Mid-Nineteenth Century City*, Harvard University Press, Cambridge, Mass.
Keeble, D. and Wever, E. (eds.) (1986) *New Firms and Regional Development in Europe*, Croom Helm, London.
Kelson, M. and Heller, R. (1983) The effect of death certification and coding practices on observed differences in respiratory disease mortality in EC countries, *Rev. Epidemiol. Santé Publ.*, Vol. 31, pp. 423–32.
Kennett, S. (1983) Migration within and between labour markets, in Goddard and Champion (eds.), pp. 215–38.
Kiernan, K. (1989) The family: formation and fission, in H. Joshi (ed.) *The Changing Population of Britain*, Blackwell, Oxford, pp. 27–55.
King, P. (1979) Problems of spatial analysis in geographical epidemiology, in G. Pyle (ed.) *New Directions in Medical Geography*, Pergamon, Oxford, pp. 249–52.
King, R. (1976) The evolution of international labour migration movements concerning the EEC, *Tijdschrift voor Economische en Sociale Geografie*, Vol. 67, pp. 66–82.
King, R. (ed.) (1986) *Return Migration and Regional Economic Problems*, Croom Helm, London.
Kirk, D. (1942) The relationship of employment levels to births in Germany, *Milbank Memorial Fund Quarterly*, Vol. 28, pp. 126–38.
Kirk, D. (1971) A new demographic transition?, in National Academy of Sciences *Rapid Population Growth: Consequences and Policy Vol. 2 Implications*, Johns Hopkins Press, Baltimore, pp. 123–47.
Kitagawa, E. and Hauser, P. (1973) *Differential Mortality in the United States: A Study in Socioeconomic Epidemiology*, Harvard University Press, Cambridge, Mass.
Klaassen, L. and Drewe, P. (1973) *Migration Policy in Europe*, Saxon House, Farnborough.
Kleinpenning, J. and Zoomers, E. (1988) Internal colonisation as a policy instrument for changing a country's rural system: Paraguay, *Tijdschrift voor Economische en Sociale Geografie*, Vol. 79, pp. 257–65.
Kliot, N. (1987) The era of homeless man, *Geography*, Vol. 72, pp. 109–21.
Kloos, H. and Thompson, K. (1979) Schistosomiasis in Africa: an ecological perspective, *Journal of Tropical Geography*, Vol. 48, pp. 31–46.
Knights, P. (1971) *The Plain People of Boston 1830–60*, Oxford University Press, New York.
Knodel, J., Chamratrithirong, A. and Debavalya, N. (1987) *Thailand's Reproductive Revolution*, University of Wisconsin Press, Madison.
Knox, P. (1978) The intraurban ecology of primary medical care, *Environment and Planning A*, Vol. 10, pp. 415–35.

Köllmann, W. and Rudenhausen, C. (1982) Past and present policy reactions to fertility decline in Germany, in Eversley and Köllmann (eds.), pp. 414–25.

Kono, S. (1971) Evaluation of the Japanese population register data on internal migration, in International Union for the Scientific Study of Population, *International Population Conference London 1969*, Vol. 4, Liège, pp. 2766–75.

Korner, E. (1980) *The First Report to the Secretary of State on the Collection and Use of Information about Hospital Clinical Activity in the NHS*, HMSO, London.

Kosiński, L. (1980) Population geography and the IGU, *Population Geography*, Vol. 2, pp. 1–20.

Kosiński, L. (1984) The roots of population geography, in Clarke (ed.), pp. 11–24.

Kosiński, L. and Elahi, K. M. (eds.) (1985) *Population Redistribution and Development in South Asia*, Reidel, Dordrecht.

Kosiński, L. and Prothero, R. M. (eds.) (1975) *People on the Move: Studies on Internal Migration*, Methuen, London.

Kosiński, L. and Webb, J. (eds.) (1976) *Population at Microscale*, IGU Commission on Population Geography and New Zealand Geographical Society, Christchurch.

Kraeger, P. (1982) Demography *in situ, Population and Development Review*, Vol. 8, pp. 237–66.

Krause, J. (1958) Changes in English fertility and mortality 1781–1850, *Economic History Review*, Vol. 11, pp. 52–70.

Krause, J. (1965) The changing adequacy of English registration 1690–1837, in Glass and Eversley (eds.), pp. 379–93.

Kuhn, W. (1978) Guest workers as an automatic stabilizer of cyclical unemployment in Switzerland and Germany, *International Migration Review*, Vol. 12, pp. 210–24.

Kulldorff, G. (1955) *Migration Probabilities*, Lund Studies in Geography B No. 14, Gleerup, Lund.

Kunitz, S. (1984) Mortality change in America, 1620–1920, *Human Biology*, Vol. 56, pp. 559–82.

Kunitz, S. (1986) Mortality since Malthus, in Coleman and Schofield (eds.), pp. 279–302.

Kuznets, S. (1973) *Population, Capital and Growth*, Heinemann, London.

Kwofie, K. (1976) A spatio-temporal analysis of cholera diffusion in western Africa, *Economic Geography*, Vol. 52, pp. 127–35.

Lacey, R. and Shaper, A. (1984) Changes in water hardness and cardiovascular death-rates, *International Journal of Epidemiology*, Vol. 13, pp. 18–24.

Ladinsky, J. (1967) The geographical mobility of professional and technical manpower, *Journal of Human Resources*, Vol. 2, pp. 475–94.

Land, K. (1969) Duration of residence and prospective migration, *Demography*, Vol. 6, pp. 133–40.

Langer, W. (1963) Europe's initial population explosion, *American Historical Review*, Vol. 69, pp. 1–17.

Langer, W. (1974) Infanticide: a historical survey, *History of Childhood Quarterly*, Vol. 1, pp. 353–69.

Langer, W. (1975) American foods and Europe's population growth 1750–1850, *Journal of Social History*, Vol. 5, pp. 51–66.

Langford, C. (1981) Fertility change in Sri Lanka since the War: an analysis of the experience of different districts, *Population Studies*, Vol. 35, pp. 285–306.

Lapham, R. and Mauldin, W. P. (1984) Family planning program effort and birthrate decline in developing countries, *International Family Planning Perspectives*, Vol. 10, pp. 109–18.

Lapham, R. and Mauldin, W. P. (1985) Contraceptive prevalence: the influence of organised family planning programs, *Studies in Family Planning*, Vol. 16, pp. 117–37.

Laslett, R. and Wall, R. (eds.) (1972) *Household and Family in Past Time*, Cambridge University Press.

Lawton, R. (1968) Population changes in England and Wales in the late nineteenth century, *Transactions of Institute of British Geographers*, Vol. 44, pp. 55–74.

Lawton, R. (ed.) (1978) *The Census and Social Structure: An Interpretative Guide to Nineteenth Century Censuses for England and Wales*, Cass, London.
Lawton, R. and Pooley, C. (1976) *The Social Geography of Merseyside in the Nineteenth Century*, Report to Social Science Research Council.
Learmonth, A. (1977) Malaria, in G. M. Howe (ed.) *A World Geography of Human Diseases*, Academic Press, London, pp. 61–108.
Lee, E. (1966) A theory of migration, *Demography*, Vol. 3, pp. 47–57.
Lee, R. (1987) Population dynamics of humans and other animals, *Demography*, Vol. 24, pp. 443–65.
Lee, R. B. and DeVore, I. (eds.) (1976) *Kalahari Hunter Gatherers*, Harvard University Press, Cambridge, Mass.
Lee, R. D. (ed.) (1977) *Population Patterns in the Past*, Academic Press, New York.
Leete, R. and Fox, J. (1977) Registrar General's social classes: origins and uses, *Population Trends*, Vol. 8, pp. 1–7.
Leff, N. (1982) *Underdevelopment and Development in Brazil*, Allen & Unwin, London.
Lesthaeghe, R. (1977) *The Decline of Belgian Fertility, 1800–1970*, Princeton University Press.
Lesthaeghe, R. (1980) On the social control of human reproduction, *Population and Development Review*, Vol. 6, pp. 527–48.
Lesthaeghe, R. (1986) On the adaptation of sub-Saharan systems of reproduction, in Coleman and Schofield (eds.), pp. 212–38.
Lesthaeghe, R. and Surkyn, J. (1988) Cultural dynamics and economic theories of fertility change, *Population and Development Review*, Vol. 14, pp. 1–45.
Lesthaeghe, R. and Wilson, C. (1986) Modes of production, secularisation, and the pace of the fertility decline in western Europe, 1870–1930, in Coale and Watkins (eds.), pp. 261–92.
Levine, D. (1977) *Family Formation in an Age of Nascent Capitalism*, Academic Press, New York.
Levine, D. (1984) Production, reproduction and the proletarian family in England, 1500–1851, in D. Levine (ed.) *Proletarianization and Family History*, Academic Press, London, pp. 87–127.
Levine, D. (1987) *Reproducing Families: The Political Economy of English Population History*, Cambridge University Press.
Lewis, G. (1982) *Human Migration: A Geographical Perspective*, Croom Helm, London.
Lewis, J. (1986) International labour migration and uneven regional development in labour exporting countries, *Tijdschrift voor Economische en Sociale Geografie*, Vol. 77, pp. 27–41.
Lewis, J. and Townsend, A. (eds.) (1989) *The North–South Divide: Regional Change in Britain in the 1980s*, Paul Chapman, London.
Lewis, O. (1959) *Five Families: A Mexican Case Study in the Culture of Poverty*, Basic Books, New York.
Lipton, M. (1977) *Why Poor People Stay Poor: A Study of Urban Bias in World Development*, Temple Smith, London.
Livi-Bacci, M. (1971) *A Century of Portuguese Fertility*, Princeton University Press.
Livi-Bacci, M. (1972) Fertility and population growth in Spain in the eighteenth and nineteenth centuries, in Glass and Revelle (eds.), pp. 173–84.
Livi-Bacci, M. (1977) *A History of Italian Fertility during the Last Two Centuries*, Princeton University Press.
Livi-Bacci, M. (1983) The nutrition–mortality link in past times, *Journal of Interdisciplinary History*, Vol. 13, pp. 293–8.
Livi-Bacci, M. (1986) Social-group forerunners of fertility control in Europe, in Coale and Watkins (eds.), pp. 182–200.
Lloyd, O., Williams, F., Berry, W. and Florey, C. (1987) *An Atlas of Mortality in Scotland*, Croom Helm, London.
Lloyd, P. (1979) *Slums of Hope*, Penguin, Harmondsworth.

Lockhart, D. (1982) Patterns of migration and movement of labour to the planned villages of N.E. Scotland, *Scottish Geographical Magazine*, Vol. 98, pp. 35–47.

Lockridge, K. (1983) *The Fertility Transition in Sweden: A Preliminary Look at Smaller Geographic Units, 1855–1890*, Demographic Data Base Report 3, Umea.

Long, L. (1988) *Migration and Residential Mobility in the United States*, Russell Sage, New York.

Long, L. and Hansen, K. (1979) *Reasons for Interstate Migration*, US Bureau of the Census, Washington.

Long, L., Tucker, J. and Urton, W. (1988) Migration distances: an international comparison, *Demography*, Vol. 25, pp. 633–40.

Lovett, A. and Gatrell, A. (1988) The geography of spina bifida in England & Wales, *Transactions of Institute of British Geographers*, Vol. 13, pp. 288–302.

Lowry, I. (1966) *Migration and Metropolitan Growth: Two Analytical Models*, University of California Press, Los Angeles.

Luling, V. (1989) Wiping out a way of life, *Geographical Magazine*, Vol. 61, no. 7, pp. 34–37.

Mabogunje, A. (1970) Systems approach to a theory of rural–urban migration, *Geographical Analysis*, Vol. 2, pp. 1–18.

MacDonald, E. (1976) Demographic variation in cancer in relation to industrial and environmental influence, *Environmental Health Perspectives*, Vol. 17, pp. 153–66.

McCay, B. and Acheson, J. (eds.) (1987) *The Question of the Commons: The Culture and Ecology of Communal Resources*, University of Arizona Press, Tucson.

McDowall, M. (1983) William Farr and the study of occupational mortality, *Population Trends*, Vol. 31, pp. 12–14.

McElroy, J. and Albuquerque, K. de (1988) Migration transition in small northern and eastern Caribbean states, *International Migration Review*, Vol. 22, pp. 30–58.

McGinnis, R. (1974) Review of: The Limits to Growth, *Demography*, Vol. 10, pp. 295–9.

McGlashan, N. (1967) Geographical evidence on medical hypotheses, *Tropical and Geographical Medicine*, Vol. 19, pp. 333–43.

McGlashan, N. (ed.) (1972) *Medical Geography, Techniques and Analysis*, Methuen, London.

McGlashan, N. (1977) Spatial variations in cause-specific mortality in Australia, in N. McGlashan (ed.) *Studies in Australian Mortality*, University of Tasmania, Hobart, pp. 1–28.

McInnis, R. M. (1977) Childbearing and land availability, in Lee (ed.), pp. 201–7.

McIntosh, C. (1983) *Population Policy in Western Europe*, Sharpe, New York.

McIntosh, C. (1987) Recent pronatalist policies in Western Europe, in Davis *et al.* (eds.), pp. 318–34.

McKay, J. and Whitelaw, J. (1977) The role of large private and government organizations in generating flows of inter-regional migrants: the case of Australia, *Economic Geography*, Vol. 53, pp. 28–44.

McKeown, T. (1976) *The Modern Rise of Population*, Arnold London.

McKeown, T. and Brown, R. (1955) Medical evidence related to English population changes in the eighteenth century, *Population Studies*, Vol. 9, pp. 119–41.

McKeown, T., Brown, R. and Record, R. (1972) An interpretation of the modern rise of population in Europe, *Population Studies*, Vol. 26, pp. 345–82.

McNabb, R. (1979) A socio-economic model of migration, *Regional Studies*, Vol. 13, pp. 297–304.

McNamara, R. (1977) Possible interventions to reduce fertility, *Population and Development Review*, Vol. 3, pp. 163–76.

McNicoll, G. (1978) Population and development: outlines for a structuralist approach, in Hawthorn (ed.), pp. 79–99.

McNicoll, G. (1987) Economic growth with below-replacement fertility, in Davis *et al.* (eds.), pp. 217–38.

Madeley, J. (1988) Malaria in the Solomons, *World Health*, June, pp. 14–15.

Mamdani, M. (1972) *The Myth of Population Control: Family, Caste and Class in an Indian Village*, Monthly Review Press, New York.
Mandle, J. (1970) The decline in mortality in British Guiana 1911–60, *Demography*, Vol. 7, pp. 301–15.
Mann, M. (1973) *Workers on the Move: The Sociology of Relocation*, Cambridge University Press.
Margolis, J. (1977) Internal migration: measurement and models, in A. Brown and E. Neuberger (eds.) *Internal Migration: A Comparative Perspective*, Academic Press, New York.
Marmot, M. and McDowall, M. (1986) Mortality decline and widening social inequalities, *Lancet*, Vol. ii, pp. 274–6.
Martin, L. (1988) The aging of Asia, *Journal of Gerontology: Social Sciences*, Vol. 43, S99–S113.
Martin, L. (1989) The greying of Japan, *Population Bulletin*, Vol. 44, no. 2.
Martin, R. (1988) The political economy of Britain's north–south divide, *Transactions of Institute of British Geographers*, Vol. 13, pp. 389–418.
Marx, K. (1976 edn) *Capital Vol. 1*, Penguin, London.
Mason, T. et al. (1975) *Atlas of Cancer Mortality of US Counties 1950–1969*, Government Printing Office, Washington D.C.
Massey, D. (1984) *Spatial Divisions of Labour*, Macmillan, London.
Massey, D. (1985) New directions in space, in D. Gregory and J. Urry (eds.) *Social Relations and Spatial Structures*, Macmillan, London, pp. 9–19.
Matras, J. (1965) The social strategy of family formation: some variations in time and space, *Demography*, Vol. 2, pp. 349–62.
Mauldin, W. P. (1978) Patterns of fertility decline in developing countries 1950–75, *Studies in Family Planning*, Vol. 9, pp. 75–84.
Mauldin, W. P. and Berelson, B. (1978) Conditions of fertility decline in developing countries 1965–75, *Studies in Family Planning*, Vol. 9, pp. 89–147.
May, J. (1961) *Studies in Disease Ecology*, Hafner, New York.
Mayer, J. (1982) Relations between two traditions of medical geography, *Progress in Human Geography*, Vol. 6, pp. 216–30.
Meade, M. (1977) Medical geography as human ecology, *Geographical Review*, Vol. 67, pp. 377–93.
Meadows, D. H., Meadows, D. C., Randers, J. and Behrens, W. (1972) *The Limits to Growth, a Report for the Club of Rome's Project on the Predicament of Mankind*, Universe Books, New York.
Meillassoux, C. (1972) From reproduction to production: a Marxist approach to economic anthropology, *Economy and Society*, Vol. 1, pp. 93–105.
Meillassoux, C. (1981) *Maidens, Meal and Money: Capitalism and the Domestic Community*, Cambridge University Press.
Mellor, J. (1987) Food aid for food security and economic development, in E. Clay and J. Shaw (eds.) *Poverty, Development and Food*, Macmillan, London, pp. 173–91.
Melrose, D. (1981) *The Great Health Robbery: Baby Milk and Medicines in Yemen*, Oxfam, Oxford.
Mendels, F. (1972) Proto-industrialisation: the first phase in the industrialisation process, *Journal of Economic History*, Vol. 32, pp. 241–61.
Mercer, A. (1985) Smallpox and epidemiological-demographic change in Europe, *Population Studies*, Vol. 39, pp. 287–307.
Mercer, J. (1989), Asian migrants and residential location in Canada, *New Community*, Vol. 15, pp. 185–202.
Merrick, T. and Berquo, E. (1983) *The Determinants of Brazil's Recent Rapid Decline in Fertility*, National Academy Press, Washington, D.C.
Mesarovic, M. and Pestel, E. (1974) *Mankind at the Turning Point*, Hutchinson, London.
Mincer, J. (1978) Family migration decisions, *Journal of Political Economy*, Vol. 86, pp. 749–73.

Mitchell, J. C. (1969) Structural plurality, urbanization and labour circulation in Southern Rhodesia, in Jackson (ed.), pp. 156–80.
Mohan, J. (1988) Restructuring, privatization and the geography of health care provision in England 1983–87, *Transactions of Institute of British Geographers*, Vol. 13, pp. 449–65.
Monbiot, G. (1989) The transmigration fiasco, *Geographical Magazine*, Vol. 61, no. 5, pp. 26–30.
Morrill, R. (1965) The Negro ghetto: problems and alternatives, *Geographical Review*, Vol. 55, pp. 339–61.
Morrison, P. (1967) Duration of residence and prospective migration, *Demography*, Vol. 4, pp. 553–61.
Morrison, P. (1973) Theoretical issues in the design of population mobility models, *Environment and Planning*, Vol. 5, pp. 125–34.
Morrison, P. (ed.) (1983) *Population Movements: Their Forms and Functions in Urbanisation and Development*, IUSSP, Ordina Editions, Liège.
Moseley, M. (1979) *Accessibility: The Rural Challenge*, Methuen, London.
Moseley, W. H. (1984) Child survival: research and policy, in W. H. Moseley and L. Chen (eds.) *Child Survival: Strategies for Research*, Supplement to *Population and Development Review*, Vol. 10, pp. 3–23.
Mosher, W. and Hendershot, G. (1984) Religion and fertility, *Demography*, Vol. 21, pp. 185–91.
Mosk, C. and Johansson, S. (1986) Income and mortality: evidence from modern Japan, *Population and Development Review*, Vol. 12, pp. 415–40.
Mourant, A. (1978) *Blood Groups and Diseases*, Oxford University Press.
Mukherji, S. (1985) The syndrome of poverty and wage labour circulation: the Indian scene, in Prothero and Chapman (eds.), pp. 279–98.
Muller, M. (1982) *The Health of Nations: A North–South Investigation*, Faber, London.
Murray, M. (1967) The geography of death in the United States and the United Kingdom, *Annals of Association of American Geographers*, Vol. 57, pp. 301–14.
Myrdal, A. (1945) *Nation and Family: The Swedish Experiment in Democratic Family and Population Policy*, Routledge, London.
Myrdal, G. (1940) *Population: A Problem for Democracy*, Harvard University Press, Cambridge Mass.
Myrdal, G. (1957) *Economic Theory and Under-Developed Regions*, Duckworth, London.
Nag, M. (1971) The pattern of mating behaviour, emigration and contraceptives as factors affecting human fertility in Barbados, *Social and Economic Studies*, Vol. 20, pp. 111–33.
Nag, M. (1980) How modernisation can also increase fertility, *Current Anthropology*, Vol. 21, no. 5.
Nag, M. and Kak, N. (1984) Demographic transition in a Punjab village, *Population and Development Review*, Vol. 10, pp. 661–78.
Najafizadeh, M. and Mermerick, L. (1988) Worldwide educational expansion from 1950 to 1980: the failure of the expansion of schooling in developing countries, *Journal of Developing Areas*, Vol. 22, pp. 333–58.
Nalson, J. (1968) *Mobility of Farm Families*, Manchester University Press.
Nam, C. (1979) The progress of demography as a scientific discipline, *Demography*, Vol. 16, pp. 485–92.
Nayar, P. (1984) Factors in fertility decline in Kerala, in K. Mahadevan (ed.) *Fertility and Mortality*, Sage, New Delhi, pp. 155–67.
Nelson, J. (1976) Sojourners versus new urbanites, *Economic Development and Cultural Change*, Vol. 24, pp. 721–57.
Nelson, J. (1983) Population redistribution policies and migrants' choices, in Morrison (ed.), pp. 281–312.
Nelson, P. (1959) Migration, real income and information, *Journal of Regional Science*, Vol. 1, pp. 43–73.

Newell, C. (1988) *Methods and Models in Demography*, Belhaven, London.
Newman, J. and Lura, R. (1983) Fertility control in Africa, *Geographical Review*, Vol. 73, pp. 396–406.
Newsholme, A. (1929) *The Story of Modern Preventive Medicine*, Williams and Wilkins, Baltimore.
Nikolinakos, M. (1975) Notes towards a general theory of migration in late capitalism, *Race and Class*, Vol. 7, pp. 5–16.
Noin, D. (1979) *Géographie de la Population*, Masson, Paris.
Nortman, D. (1985) *Population and Family Planning Programs 1983*, Population Council, New York.
Oberai, A. (1982) An overview of migration-influencing policies and programmes, in A. Oberai (ed.) *State Policies and Internal Migration*, Croom Helm, London, pp. 11–26.
OECD (1989) *SOPEMI Annual Report 1988*, Directorate for Social Affairs, Manpower and Education, OECD, Paris.
Oechsli, F. and Kirk, D. (1974–5) Modernization and the demographic transition in Latin America and the Caribbean, *Economic Development and Cultural Change*, Vol. 23, pp. 391–419.
Office of Population Censuses and Surveys (1978a) *Trends in Mortality 1951–1975*, Series DHI no. 3, HMSO, London.
Office of Population Censuses and Surveys (1978b) *Occupational Mortality*, Series DS no. 1, HMSO, London.
Ogden, P. (1985) France: recession, politics and migration policy, *Geography*, Vol. 70, pp. 158–62.
Ogden, P. (1989) International migration in the nineteenth and twentieth centuries, in P. Ogden and P. White (eds.) *Migrants in Modern France*, Unwin Hyman, London.
Ohlin, P. (1961) Mortality, marriage and growth in pre-industrial populations, *Population Studies*, Vol. 14, pp. 190–7.
Okpara, E. (1986) Rural–urban migration and urban employment opportunities in Nigeria, *Transactions of Institute of British Geographers*, Vol. 11, pp. 67–74.
Olsson, G. (1965) Distance and human interaction: a migration study, *Geografiska Annaler*, Vol. 47B, pp. 3–43.
Omran, A. (1977) Epidemiologic transition in the US, *Population Bulletin*, Vol. 32, no. 2.
Omvedt, G. (1980) Migration in colonial India: the articulation of feudalism and capitalism by the colonial state, *Journal of Peasant Studies*, Vol. 7.
OPCS (1986) *Occupational Mortality*, Series DS no. 6, HMSO, London.
OPCS (1988) *Occupational Mortality, Childhood Supplement*, HMSO, London.
Openshaw, S. (1988) Developments in geographical information systems, *ESRC Newsletter*, Vol. 63, pp. 11–14.
Openshaw, S. and Taylor, P. (1981) The modifiable areal unit problem, in N. Wrigley and R. Bennett (eds.) *Quantitative Geography*, Routledge, London, pp. 60–69.
Orubuloye, I. and Caldwell, J. (1975) The impact of public health services on mortality differentials in a rural area of Nigeria, *Population Studies*, Vol. 29, pp. 259–72.
Osborne, R. (1958) The movements of people in Scotland 1851–1951, *Scottish Studies*, Vol. 2, pp. 1–46.
Owen, D. and Green, A. (1989) Spatial aspects of labour mobility in the 1980s, *Geoforum*, Vol. 20, pp. 107–26.
Pacione, M. (ed.) (1986) *Population Geography: Progress and Prospect*, Croom Helm, London.
Page, H. and Lesthaeghe, R. (eds.) (1981) *Child-spacing in Tropical Africa*, Academic Press, London.
Pahl, R. (1984) *Divisions of Labour*, Blackwell, Oxford.
Paine, S. (1979) Replacement of the West European migrant labour system by investment in the European periphery, in Seers *et al.* (eds.) pp. 65–95.
Palloni, A. (1981) Mortality in Latin America: emerging patterns, *Population and Development Review*, Vol. 7, pp. 623–49.

Pamuk, E. (1985) Social class inequality in mortality from 1921 to 1972 in England & Wales, *Population Studies*, Vol. 39, pp. 17–31.
Pariser, E. (1987) Post-harvest food losses in developing countries, in Gittinger *et al.* (eds.), pp. 309–25.
Peach, C. (1975) Immigrants in the inner city, *Geographical Journal*, Vol. 141, pp. 372–9.
Penninx, R. (1984) *Immigrant Populations and Demographic Development*, Population Studies 12, Council of Europe, Strasbourg.
Perry, R., Dean, K. and Brown, B. (1986) *Counterurbanisation: International Case Studies of Socioeconomic Change in Rural Areas*, Geo Books, Norwich.
Philip, G., Taylor, P. and Hutton, A. (1982) Oil-related construction workers: travelling and migration, in H. Jones (ed.) *Recent Migration in Northern Scotland: Pattern, Process, Impact*, SSRC North Sea Oil Panel Occasional Paper 13, Redhill, pp. 27–60.
Phillips, J., Simmons, R., Koenig, M. and Chakraborty, J. (1988) Determinants of reproductive change in a traditional society: evidence from Matleb, Bangladesh, *Studies in Family Planning*, Vol. 19, pp. 313–34.
Pinstrup-Anderson, P. and Hazell, P. (1987) The impact of the Green Revolution and prospects for the future, in Gittinger *et al.*, (eds.), pp. 106–18.
Plant, J. (1987) Regional geochemical maps of the UK, *Natural Environment Research Council Newsjournal*, Vol. 3, no. 4, pp. 5–7.
Pocock, S., Shaper, A. and Cook, D,. (1980) British Regional Heart Study: geographic variations in cardiovascular mortality and the role of water quality, *British Medical Journal*, Vol. 280, pp. 1243–9.
Pooley, C. (1977) The residential segregation of migrant communities in mid-Victorian Liverpool, *Transactions of Institute of British Geographers*, Vol. 2, pp. 364–82.
Pooley, C. (1979) Residential mobility in the Victorian city, *Transactions of Institute of British Geographers*, Vol. 4, pp. 258–77.
Pooley, M. and Pooley, C. (1984) Health, society and environment in Victorian Manchester, in Woods and Woodward (eds.), pp. 148–75.
Portes, A. and Walton, J. (1981) *Labor, Class and the International System*, Academic Press, New York.
Power, J. (1976) *Western Europe's Migrant Workers*, Minority Rights Group Report No. 28, London.
Preston, S. (ed.) (1978) *The Effects of Infant and Child Mortality on Fertility*, Academic Press, New York.
Preston, S. (1979) Urban growth in developing countries: a demographic reappraisal, *Population and Development Review*, Vol. 5, pp. 195–215.
Preston, S. (1980) Causes and consequences of mortality declines in less developed countries during the twentieth century, in R. Easterlin (ed.) *Population and Economic Change in Developing Countries*, University of Chicago Press, pp. 289–360.
Preston, S. and Nelson, V. (1974) Structure and change in causes of death: an international summary, *Population Studies*, Vol. 28, pp. 19–51.
Preston, S. and Van de Walle, E. (1978) Urban French mortality in the nineteenth century, *Population Studies*, Vol. 32, pp. 275–97.
Pritchard, R. (1976) *Housing and the Spatial Structure of the City*, Cambridge University Press.
Prothero, R. M. and Chapman, M. (eds.) (1985) *Circulation in Third World Countries*, Routledge, London.
Proudfoot, B. (1990) The setting of immigration levels in Canada since the Immigration Act, 1976, *British Journal of Canadian Studies*.
Pryce, W. R. (1982) Migration in pre-industrial and industrial societies, in *Patterns and Processes of Internal Migration*, D301 Units 9–10, Open University Press, Milton Keynes.
Pryor, R. (1976) Conceptualising migration behaviour: a problem in micro-demographic analysis, in Kosiński and Webb (eds.), pp. 105–19.
Pyle, G. (1971) *Heart Disease, Cancer and Stroke in Chicago*, Geography Research Paper 134, University of Chicago.
Pyle, G. (1979) *Applied Medical Geography*, Wiley, New York.

Pyle, G. and Rees, P. (1971) Modeling patterns of death and disease in Chicago, *Economic Geography*, Vol. 47, pp. 475–88.
Quigley, J. and Weinberg, D. (1977) Intra-urban residential mobility: a review and synthesis, *International Regional Science Review*, Vol. 2, pp. 41–66.
Ramasubban, R. (1984) The develoment of health policy in India, in Dyson and Crook (eds.), pp. 97–116.
Ravenstein, E. (1885) The laws of migration, *Journal of the Royal Statistical Society*, Vol. 48, pp. 167–227.
Razzell, P. (1974) An interpretation of the modern rise of population in Europe – a critique, *Population Studies*, Vol. 28, pp. 5–17.
Razzell, P. (1977) *The Conquest of Smallpox: The Impact of Inoculation on Smallpox Mortality in Eighteenth Century Britain*, Caliban Books, Firle, Sussex.
Reddaway, W. (1939) *The Economics of Declining Population*, Macmillan, London.
Redford, A. (1926) *Labour Migration in England 1800–50*, University of Manchester.
Rees, G. and Rees, T. (1977) Alternatives to the census: the example of sources of internal migration data, *Town Planning Review*, Vol. 48, pp. 123–40.
Rees, G. and Rees, T. (1981) *Migration, Industrial Restructuring and Class Relations: The Case of South Wales*, Papers in Planning Research 22, UWIST, Cardiff.
Rees, P. (1977) The measurement of migration from census data and other sources, *Environment and Planning A*, Vol. 9, pp. 247–72.
Rees, P., Stillwell, J. and Boden, P. (1989) Migration trends and population projections for the elderly, in Congdon and Batey (eds.), pp. 205–26.
Rees, P. and Wilson, A. (1977) *Spatial Population Analysis*, Arnold, London.
Reichert, J. and Massey, D. (1979) Migration from a rural Mexican town, *Intercom*, Vol. 7, no. 6, pp. 6–7.
Reid, D. (1973) Arteriosclerotic disease in relation to the environment, in Howe and Lorraine (eds.), pp. 145–53.
Reid, J. (1982) Black America in the 1980s, *Population Bulletin*, Vol. 37, no. 4.
Rele, J. (1987) Fertility levels and trends in India, 1951–81, *Population and Development Review*, Vol. 13, pp. 513–30.
Repetto, R. (1979) *Economic Equality and Fertility in Developing Countries*, Johns Hopkins University Press, Baltimore.
Repetto, R. (1987) Population, resources, environment: an uncertain future, *Population Bulletin*, Vol. 42, no. 2.
Retherford, R. and Cho, L. (1973) Comparative analysis of recent fertility trends in East Asia, in *Proceedings of the International Population Conference*, Vol. 2, IUSSP, Liège.
Rich, W. (1973) *Smaller Families through Social and Economic Progress*, Overseas Development Council, Washington D.C.
Richmond, A. and Kubat, D. (eds.) (1976) *Internal Migration: The New World and the Third World*, Sage, London.
Richmond, A. and Verma, R. (1978) The economic adaptation of immigrants: a new theoretical perspective, *International Migration Review*, Vol. 12, pp. 3–39.
Rigg, J. (1989) The Green Revolution and equity, *Geography*, Vol. 74, pp. 144–9.
Rindfuss, R. and Sweet, J. (1977) *Postwar Fertility Trends and Differentials in the United States*, Academic Press, New York.
Roberts, B. (1978) *Cities of Peasants: The Political Economy of Urbanization in the Third World*, Arnold, London.
Roberts, G. (1955) Some aspects of fertility and mating in the West Indies, *Population Studies*, Vol. 8, pp. 199–227.
Roberts, J. (1987) Geographical mobility and housing, *Employment Gazette*, Vol. 95, pp. 125–9.
Robertson, I. M. (1972) Population distribution and location problems, *Regional Studies*, Vol. 6, pp. 237–45.
Robertson, I. M. (1974) Scottish population distribution: implications for locational decisions, *Transactions of Institute of British Geographers*, Vol. 63, pp. 111–24.

Robinson, V. (1989) Up the creek without a paddle? Britain's boat people ten years on, *Geography*, Vol. 74, pp. 332–38.
Robinson, W. S. (1950) Ecological correlations and the behaviour of individuals, *American Sociological Review*, Vol. 15, pp. 351–7.
Rogers, A. (1966) A Markovian policy model of interregional migration, *Papers of Regional Science Association*, Vol. 17, pp. 205–24.
Rogers, A. (1967) A regression analysis of interregional migration in California, *Review of Economics and Statistics*, Vol. 49, pp. 262–7.
Rogers, A. (1968) *Matrix Analysis of Interregional Population Growth and Distribution*, University of California Press, Los Angeles.
Rogers, A. (1984) *Migration, Urbanisation and Spatial Population Dynamics*, Westview Press, Boulder, Colorado.
Rogers, A. (1988) Age patterns of elderly migration: an international comparison, *Demography*, Vol. 25, pp. 355–70.
Rogerson, P. (1987) Changes in United States national mobility levels, *Professional Geographer*, Vol. 39, pp. 344–51.
Romaniuk, A. (1980) Increase in natural fertility during the early stages of modernization: evidence from an African case study, Zaire, *Population Studies*, Vol. 34, pp. 293–310.
Romaniuk, A. (1984) *Current Demographic Analysis: Fertility in Canada*, Statistics, Canada, Ottawa.
Rose, H. (1971) *The Black Ghetto: A Spatial Behavioral Perspective*, McGraw-Hill, New York.
Ross, C. (Chairman) (1973) *Report of the Population Panel*, Cmnd 5258, HMSO, London.
Rossi, P. (1955) *Why Families Move: A Study in the Social Psychology of Urban Residential Mobility*, Free Press, Glencoe, Ill.
Roth, E. (1985) A note on the demographic concomitants of sedentism, *American Anthropologist*, Vol. 87, pp. 380–2.
Rostow, W. (1960) *The Stages of Economic Growth*, Cambridge University Press.
Ruzicka, L. and Hansluwka, H. (1982) Mortality transition in South and East Asia: technology confronts poverty, *Population and Development Review*, Vol. 8, pp. 567–88.
Salt, J. (1984) *Labour Migration within Multi-Locational Organisations in Britain*, Final Report to ESRC, London.
Salt, J. (1985) West German dilemma: little Turks or young Germans?, *Geography*, Vol. 70, pp. 162–8.
Salt, J. (1988) Highly-skilled international migrants, careers and internal labour markets, *Geoforum*, Vol. 19, pp. 387–99.
Sandbach, F. (1978) Ecology and the 'Limits to Growth' debate, *Antipode*, Vol. 10, no. 2, pp. 22–32.
Sandell, S. (1977) Women and the economics of family migration, *Review of Economics and Statistics*, Vol. 59, pp. 406–14.
Sarda, R. (1987) Schisto comes to town, *World Health*, June, pp. 27–9.
Sathar, Z., Crook, N., Callum, C. and Kazi, S. (1988) Women's status and fertility change in Pakistan, *Population and Development Review*, Vol. 14, pp. 415–32.
Sassen-Koob, S. (1987) Issues of core and periphery: labour migration and global restructuring, in J. Henderson and M. Castells (eds.) *Global Restructuring and Territorial Development*, Sage, London, pp. 60–87.
Sauer, R. (1978) Infanticide and abortion in nineteenth century Britain, *Population Studies*, Vol. 32, pp. 81–93.
Schmink, M. and Wood, C. (eds.) (1984) *Frontier Expansion in Amazonia*, University of Florida Press.
Schroeder, H. (1960) Relation between mortality from cardiovascular disease and treated water supplies, *Journal of American Medical Association*, Vol. 172, pp. 1902–8.
Schuh, G. (1987) The changing context of food and agricultural development policy, in Gittinger *et al.* (eds.), pp. 72–88.

Scottish Economic Planning Board (1970) *Migration to and from Scotland*, Edinburgh.
Scrimshaw, S. (1978) Infant mortality and behaviour in the regulation of family size, *Population and Development Review*, Vol. 4, pp. 383–403.
Seccombe, I. (1986) Immigrant workers in an emigrant economy, *International Migration*, Vol. 24, pp. 377–96.
Seccombe, W. (1983) Marxism and demography, *New Left Review*, Vol. 137, pp. 22–47.
Seers, D., Schaffer, B. and Kiljunen, M. (eds.) (1979) *Underdeveloped Europe: Studies in Core-Periphery Relations*, Harvester Press, Hassocks.
Sell, R. (1988) Egyptian international labor migration and social processes, *International Migration Review*, Vol. 22, pp. 87–108.
Sen, A. (1981) *Poverty and Famines*, Clarendon Press, Oxford.
Shannon, G. and Dever, G. (1974) *Health Care Delivery: Spatial Perspectives*, McGraw-Hill, New York.
Shaper, A. (1984) Geographic variations in cardiovascular mortality in Great Britain, *British Medical Bulletin*, Vol. 40, pp. 366–73.
Shaper, A., Pocock, S., Packham, R., Lacey, R. and Powell, P. (1983) Softness of drinking water and cardiovascular disease – practical implications of recent research, *Health Trends*, Vol. 15, pp. 22–4.
Shaw, R. P. (1979) Migration and employment in the Arab world: construction as a key policy variable, *International Labour Review*, Vol. 118, pp. 589–605.
Short, J. (1978) Residential mobility, *Progress in Human Geography*, Vol. 2, pp. 419–47.
Short, J. (1989) Yuppies, yuffies and the new urban order, *Transactions of Institute of British Geographers*, Vol. 14, pp. 173–88.
Short, R. (1986) The evolution of human reproduction, *Proceedings of the Royal Society of London B*, Vol. 195, pp. 3–24.
Shorter, E. (1976) *The Making of the Modern Family*, Collins, London.
Shrestha, N. (1988) A structural perspective on labour migration in underdeveloped countries, *Progress in Human Geography*, Vol. 12, pp. 179–207.
Shrewsbury, J.(1970) *A History of Bubonic Plague in the British Isles*, Cambridge University Press.
Shryock, H. and Siegel, J. (1976) *The Methods and Materials of Demography* (condensed edn by E. Stockwell), Academic Press, New York.
Silk, J. (1971) *Search Behaviour*, Geography Paper 7, University of Reading.
Silver, M. (1965) Births, marriages and business cycle in the US, *Journal of Political Economy*, Vol. 74, pp. 237–55.
Simon, J. (1981) *The Ultimate Resource*, Princeton University Press.
Simon, J. (1986) *Theory of Population and Economic Growth*, Blackwell, Oxford.
Simon, J. and Kahn, H. (eds.) (1984) *The Resourceful Earth: A Response to Global 2000*, Blackwell, Oxford.
Sivamurthy, M. (1981) The deceleration of mortality decline in Asian countries, in International Union for the Scientific Study of Population, *International Population Conference, Manila 1981*, Vol. 2, Liège, pp. 51–76.
Sjaastad, L. (1962) The costs and returns of human migration, *Journal of Political Economy*, Vol. 70, pp. 80–93.
Skeldon, R. (1985) Circulation: a transition in mobility in Peru, in Prothero and Chapman (eds.), pp. 100–20.
Smith, D. S. (1977) A homeostatic demographic regime: patterns in West European family reconstitution studies, in Lee (ed.), pp. 19–51.
Soldo, B. and Agree, E. (1988) America's elderly, *Population Bulletin*, Vol. 43, no. 3.
Spengler, J. (1938) *France Faces Depopulation*, Duke University Press, Durham.
Srinivasan, K., Reddy, P. and Raju, K. (1978) From one generation to the next: changes in fertility, family size preferences, and family planning in an Indian State between 1951 and 1975, *Studies in Family Planning*, Vol. 9, pp. 258–71.
Stamp, L. D. (1964) *The Geography of Life and Death*, Collins, London.

Stevens, G. (1981) Social mobility and fertility, *American Sociological Review*, Vol. 56, pp. 573–85.

Stewart, C. (1960) Migration as a function of population and distance, *American Sociological Review*, Vol. 25, pp. 347–56.

Stolnitz, G. (1975) International mortality trends: some main trends and implications, in United Nations, *The Population Debate: Dimensions and Perspectives*, Papers of the World Population Conference 1974, Vol. 2, New York, pp. 151–9.

Stone, P. (1978) The implications for the conurbations of population changes, *Regional Studies*, Vol. 12, pp. 95–123.

Stouffer, S. (1940) Intervening opportunities: a theory relating mobility and distance, *American Sociological Review*, Vol. 5, pp. 845–67.

Stouffer, S. (1960) Intervening opportunities and competing migrants, *Journal of Regional Science*, Vol. 2, pp. 1–26.

Streatfield, K. (1986) *Fertility Decline in a Traditional Society: The Case of Bali*, Australian National University, Canberra.

Strodbeck, F. (1949) Equal opportunity intervals: a contribution to the method of intervening opportunity analysis, *American Sociological Review*, Vol. 14, pp. 490–7.

Stycos, J. M. (1968) *Human Fertility in Latin America*, Cornell University Press, Ithaca.

Stycos, J. M. (1971a) Family planning and American goals, in D. Chaplin (ed.), *Population Policies and Growth in Latin America*, Heath, Lexington, pp. 111–31.

Stycos, J. M. (1971b) *Ideology, Faith and Family Planning in Latin America*, McGraw-Hill, New York.

Stycos, J. M. (1977) Population policy and development, *Population and Development Review*, Vol. 3, pp. 103–12.

Sublett, M. (1975) *Farmers on the Road*, Geography Research Paper 168, University of Chicago.

Sundquist, J. (1975) *Dispersing Population: What Americans can Learn from Europe*, Brookings Institute, Washington, D.C.

Sutton, K. and Lawless, R. (1978) Population regrouping in Algeria, *Transactions of Institute of British Geographers*, Vol. 3, pp. 331–50.

Swerdlow, A. (1987) 150 years of Registrar Generals' medical statistics, *Population Trends*, Vol. 48, pp. 20–6.

Swindell, K. (1985) Seasonal agricultural circulation: the Strange Farmers of the Gambia, in Prothero and Chapman (eds.), pp. 178–201.

Taeuber, I. (1958) *The Population of Japan*, Princeton University Press.

Taeuber, K., Chiazze, L. and Haenszel, W. (1968) *Migration in the United States: An Analysis of Residence Histories*, US Dept. of Health, Education and Welfare, Washington, D.C.

Tarrant, J. (1980) The geography of food aid, *Transactions of Institute of British Geographers*, Vol. 5, pp. 125–40.

Taylor, A. J. (ed.) (1975) *The Standard of Living in Britain in the Industrial Revolution*, Methuen, London.

Taylor, J. (1986) Some consequences of recent reductions in mine labour recruitment in Botswana, *Geography*, Vol. 71, pp. 34–46.

Taylor, P. J. (1979) 'Difficult-to-let', 'difficult-to-live-in', and sometimes 'difficult-to-get-out-of': an essay on the provision of council housing, *Environment and Planning*, Vol. 11A, pp. 1305–20.

Taylor, R. C. (1969) Migration and motivation, in Jackson (ed.), pp. 99–133.

Taylor, R. C. (1979) Migration and the residual community, *Sociological Review*, Vol. 27, pp. 475–89.

Teitelbaum, M. (1984) *The British Fertility Decline: Demographic Transition in the Crucible of the Industrial Revolution*, Princeton University Press.

Teitelbaum, M. and Winter, J. (1985) *The Fear of Population Decline*, Academic Press, London.

Teller, C. (1981) The demography of malnutrition in Latin America, *Intercom*, Vol. 9, no. 8, pp. 8–11.

Thomas, B. (1954) *Migration and Economic Growth: A Study of Great Britain and the Atlantic Economy*, Cambridge University Press.

Thomas, B. (1972) *Migration and Urban Development: A Re-appraisal of British and American Long Cycles*, Methuen, London.

Thomas, C. and Phillips, D. (1978) An ecological analysis of child medical emergency admissions to hospitals in West Glamorgan, *Social Science and Medicine*, Vol. 12D, pp. 183–92.

Thomas, D. S. (1938) *Research Memorandum on Migration Differentials*, Social Science Research Council, New York.

Thomas, D. S. (1941) *Social and Economic Aspects of Swedish Population Movements 1740–1933*, Macmillan, News York.

Thomas, I. (1982) Villagization in Tanzania, in Clarke and Kosiński (eds.), pp. 182–91.

Thomlinson, R. (1965) *Population Dynamics*, Random House, New York.

Thunhurst, C. (1985) The analysis of small area statistics and planning for health, *Statistician*, Vol. 34, pp. 93–106.

Tien, Y. (1989) Second thoughts on the second child, *Population Today*, Vol. 17, no. 4, pp. 6–9.

Tilakaratne, W. (1978) Economic change, social differentiation and fertility: Aluthgama, in Hawthorn (ed.), pp. 186–97.

Tilly, C. (1978) The historical study of vital processes, in Tilly (ed.), pp. 3–56.

Tilly, C. (ed.) (1978) *Historical Studies of Changing Fertility*, Princeton University Press.

Toney, M. (1976) Length of residence, social ties, and economic opportunities, *Demography*, Vol. 13, pp. 297–309.

Tout, K. (1989) *Ageing in Developing Countries*, Oxford University Press.

Townsend, P. and Davidson, N. (1982) *Inequalities in Health*, Penguin, Harmondsworth.

Townsend, P., Phillimore, P. and Beattie, A. (1988) *Health and Deprivation: Inequality and the North*, Croom Helm, London.

Townsend, P., Simpson, D. and Tibbs, N. (1985) Inequalities in health in the city of Bristol, *International Journal of Health Services*, Vol. 15, pp. 637–63.

Trewartha, G. (1953) A case for population geography, *Annals of Association of American Geographers*, Vol. 43, pp. 71–97.

Trewartha, G. (1969) *A Geography of Population*, Wiley, New York.

Tromp, S. (1973) The relationship of weather and climate to health and disease, in Howe and Lorraine (eds.), pp. 72–99.

Tsui, A. and Bogue, D. (1978) Declining world fertility: trends, causes, implications, *Population Bulletin*, Vol. 33, no. 4.

UN Dept. of Economic and Social Affairs (1970) *Methods of Measuring Internal Migration*, Population Studies 47, New York.

US Bureau of Census (1960) *Historical Statistics of the United States*, Washington D.C.

Valentine, C. and Revson, J. (1979) Cultural traditions, social change and fertility in sub-Saharan Africa, *Journal of Modern African Studies*, Vol. 17, pp. 453–72.

van de Kaa, D. (1987) Europe's second demographic transition, *Population Bulletin*, Vol. 42, no. 1.

Van de Walle, E. (1972) Marriage and marital fertility, in Glass and Revelle (eds.), pp. 137–51.

Van de Walle, E. (1978) Alone in Europe: the French fertility decline until 1850, in Tilly (ed.), pp. 257–88.

Van de Walle, E. and Knodel, J. (1980) Europe's fertility transition: new evidence and lessons for today's developing world, *Population Bulletin*, Vol. 34, no. 6.

Van Heek, F. (1956–7) Roman Catholicism and fertility in the Netherlands, *Population Studies*, Vol. 10, pp. 125–38.

Vanhove, N. and Klaassen, L. (1980) *Regional Policy: A European Approach*, Saxon House, Farnborough.

Vedder, R. and Gallaway, L. (1970) Settlement patterns of Canadian emigrants to the United States 1850–1960, *Canadian Journal of Economics*, Vol. 3, pp. 476–86.

Verhoef, R. and van de Kaa, D. (1987) Population registers and population statistics, *Population Index*, Vol. 53, pp. 633–42.
Vinovskis, M. (1979) *Studies in American Historical Demography*, Academic Press, New York.
Visaria, P. and Visaria, L. (1981) India's population: second and growing, *Population Bulletin*, Vol. 36, no. 4.
Visvalingam, M. (1978) The signed Chi-square measure for mapping, *Cartographic Journal*, Vol. 15, pp. 93–8.
Wade, R. (1979) Fast growth and slow development in southern Italy, in Seers *et al.* (eds.), pp. 197–221.
Walker, C. (1985) *Tayside Infant Morbidity and Mortality Study Final Report*, Tayside Health Board, Dundee.
Wallerstein, I. (1979) *The Capitalist World Economy*, Cambridge University Press.
Warnes, A. (1986) The elderly in less-developed world regions, *Ageing and Society*, Vol. 6, pp. 373–80.
Warnes, A. and Law, C. (1984) The elderly population of Great Britain: locational trends and policy implications, *Transactions of Institute of British Geographers*, Vol. 9, pp. 37–59.
Warren, H. (1989) Geology, trace elements and health, *Social Science and Medicine*, Vol. 29, pp. 923–26.
Warwick, D. (1986) The Indonesian family planning program, *Population and Development Review*, Vol. 12, pp. 453–90.
Warwick, D. (1988) Culture and the management of family planning programs, *Studies in Family Planning*, Vol. 19, pp. 1–18.
Watkins, S. and Menken, J. (1985) Famines in historical perspective, *Population and Development Review*, Vol. 11, pp. 647–75.
Watson, A. and Moss, R. (1970) Dominance, spacing behaviour and aggression in relation to population limitation in vertebrates, in A. Watson (ed.), *Animal Populations in Relation to their Food Resources*, Blackwell, Oxford.
Watson, W. (1964) Social mobility and social class in industrial communities, in M. Gluckman (ed.), *Closed Systems and Open Minds*, Oliver and Boyd, Edinburgh, pp. 129–57.
Watts, M. (1989) The agrarian question in Africa: debating the crisis, *Progress in Human Geography*, Vol. 13, pp. 1–41.
Weaver, C. (1987) Social security in ageing societies, in Davis *et al.* (eds.), pp. 273–94.
Webber, R. and Craig, J. (1976) Which local authorities are alike?, *Population Trends*, Vol. 5, pp. 13–19.
Weeden, R. (1973) *Interregional Migration Models and their Application to Great Britain*, Cambridge University Press.
Weeks, J. (1988) The demography of Islamic nations, *Population Bulletin*, Vol. 43, no. 4.
Weil, C. and Kvale, K. (1985) Current research on geographical aspects of Schistosomiasis, *Geographical Review*, Vol. 75, pp. 186–216.
Weintraub, R. (1962) The birth rate and economic development: an empirical study, *Econometrica*, Vol. 15, pp. 182–217.
Werner, B. (1985) Fertility trends in different social classes: 1970–83, *Population Trends*, Vol. 41, pp. 5–13.
White, M. and Mueser, P. (1988) Implications of boundary choice for the measurement of residential mobility, *Demography*, Vol. 25, pp. 443–59.
White, P. (1980) Migration loss and the residual community: a study in rural France 1962–75, in White and Woods (eds.), pp. 198–222.
White, P. (1986) International migration in the 1970s: revolution or evolution, in Findlay and White (eds.), pp. 50–80.
White, P. and Woods, R. (eds.) (1980) *The Geographical Impact of Migration*, Longman, London.
Whitehead, M. (1988) *The Health Divide*, Penguin, Harmondsworth.

Wilkinson, T. (1967) Japan's population problem, in S. Chandrasekhar (ed.), *Asia's Population Problems*, Allen, London, pp. 100–18.
Williams, P. (1978) Building societies and the inner city, *Transactions of Institute of British Geographers*, Vol. 3, pp. 23–34.
Willigan, J. and Lynch, K. (1982) *Sources and Methods of Historical Demography*, Academic Press, London.
Willis, K. (1972) The influence of spatial structure and socio-economic factors on migration rates, *Regional Studies*, Vol. 6, pp. 69–82.
Wilson, F. (1972) *Labour in the South African Gold Mines 1911–69*, Cambridge University Press.
Wilson, M. G. (1968) *Population Geography*, Nelson, London.
Wilson, M. G. (1971) The spatial dimension of human reproduction in Victoria, *Proceedings, Sixth New Zealand Geography Conference*, NZ Geographical Society, pp. 258–64.
Wilson, M. G. (1978a) A spatial analysis of human fertility in Scotland: reappraisal and extension, *Scottish Geographical Magazine*, Vol. 94, pp. 130–43.
Wilson, M. G. (1978b) The pattern of fertility in a medium-sized industrial city, *Tijdschrift voor Economische en Sociale Geografie*, Vol. 69, pp. 225–32.
Wilson, M. G. (1979) Infant death in metropolitan Australia 1970–73, *Canadian Studies in Population*, Vol. 6, pp. 127–42.
Wilson, M. G. (1984) The changing pattern of urban fertility in eastern Australia, 1966–76, *Australian Geographical Studies*, Vol. 22, pp. 202–20.
Wilson, M. G. (1988) The end of an affair? Geography and fertility in late post-transitional societies, paper delivered at IGU Population Geography Commission Symposium, Macquarie University, Sydney. *Australian Geographer*, Vol. 21, pp. 53–67.
Wolpert, J. (1965) Behavioural aspects of the decision to migrate, *Papers of the Regional Science Association*, Vol. 15, pp. 159–69.
Wolpert, J. (1966) Migration as an adjustment to environmental stress, *Journal of Social Issues*, Vol. 22, pp. 92–102.
Wood, C. and Carvalho, J. (1988) *The Demography of Inequality in Brazil*, Cambridge University Press.
Woods, R. (1979) *Population Analysis in Geography*, Longman, London.
Woods, R. (1982a) *Theoretical Population Geography*, Longman, London.
Woods, R. (1982b) Population Studies, *Progress in Human Geography*, Vol. 6, pp. 247–53.
Woods, R. (1986) The spatial dynamics of the demographic transition in the West, in Woods and Rees (eds.), pp. 21–44.
Woods, R. (1987) Approaches to the fertility transition in Victorian England, *Population Studies*, Vol. 41, pp. 283–311.
Woods, R. and Rees, P. (eds.) (1986) *Population Structures and Models*, Allen & Unwin, London.
Woods, R. and Woodward, J. (eds.) (1984) *Urban Disease and Mortality in Nineteenth-Century England*, Batsford, London.
Woodward, J. (1984) Medicine and the city, in Woods and Woodward (eds.), pp. 65–78.
Wooldridge, S. W. and East, W. G. (1951) *The Spirit and Purpose of Geography*, Hutchinson, London.
World Bank (1984) *World Development Report 1984*, Oxford University Press.
World Health Organization (1976) Health care in rural areas, *WHO Chronicle*, Vol. 30, pp. 11–17.
World Health Organization (1978) *Primary Health Care: Report of International Conference on Primary Health Care, Alma Ata 1978*, WHO, Geneva.
World Health Organization (1987) *World Health Statistics*, WHO, Geneva.
Wrigley, E. A. (1961) *Industrial Growth and Population Change*, Cambridge University Press.
Wrigley, E. A. (ed.) (1966) *An Introduction to English Historical Demography*, Weidenfeld and Nicolson, London.

Wrigley, E. A. (1967) Demographic models and geography, in R. Chorley and P. Haggett (eds), *Socio-Economic Models in Geography*, Methuen, London, pp. 189–215.
Wrigley, E. A. (ed.) (1972) *Nineteenth Century Society: Essays in the Use of Quantitative Methods for the Study of Social Data*, Cambridge University Press.
Wrigley, E. A. (ed.) (1973) *Identifying People in the Past*, Arnold, London.
Wrigley, E. A. and Schofield, R. (1981) *The Population History of England 1541–1871*, Arnold, London.
Wrigley, N., Morgan, K. and Martin, D. (1988) Geographical information systems and health care: the Avon Project, *ESRC Newsletter*, Vol. 63, pp. 8–11.
Wynne-Edwards, V. (1962) *Animal Dispersion in Relation to Social Behaviour*, Hafner, New York.
Wynne-Edwards, V. (1965) Self-regulating systems in populations of animals, *Science*, Vol. 147, pp. 1543–8.
Yasuba, Y. (1961) *Birth Rates of the White Population in the United States 1800–1860*, Johns Hopkins University Press, Baltimore.
Young, J. (1974) *Suspected Food Poisoning in Consett, County Durham*, Occasional Publication 4, Department of Geography, University of Durham.
Zachariah, K. (1984) *The Anomaly of the Fertility Decline in India's Kerala State*, World Bank Staff Working Paper, Washington, D.C.
Zelinsky, W. (1966) *A Prologue to Population Geography*, Prentice-Hall, Englewood Cliffs, N.J.
Zelinsky, W. (1971) The hypothesis of the mobility transition, *Geographical Review*, Vol. 61, pp. 219–49.
Zelinsky, W. (1983) The impasse in migration theory: a sketch map for potential escapees, in Morrison (ed.), pp. 19–46.
Zimmer, B. (1955) Participation of migrants in urban structures, *American Sociological Review*, Vol. 20, pp. 218–24.
Zipf, G. (1949) *Human Behavior and the Principle of Least Effort*, Hafner, New York.
Zolberg, A., Suhrke, A. and Aguayo, S. (1986) International factors in the formation of refugee movements, *International Migration Review*, Vol. 20, pp. 151–69.

INDEX